THE RISE OF THE
NEW PHYSICS

ITS MATHEMATICAL AND PHYSICAL THEORIES

formerly titled 'DECLINE OF MECHANISM'

A. d'Abro

in two volumes

VOL

1

DOVER PUBLICATIONS, INC.

NEW YORK

Published in Canada by General Publishing Company, Ltd., 30 Lesmill Road, Don Mills, Toronto, Ontario.
Published in the United Kingdom by Constable and Company, Ltd., 10 Orange Street, London WC 2.

This Dover edition, first published in 1952, is an unabridged republication of the work originally published in 1939 by D. Van Nostrand Co., Inc. under the title *Decline of Mechanism*. This Dover edition contains corrections by the author and a new section of portraits.

International Standard Book Number: 0–486–20003–5
Library of Congress Catalog Card Number: 52–8895

Manufactured in the United States of America
Dover Publications, Inc.
180 Varick Street
New York, N. Y. 10014

PREFACE

The present century has witnessed the emergence of two grand theories of mathematical physics: the Theory of Relativity and the Quantum Theory. Both theories were conceived for the purpose of coordinating certain bodies of facts which the classical theories were unable to interpret; and neither theory would have seen the day had it not been for the increased refinement of experimental measurements which rendered the disclosure of these facts possible. But although the two theories were born under similar circumstances, they soon branched out in opposite directions. The theory of relativity has developed into a doctrine whose principal field of application is found in the world of large-scale phenomena, whereas the quantum theory has become identified with the microscopic processes of the atomic and subatomic worlds. To this extent the two theories are complementary. In this book we shall be concerned more especially with the quantum theory and with the researches undertaken by physicists in the subatomic world.

Probably the most remarkable discovery that has issued from these researches is that subatomic phenomena do not appear to be mere repetitions on a microscopic scale of the macroscopic processes with which we have become familiar from our daily experience. In the subatomic world mechanical representations and classical concepts are no longer of much avail, except as props to a bewildered imagination which is unable to feel at ease in its new surroundings. But when classical concepts are utilized in the interpretation of subatomic phenomena, we find ourselves confronted with unexpected difficulties: waves and particles seem to dissolve one into the other as though they were the same and yet not the same.

If the subatomic world were neatly separated from the better-known macroscopic one, we might view the new discoveries as peculiar to the subatomic world. There would then be two physics: the classical (or relativistic) physics of the macroscopic world and the quantum physics of the microscopic world. However, the two worlds exhibit no clear-cut separation; and so we must suppose that the strange characteristics of the new world are present also in the world of our common experience. If this be the case classical physics, even in the macroscopic world, must be superseded by quantum physics. No inconsistency with the facts of observation is involved in this conclusion, for the quantum theory shows that in our ordinary world the strange quantum characteristics should become obscured to such an extent that observations of far greater refinement than we are able to perform in practice would be required to detect them.

The philosophical implications of the new discoveries are still controversial, the main disagreements centering around the status of the principle of rigorous causality. Classical scientists assumed the validity of this principle in connection with the evolution of inorganic systems. Doubts on its validity arose in the earlier years of this century, when the difficulty of accounting for the radiation law led Planck to propose the initial form of the quantum theory. But at that time it was thought that causal connections would be reestablished subsequently as a better understanding of quantum phenomena was obtained, and so no special prominence was given to discussions on the merits of the causal doctrine. Then, in 1927, Heisenberg discovered his celebrated "Uncertainty Relations," as a result of which he contended that a rigorously deterministic scheme could no longer be entertained in physics. Heisenberg's contention occasioned a split between the leading natural philosophers. The quantum theorists, represented by Born, Bohr, and Dirac, supported Heisenberg; Bohr, in particular, suggested his "Principle of Complementarity" as a substitute for the discarded causal doctrine. But the attitude of the quantum theorists was resisted by Planck and by Einstein, who expressed the opinion that the apparent breakdown of the causal doctrine was due to the incompleteness of our present knowledge and that it did not betray any fundamental indeterminism in natural processes.

In the final analysis the quantum theory, like the theory of relativity, is a mere refinement of the classical theories, and to this extent an understanding of the quantum theory requires that we take into account the historical background. Our aim has been to give a semi-popular presentation of the progressive growth of the new ideas, starting from the most elementary notions and giving due consideration to the mathematical development without which the new theories could never have been constructed.

We have divided the book into three parts. Part I deals with the fundamentals of natural philosophy (in relation to physics). Part II contains chapters on mathematics and a review of the more important physical theories of the classical period. In Part III we have discussed the various quantum theories, and have included whatever preliminary mathematical information seemed necessary to their understanding. In the last chapter we have attempted to examine the present status of the doctrine of causality in the light of the new discoveries.

Some of the subject matter might have been omitted: for example, Galois' theory of groups (Chapter XIV), the controversies on the nature of mathematics (Chapter XVI), the Problem of Three Bodies (Chapter XVII), and the theory of contact transformations (Chapter XX). Also Bohr's theory of the atom (Chapters XXVI and XXVII) might have been shortened. But we have preferred to retain these chapters for various reasons. Our excuse for devoting a few pages to Galois' theory of groups and to contact transformations is that these theories are of sufficient interest in themselves to merit mention, and we believe that no elementary presentation of them has yet been given. We have included a brief discussion of the Problem of Three Bodies because the tremendous mathematical difficulties which beset this seemingly simple mechanical problem show how the progress of theoretical physics may be delayed by purely mathematical obstacles. For a somewhat different reason we have given a summary of the controversies on the nature of mathematics. In our opinion these controversies originate from the same psychological differences which appear to be responsible for the current controversy concerning the principle of causality in physics. As such, the controversies in mathematics help to shed light on the nature of the dispute between the quantum theorists and their opponents. Finally we have treated Bohr's theory of the atom at some length because this theory illustrates the difficulty of providing mechanical interpretations for subatomic phenomena. Indeed it was the insufficiency of Bohr's theory that was primarily responsible for the development of the newer quantum theories, and the justification for these newer theories can be appreciated only when the limitations of Bohr's theory are understood.

The non-mathematical reader should experience no serious difficulty in following the presentation, especially so since the more technical chapters (Chapters XXXI to XL) are not essential to a general understanding of the new ideas, and may be omitted in consequence.

New York, N. Y.,
January 1939

CONTENTS

PORTRAITS

MATHEMATICIANS

following page 268

EUCLID, ARCHIMEDES, GALILEO, SIR ISAAC NEWTON, LEONARD EULER, JOSEPH LOUIS LAGRANGE, PIERRE SIMON LAPLACE, AUGUSTIN CAUCHY, WILLIAM ROWAN HAMILTON, CARL G. J. JACOBI, EVARISTE GALOIS, CHARLES HERMITE, MARIUS SOPHUS LIE, HENRI POINCARÉ, DAVID HILBERT

PRE-QUANTUM PHYSICISTS

following page 426

AUGUSTIN FRESNEL, LORD RAYLEIGH, VIKTOR MEYER, SADI CARNOT, RUDOLF JULIUS EMANUEL CLAUSIUS, J. WILLARD GIBBS, WALTER HERMANN NERNST, JAMES CLERK MAXWELL, LUDWIG BOLTZMANN

PART I

GENERAL CONSIDERATIONS

CHAPTER I

THE HISTORICAL BACKGROUND OF THE SCIENTIFIC METHOD

THE craving to understand appears to be one of the characteristic incentives of the human species. With primitive races the form of understanding sought is of a kind that is useful to man in his daily struggle for existence. As such, the motive underlying the urge is a practical one. In later periods sheer curiosity becomes the main driving force, and we find men seeking knowledge for its own sake. But though the urge to acquire knowledge is as old as the human race, the method whereby the urge may be satisfied is a comparatively recent discovery. This method, devised by Galileo and Newton, is known as the scientific method, and the knowledge obtained from its application is called science. Natural philosophy embodies the general outlook that seems best suited to account for the various discoveries. In the case of physics, with which we shall be especially concerned, natural philosophy finds its chief mode of expression in the theories of mathematical physics. To the abstruseness of these theories is due the esoteric character of natural philosophy and the difficulty of weighing its conclusions, unless the critic be well versed in general physics and in mathematics.

Prior to the discovery of the scientific method little progress was made; and though art, literature, and geometry * flourished, man had practically no understanding of natural phenomena. In considering the scientific method, it will be advantageous to stress the successive stages in its application. These may be called:

(a) The observational stage.
(b) The experimental stage.
(c) The theoretical and mathematical stage (in physics).

The order in which these stages have been listed is the order in which they arise in the study of any group of physical phenomena. It is also the chronological order in which they were discovered.

---

* We are not viewing mathematics as a science but as a form of human reasoning bearing a strong resemblance to logic. Furthermore, throughout this book the stress will be laid on physical science.

(a) *The observational stage.* This first stage consists in observing phenomena with precision. For instance, we may observe that plants die unless supplied with moisture, or that rainbows are associated with storms, or that the stars move around the Pole Star. Observations of this sort, being purely qualitative, are necessarily lacking in definiteness; but we may often attain greater precision by discerning quantitative relationships. Thus the last observation is improved when we note that, in their motions around the Pole Star, the stars describe circles with uniform speed, as though they were rigidly connected. We need scarcely mention that the more precise the information, the better equipped we are to progress; and it is thus apparent that the search for quantitative relationships arose quite naturally from a desire to attain greater accuracy in the observations. Other reasons also favor quantitative observations, reasons which may not have been obvious to the earlier investigators but which become clear when the third, or mathematical, stage comes into consideration.

Quantitative observations and measurements often reveal remarkably simple relationships among the various magnitudes measured. Such relationships are called natural laws and are referred to specifically as "empirical laws." Kepler's laws for the planetary motions are probably the most famous laws of this type. Other kinds of natural laws will be examined as we proceed.

Properly speaking, we should restrict the appellation "scientific method" to quantitative observations, because the mere noting that rainbows are associated with showers or that plants die unless supplied with moisture constitutes an observation of so commonplace a nature that it can scarcely be called scientific. Even with this reservation, the application of the first step in the scientific method must have arisen many centuries ago, for we know that the Chaldaeans and ancient Egyptians had acquired a considerable knowledge of observational astronomy. The Greek astronomers, Hipparchus in particular, established quantitative relationships with a degree of precision that has elicited the admiration of later investigators. To mention but one instance, we may recall Hipparchus's discovery of the precession of the equinoxes.

(b) *The experimental stage.* The second step in the application of the scientific method consists in supplementing the first, or purely observational, step by experiment. Men no longer content themselves with observing phenomena as and when they happen to occur; they go out of their way to produce these phenomena artificially, so as to observe them under different conditions and with the increased accuracy permitted by their repetition. Aside from this important difference, the same requirements of accurate observation of a quantitative nature (whenever possible) are imposed. The discovery of the experimental method is usually attributed to Galileo, and, for this reason among others, he is regarded as the father of modern science. Some injustice seems to be done to Archimedes, since in all truth it was he who inaugurated the experimental method in his investigations in hydrostatics. Unfortunately for the glory of Greece, Archimedes came too late to found a school. Rome was then rising in power, and the interest of the new civilization lay in conquest and in organization rather than in scientific research. The great lesson taught by Archimedes was soon forgotten, and when in the middle ages the study of Hellenic culture was revived under the scholastics, it was Aristotle and the mystic Plato who were regarded as the most eminent representatives of Greek thought. Galileo lived in a more propitious epoch, and he had the remarkably good fortune of being followed by Newton. Newton, by his discovery of the third (or theoretical) step, to which reference will be made presently, completed the philosophy of the scientific method inaugurated by Galileo. Natural philosophy as pertaining to physics was thus born and has endured without any essential change in its methods to the present day.

Aside from having rendered possible the introduction of the third (or theoretical) step, the second (or experimental) stage has an enormous importance on its own account. The majority of empirical laws in science were obtained by the application of the experimental method. Galileo measured the accelerations of falling bodies and thereby obtained the empirical law of freely falling bodies. Boyle's law for gases and Descartes' law of refraction are other examples of empirical laws derived from experiment. In another department of science we may mention Mendel's observations on the cross-breeding of different varieties of sweet peas; the empirical law obtained is known as Mendel's law.

The rôle played by experiment is less conspicuous in astronomy and in geology, for in these sciences we must be content with observation. In modern astronomy and in astrophysics, however, experiment plays an important, though indirect, rôle. Firstly, it is thanks to physical experiment (supplemented by theory) that telescopes and spectroscopes can be constructed. Secondly, it is experiment that enables us to match in the laboratory the spectral lines observed in the spectra of the stars and thereby to establish their significance. In a broad sense, the successful performance of the first and second steps requires care, precision, and ingenuity in the construction of the experimental apparatus Very little in the nature of speculative assumptions is needed.

(c) *The theoretical stage.* The third step was taken by Newton when he furnished a mathematical theory of the planetary motions, obtaining thereby his law of gravitation. It differs from the experimental step in that it is mathematical and speculative. Furthermore, physics is the only one of the sciences to which the mathematical procedure of the third stage has been applied with outstanding success. Owing to this circumstance physics has developed far more rapidly than the other sciences.

The incentives which prompted the third step are readily accounted for. In the experimental stage facts were discovered, their relations were established, and empirical laws were obtained. But the relations disclosed by experiment were the most obvious ones, and it seemed permissible to suppose that hidden relations also existed. Could these additional relations be revealed, a wider array of facts would be connected than was possible by the experimental method alone, and our understanding of natural processes would be correspondingly increased. To proceed beyond the experimental stage, investigators were compelled, however, to introduce assumptions of a more or less speculative nature. In the majority of cases the implications of these assumptions were too complicated to be deduced by commonplace reasoning and logic. A far subtler method of approach was needed; it was found in the mathematical instrument. By this means the third stage in the scientific method (in physics) was attained.

Thanks to the mathematical instrument, formerly unsuspected relations were revealed, and additional laws were derived. These mathematical laws are no longer called "empirical," for they differ entirely in their mode of derivation from the laws that may have been established in the preceding steps. Newton's law of gravitation is an illustration of a law obtained through the medium of mathematics. The doctrines developed in physics as a result of the application of the third step are called "theories of mathematical physics," or more briefly "theoretical physics." The theory of relativity and the quantum theory are of this type.

In any historical discussion of the development of the scientific method in physics, we must realize that the theoretical stage could not have arisen before mathematics had attained at least a moderate degree of development. The fact is that all physical theories involve the differential calculus, the calculus of probabilities, or still more advanced parts of analysis. Since none of these mathematical developments were known before Newton's day, it would be unfair to criticize Newton's predecessors for their failure to apply the theoretical method. On the other hand, we may well wonder why the experimental step should have been so tardy in making its appearance, especially when we realize that the transition from the observational to the experimental step is a natural one. Both steps rely essentially on observation; they differ only in that, in the first step, we observe what happens to occur, whereas in the experimental step we attempt to generate the occurrence. The child who observes the contents of an open box is applying the first step. And when the box is closed and the child deliberately opens it to view its contents, he is unwittingly applying the second step. The experimental method appears so natural today that there has always been something of a paradox, in the failure of the Greeks to have applied it, other than in the most trivial cases. Are we then to believe that the gifted Greek race, that gave to the world such varied geniuses as Homer, Phidias, Sophocles, and Euclid, produced only one thinker, Archimedes, who realized the advisability of opening the box rather than of speculating endlessly on its hidden contents? A possible answer to this question will be considered presently. At all events, the failure of the ancients to rely on experiment was responsible for the insignificance of their contributions to natural philosophy. Indeed, if we except Archimedes and the geometricians, we find that the Greeks confined themselves to wild guesses, supported by metaphysical arguments so transparently irrelevant that even a modern schoolboy could refute them with ease. In these speculations, matter, space, motion, atoms with hooks and eyes, and other notions drawn from commonplace experience played a prominent part. The procedure of the ancients has not even the merit of being consistent, for if we choose to ignore the information revealed by intelligent experimentation, we should in all logic disregard likewise the cruder knowledge we have gained from commonplace observation, and we should refrain from taking this knowledge as the basis of our speculations.

A. H. Compton attributes the strange lack of understanding of the Greeks to the influence of Socrates and of the Persian magicians. In our opinion Compton's explanation cannot contain the whole truth, for we may safely presume that not all the Greeks were mystics. And who aside from a mystic could be swayed by the arguments of the magicians or of Socrates and Plato?

Another explanation, which appears to us more plausible, is that the Greeks were perfectly capable of understanding the value of experiment but that they did not wish to recognize it. Plainly, the application of the experimental method is not a matter of mere intelligence; it involves in a less conspicuous form other qualities; namely, character, sincerity, and modesty. It requires character to seek the truth even when we have reason to fear that it will not be to our liking. It requires sincerity to accept the truth when this truth happens to contradict all that we have previously professed. Finally, it requires modesty to recognize that man cannot, by his inner vision alone, attain to truth and that he must stoop to experiment. Experiment has always been anathema to the egotist, not necessarily because it involves manual labor, but because it belittles man by placing him in the position of a humble student of Nature instead of revealing him as the Lord of all Creation.

These qualities of courage, sincerity, and modesty, which the ancients seem to have lacked, prevented them from creating a science. Thus we find Pythagoras concealing his discovery of irrationals because it upset his doctrine of numbers. In Galileo's day the same spirit is shown by the metaphysician who refused to look at the heavens through a telescope for fear of being convinced of the errors in Aristotle's teachings. And, if Plato by opening a box could have tested his theory of universals, we may be certain that he would have destroyed the box rather than run the risk of being refuted.

Of course, when we speak of the experimental method as exemplified by the opening of a box to see what it contains, we are taking a trivial illustration, and we may find plenty of trivial instances in which even the Greek metaphysicians resorted to experiment. But they proceeded in a half-hearted, inconsistent way. We shall not dwell on the speculations of the Pythagoraeans and the early Atomists, for at the dawn of any civilization guesses are all that can be expected. Plato may also be disregarded since he was avowedly a mystic seeking truth from his inner soul. A study of Aristotle's writings will be more instructive from our present standpoint.

Aristotle achieved well-deserved fame as the founder of logic; but he was also a student of natural phenomena, and it is with his investigations in this latter field of inquiry that we shall here be concerned. The annual overflow of the Nile excited his curiosity; and to determine its origin he proceeded in a thoroughly commendable way by sending an expedition to the sources of the river. In other words he opened the box and looked inside. Yet the same Aristotle informs us that the increase in weight of a growing plant is due to the matter it absorbs from the soil. Of course Aristotle, like every one else, had observed that plants grow from the ground and do not spring from a bare slab of stone; accordingly, his guess was plausible. But never did he attempt to test it by weighing a pot containing a growing plant at successive periods of its growth. It is significant that, not until the eighteenth century, was this simple test made and the weight of the pot found to increase. The foregoing elementary experiment, which might have convinced Aristotle of his mistake, would surely have been less costly and more easily accomplished than the sending of an expedition to the sources of the Nile.

Elsewhere Aristotle tells us that women have less teeth than men and that they have fewer ribs than men. We are also told that a body dropped from the mast of a moving ship falls behind the mast. Obviously the experiment was not performed. In all cases we find guesses, some of which, such as the prevalence of earthquakes in the coastal regions, have turned out to be correct. But whether the guesses were correct or incorrect is a minor point; the characteristic fact is that they were not even based on careful observation. As such, they could teach Aristotle nothing of the phenomena in which he was interested. The net result is that Mendel by his simple experiments contributed more to biology than did Aristotle in his massive writings.

A comparative study of the rise of the experimental method in the different civilizations would be of interest in disclosing whether the strange deficiencies of the Greeks were peculiar to them. Unfortunately such a study could be applied only to the civilizations of the white race, for the white race alone has evolved the experimental method on its own initiative. We may note, however, that in the few instances where some kind of a comparison is possible, the application of the experimental method has invariably followed the golden age of literature and art: Archimedes comes after Pericles, Galileo after Dante and Raphael, and Newton after Shakespeare. From these illustrations we might surmise that the failure of the Greeks to develop the experimental method was due to the premature eclipse of their civilization. Other considerations support this view. Homocentric tendencies have always been pernicious to the development of science; and we find that in the early stages of any civilization these tendencies predominate. The history of painting furnishes an illustration. The first subjects to be represented were men, gods fashioned like men, and animals useful to men. Presently landscapes were added, but only as a background; man was still the centre of interest. Only considerably later, with the rise of the Dutch school, were landscapes painted for their own sake. Similarly in literature, the earlier poets sang of their loves and battles but never extolled the beauty of their lands. Even in more modern times Montesquieu in his description of the Alps speaks of them as a mass of rocks which make travelling tedious. Not until the latter part of the eighteenth century, in the writings of J. J. Rousseau, is the beauty of nature considered worthy of more than passing comment. In the scientific field we find the same persistent homocentric tendencies. These tendencies are seen in the heated controversies which followed Copernicus's defense of the heliocentric system. The true importance of Copernicus's discovery lies in its having paved the way for Kepler's laws and Newton's investigations. But what caused so much excitement in Copernicus's days was the dethronement of the Earth from its central position in the Universe. Yet who but the egotist cares one whit whether the Earth on which he lives be the centre of Creation or a mere satellite of the sun? In more recent times the antagonism manifested in certain quarters to the theory that man's ancestors were apes is traceable to a similar egotism.* However natural all these homocentric tendencies may be, they have undoubtedly exerted a pernicious influence on the development of disinterested research.

With the application of the experimental method by Galileo, the effects of egotism began to lose their force, at least in scientific matters; facts were sought whether they flattered man or not. The seventeenth century witnessed several attempts to extend to other phenomena the method of Galileo. Thus Descartes measured the angles of incidence and of refraction of a ray of light when it passed from one medium into another, and he obtained from his measurements the empirical law of refraction. Pascal performed experiments in hydrostatics and established the decrease in the atmospheric pressure that accompanies increasing elevation. Boyle studied the change in volume of a gas when compressed at constant temperature. Newton supplemented Galileo's mechanical experiments with others of his own invention and thereby obtained the fundamental laws of mechanics. Experiments were conducted by Newton and by Huyghens in optics. These and a few others constitute the sole contributions of the experimenters of the seventeenth century. The results are meagre, and no more rapid advance is registered in the following century. The slow progress was due to the crudeness of the experimental apparatus and to the limited control the investigators had over the conditions under which their experiments were performed.

The truly significant development that occurred in the seventeenth century was the discovery and application by Newton of the third step in the scientific method. The facts that Newton attempted to coordinate were Kepler's laws of the planetary motions and the mechanical facts expressed by the laws of mechanics. Newton made the assumption that the planetary masses were governed by the same mechanical laws that held for terrestrial bodies. It is the deduction of the necessary mathematical consequences of this simple assumption that constitutes one of

---

* We are here referring to objections from those who complain that man's dignity is degraded by this theory, and not from those who combat it for religious or social reasons.

Newton's chief titles to fame. As is well known, a direct consequence of Newton's assumption (when consideration is given to Kepler's laws) is the law of gravitation. Several important points are connected with Newton's investigations, but we shall defer fuller consideration of them for the present. Suffice it to say, not only did Newton's theory account for Kepler's laws (and this was its original aim), but it also showed the connection between the fall of bodies and such seemingly independent phenomena as the tides and the precession of the equinoxes. Newton's findings illustrate the ability of a scientific theory to coordinate a greater array of facts than would be possible by means of the experimental method alone.

The importance of Newton's method was so obvious that it could not fail to impress his contemporaries. His success appeared all the more striking when contrasted with Descartes' rival scheme. Descartes was a creative mathematician fully conversant with the limited mathematics of his day; he also realized the importance of ascertaining facts correctly. (To him we owe the empirical laws of reflection and refraction.) Nevertheless, in the later stages of his life, he preferred to pursue the facile rationalizations of the Greek metaphysicians, creating a world to suit his fancy without the slightest regard for the known facts. Thus Descartes sought to account for the planetary motions, and for this purpose he imagined vortices sweeping the planets around the sun. But, in contradistinction to Newton, he failed to take into consideration the precise facts established by Kepler or indeed any of the mechanical facts discovered by Galileo. In Descartes' scheme, the planets might just as well have followed arbitrary curves, open or closed; whereas in the Newtonian scheme, had the courses and the motions of the planets differed from those established by Kepler, Newton's law of the inverse square would have been untenable. The net result of Descartes' faulty procedure was that his theory accounted for nothing, coordinated nothing, and predicted nothing; and his speculations have exerted no greater influence on the subsequent development of natural philosophy in physics than have his similar ones in medicine.

In this condemnation of Descartes, we must make clear that the objectionable feature in his writings is not due to his speculative spirit; indeed, many of the theories of mathematical physics which occupy a high place in the history of science make greater demands on the imagination than anything Descartes ever advanced. What condemns Descartes is his utter disregard for facts. Some excuse for him may be found when we remember that in his time the differential calculus was unknown and the proper treatment of the planetary motions impossible. Less excuse can be made for Leibnitz. The extravagant writings of Descartes and of Leibnitz did, however, exert some beneficial influence indirectly, by convincing mathematicians of the futility of baseless speculations. Never again do we find creative mathematicians constructing world systems to suit their fancy; instead, they concentrated their efforts on perfecting the mathematical instrument, thereby preparing the way for the theoretical physicist.

A point which cannot be stressed too strongly is that although mathematicians pursue their investigations in a disinterested spirit, their discoveries prove of inestimable value to the theoretical physicist in his attempts at constructing a theory. Our mention of the bearing of the differential calculus on the theory of gravitation is but one illustration among many. Even if all the facts on which the theory of relativity and the quantum theory are based had been known in Newton's time (or even a century later), it would still have been impossible, owing to mathematical limitations, to construct these theories. The general theory of relativity, for instance, could never have arisen prior to Riemann's investigations in pure mathematics on curved spaces.

Having noted the origins of the scientific method, we pass to the principal results of its application in physics. During the seventeenth century Newton effected the synthesis of the facts pertaining to the dynamics of rigid bodies and to the planetary motions; other facts were meagre and incomplete, and there was little left to coordinate. The early part of the eighteenth century did not contribute new facts of importance. Only at its close was the electric current generated by Volta, and the law of electrostatic actions formulated by Coulomb. As against this lack of experimental progress, the eighteenth century witnessed a remarkable mathematical development at the hands of Euler, d'Alembert, Lagrange, Daniel Bernoulli, and Laplace. Most of the advance arose from attempts to further the application of Newtonian dynamics to point-masses and to extended solids. Thus we find Euler obtaining his celebrated equations for solids in rotation; d'Alembert linking statics to Newtonian dynamics by means of his principle; Lagrange condensing Newtonian dynamics into a deductive form in his "Mecanique Analytique"; and Laplace contributing advances of major importance in celestial mechanics. We may also mention Euler's and Lagrange's extension of Newton's dynamics to fluids in motion. Thus statics and hydrostatics, whose laws the genius of Archimedes had established, were supplemented by dynamics and hydrodynamics. Along with this progress in the furthering of Newtonian mechanics, important branches of pure mathematics were developed, such as the calculus of variations (Lagrange), harmonic analysis (Legendre, Laplace), and partial differential equations (Lagrange).

The eighteenth century, as we have pointed out, did not contribute any physical facts of importance, so that no new physical theories were forthcoming. But during the nineteenth and twentieth centuries a whole mass of experimental data was obtained in every field of physics, thereby providing material for additional physical theories. Nevertheless, only some of the theories of the nineteenth and twentieth centuries would have seen the day, had the mathematical advance failed to keep pace with the experimental discoveries. Fortunately, the nineteenth century was also one of tremendous mathematical development, and thanks to this joint progress, theories of mathematical physics began to appear. The most important of these are thermodynamics (Carnot, Clausius, Maxwell, Gibbs, Nernst); Maxwell's theory of electromagnetism; the kinetic theory of gases (Boltzmann, Maxwell, Gibbs, Einstein); Lorentz's theory of electrons; and, finally, the theory of relativity and the quantum theory.

CHAPTER II

ASSUMPTIONS IN SCIENCE

ACCORDING to Planck, the two fundamental canons of natural philosophy are:

1. There is a real outer world which exists independently of our act of knowing.
2. The real outer world is not directly knowable.*

The first of these statements cannot be proved or disproved either by *a priori* arguments or by experiment; the stand of the solipsist is unassailable. For pragmatic reasons, however, the independent existence of an outer world must be granted.

Planck's second canon states that we have no "direct knowledge" of the real world. But first, what do we mean by direct knowledge? Obviously, knowledge that is unaccompanied by inference must be classed as direct. Thus our immediate sensations of warmth, color, etc., constitute direct knowledge. In a more extended sense direct knowledge applies to our perceptions, though here inference is not entirely lacking. Planck's second canon therefore implies that the real world, which is the cause of our sensations, is not revealed to us by these sensations alone. We might also add that this world cannot be disclosed by mere meditation and introspection. No one conversant with science would dispute Planck's contention; the direct knowledge of the real world claimed by the mystic and the modern neo-realist has no place in a scientific discussion.

In our ordinary life our attention is absorbed by our sensations and perceptions, and the real impersonal world (the reality behind the appearances) is disregarded. But the physicist cannot restrict himself to such subjective statements as "snow is white" or "sugar is sweet"; he is compelled by the nature of his studies to peer beyond and explore the real world. Only thus can he discover the hidden relations that clarify the workings of phenomena. It is the disclosure of these relations that constitutes the aim of physical science. Since direct knowledge is insufficient to reveal the real world, the physicist proceeds in a roundabout

* "Where Is Science Going?" Planck, p. 82. W. W. Norton & Co., 1932.

way, by coordinating direct knowledge (*e.g.*, the readings of his instruments), experiment, elementary inference, and rationalization. The picture he thus obtains represents the real world of physics.

The discovery of atmospheric pressure illustrates these points. The physicist has been led to believe that the atmosphere exerts a pressure of some 14 lbs. to the square inch. What evidence does he advance for this belief? He certainly does not claim that we are directly aware of this pressure on our bodies. Hence, should science be restricted to a cataloguing of immediate sense impressions, the very notion of atmospheric pressure would have no place in physics. The evidence in favor of atmospheric pressure is entirely indirect. Thus we note that a column of mercury stands at a certain height in a Torricelli tube. Or again, we exhaust the air between two hollow hemispheres and find that we are unable to tear them apart. In the second experiment direct knowledge comes into play through the effort we exert to separate the hemispheres, but only by argumentation and reasoning can we ascribe this effort to the existence of atmospheric pressure. Practically the whole of physical science is thus one mass of inference based ultimately, but not immediately, on direct knowledge.

These considerations entail important consequences. In the grouping of phenomena it is the similarities established in the impersonal world that are regarded as of particular significance; the more obvious similarities detected by our unaided senses are held to be of minor importance. For this reason the physicist claims that visible light and invisible ultra-violet light are kindred in nature, for both are electromagnetic vibrations. The accidental circumstance that the human eye detects the first kind of radiation, and not the second, is viewed as unimportant. Indeed, we need not appeal to theoretical physics to find illustrations of the same tendency. Thus, if we were judging from our immediate sensations, we should presumably claim that sugar and saccharin exhibited a striking similarity, for both are white crystals and have a sweet taste. On this basis we should be tempted to class the two substances into one family. Yet, insofar as the chemist is concerned, sugar and saccharin are entirely different: sugar is a carbohydrate containing carbon, hydrogen, and oxygen; saccharin, though sweet to the taste, has the chemical characteristics of an acid, and its molecule contains carbon, hydrogen, oxygen, and also nitrogen and sulphur. Likewise the botanist places in the same family the potato and tomato plants, or the yellow buttercup and the hooded blue aconite. No obvious similarities can justify such a classification. These examples and many others that could be mentioned bring out the important point that it is not the obvious that is necessarily credited with any deep significance in science. It is the veiled. This attitude on the part of physicists must not be attributed to any particular love for sophistication or to a failure to note the obvious. It is due to a prolonged experience which shows that the truly important similarities, the detection of which leads to the most fruitful results, are not usually those appearing on the surface.

We may now examine more fully some of the points we touched upon in the previous chapter. We mentioned that in the second, or experimental, stage of the scientific method, no *a priori* assumptions of a speculative character are introduced. More precisely, we meant that the assumptions made are of a type which any man would naturally accept in the pursuit of his ordinary activities and are not of the highly speculative kind that we shall meet with in the theories of mathematical physics. Obviously, even in the simplest experiment assumptions cannot be avoided; ordinary generalization, without which science could not proceed, rests on an assumption. Thus an experiment repeated one thousand times cannot give us any positive assurance that the result will be the same when the experiment is performed once again. On the other hand, if we are ultra cautious and refuse to generalize, experiment becomes useless since it cannot be used as a source of more general knowledge. Faced with the alternative of choosing safety with sterility or doubt with possible progress, the physicist adopts the same course as does the layman in his daily activities; he relinquishes safety and he generalizes. By so doing he is undoubtedly introducing an assumption; namely, the postulate that generalization is permissible. In this tenet consists the principle, or postulate, of scientific induction.

Since the postulate of induction constitutes a mere assumption and is therefore speculative, all empirical laws are necessarily themselves speculative. However, the postulate of induction cannot be regarded as speculative in the same sophisticated sense as are the postulates introduced in the construction of theories, for even in our daily activities we apply induction constantly. Presumably its instinctive appeal is due to a certain inertia of our minds, which find it easier to follow a pattern than to contemplate a new possibility at each step.

Though constantly utilized in science, the principle of induction must be applied with caution. For example, there is no more conclusively established fact of observation than the passage of heat by conduction from a hotter to a colder body. It takes place under the most varied conditions. Small wonder then that Clausius regarded this unilateral passage of heat as exhibiting a fundamental law of Nature (the Law of Entropy). Yet today, for theoretical reasons connected with the kinetic theory of gases, we believe that this law is only statistically valid, and we have reasons to suppose that if we could experiment long enough, heat would occasionally be found to leave the colder body of its own accord and increase the temperature of the warmer one. We have here an illustration in which the original attempt to apply the postulate of induction led to error.

There is also another difficulty which besets the application of the principle of induction. The principle states that the same experiment repeated under the same conditions will always yield the same result. When stated in this way, the principle is seen to be an expression of the law of causality. Unfortunately we cannot repeat an experiment under exactly the same conditions. The time of the day, the season of the year, the past history of the bodies operated on are subject to incessant change. No two experiments can truthfully be said to occur under exactly the same conditions.

In practice, we overcome this difficulty by disregarding some of the varying conditions, on the plea that their influence is negligible or strictly nil. Since we cannot test all conceivable influences, it is only in virtue of an assumption that we may rule out some of them. The assumptions accepted in this process of elimination are usually of the most plausible kind, but they are more or less speculative nevertheless, and further investigation may prove them to be unwarranted. A simple hypothetical illustration is the following: The first man who placed an egg in boiling water found that the egg became hard-boiled, and in view of the postulate of induction he might have been tempted to express this result as an empirical law. It would have required a mind of uncanny foresight to have felt the necessity of varying the conditions of the experiment by repeating it on a very high mountain; and yet, had this been done, the influence of elevation would have been found to be far from negligible, since the egg would no longer have become hard-boiled.

In the foregoing example our hypothetical experimenter erred in discarding as irrelevant the influence of elevation, but in the majority of cases the experimenter's selection of supposedly irrelevant influences has not been challenged at later times. As an illustration let us consider Galileo's experiments on balls rolling down inclined planes. Galileo certainly did not operate with balls of every hue and color. Hence when he concluded from his experiments that the acceleration of any ball rolling down an inclined plane would be constant, he was implicitly assuming that the color of the ball was irrelevant. Galileo's assumption, in contradistinction to that of the experimenter in the earlier example, is recognized today as fully justified. The point we wish to stress here, however, is that both assumptions were plausible. We should find quite generally that the assumptions introduced by the experimenters when they disregard certain influences as irrelevant, are always plausible assumptions, even though they may turn out to be incorrect.

Let us examine whether, in addition to the type of assumptions previously enumerated, others may not also be found in the performance of experiments, *e.g.*, in Galileo's. Galileo's measurements of the acceleration of the balls necessitated measurements of spatial distances and of intervals of time. He utilized rigid rods or marks on his inclined planes for the measurement of distances, and sand clocks or others of a similar kind for the measurement of time. We might therefore claim that Galileo was introducing assumptions on the manner in which space and time should be measured. We do not believe, however, that the choice of a system of measurement should be classed as an assumption. Rather, the use of rods and of clocks constituted a definition of the way in which Galileo proposed to measure space and time. As is well known, some kinds of measurements lead to simpler results than others, and such measurements are then claimed to be the correct ones. But so long as the measurements we propose to apply are fully stipulated, a definition and not an assumption is at stake.

In most experiments many of the assumptions merely express the results of earlier investigations. Modern experiments, in particular, rely to a considerable extent on previous experimental results. Thus today when a physicist performs experiments on cathode rays, he assumes from the start that the rays are formed of swiftly moving electrons, an assumption which is merely a conclusion derived from earlier experiments.

Our analysis of the assumptions made in the course of an experiment shows that these assumptions are either of the trivial kind, which any sane man would naturally accept, or that they are assumptions inherited as conclusions from former experiments. But if we pursue a similar analysis for theories of mathematical physics, we shall find that the assumptions introduced are of an entirely different kind. In any theory of mathematical physics one or more speculative assumptions are accepted as a basis. Some restrictions are imposed by common agreement, however. In the first place, the assumptions must not conflict with the facts we are attempting to co-ordinate. Secondly, they must not be inevitable consequences of these facts, for, if they were, they would not constitute independent assumptions. Thirdly, our assumptions must be such as to give rise to observable consequences, preferably of a quantitative nature, so that they may be checked by accurate measurement. If the assumptions do not lead to consequences which can be observed, we shall never be in a position to know whether they are justified or not, and our theory will degenerate into a set of irrelevant statements.

Newton's theory of the planetary motions is one of the first theories of mathematical physics to have been evolved,* and it will pay us to examine it more closely. As usually happens in the development of a scientific theory, the strictly logical course is not always followed; the investigator may guess at the final result before verifying it. But since our present purpose is to uncover some of Newton's more conspicuous assumptions, we shall adhere to the logical rather than to the historical order of his discovery.

Newton's aim was to construct a theory which could account for Kepler's empirical laws. Kepler's laws, derived from Tycho Brahé's astronomical observations, are as follows:

1. The radius joining the sun to a planet sweeps over equal areas in equal times.

2. The orbit described by a planet is an ellipse with the sun at one of its foci.

3. The squares of the periods of revolution of the various planets are proportional to the cubes of the major axes of the respective elliptical orbits.

Newton first assumed that the planets were massive bodies whose motions were controlled by the same laws of mechanics that were established by Galileo and by Newton himself for terrestrial bodies. In this consists Newton's fundamental assumption. We have here an example of an assumption which differs entirely from those we mentioned when discussing the assumptions of the experimenters. Though plausible, Newton's assumption is speculative and is not one which any man would inevitably make. As a case in point, Aristotle and Descartes, both of whom were concerned with the planetary motions, failed to suspect any connection between the mechanics of terrestrial bodies and that of the planets; they did not therefore make the assumption in question. In more recent times Hegel, the metaphysician, decided that the planetary motions were not governed by the mechanics of earthly bodies. Though there is no excuse for Hegel's aberration, we mention it to stress that Newton's assumption has not appeared inevitable to all men.

If we accept Newton's assumption and the laws of mechanics, Kepler's second law shows that the planets must necessarily be submitted to forces.

---

* The mathematical theory of mechanics had previously been furnished by Newton.

If there were no forces the planets would pursue straight courses according to the law of inertia. But the existence of forces acting on the planets is about all the information that can be obtained when we rely solely on commonplace logical reasoning; anything else that we might say would be a mere guess. It is noteworthy in this connection that Hooke, who made the same assumption as Newton, surmised correctly that the force acting on a planet was directed towards the sun and was inversely proportional to the square of the planet's distance. Hooke's guess was correct, but precisely because it was a mere guess, it was worthless; and Hooke is not given credit for the discovery of the law of gravitation. To solve the problem rigorously, recourse must be had to that more powerful method of investigation offered by mathematics. Newton's solution of this mathematical problem is considered his major achievement.

Let us mention briefly the results derived from the mathematical analysis. Kepler's first law requires that the planets be acted upon by forces directed towards the sun. Kepler's second law shows that the force must be inversely proportional to the square of the distance from planet to sun. Finally, Kepler's third law requires that the same permanent field of force be acting on all the planets.* These mathematical deductions may be summarized in the statement: Each planet must be attracted towards the sun by a force which is proportional to the mass of the planet, and the magnitude of this force must vary inversely to the square of the distance. Newton next assumed that the sun itself was responsible for the force. The assumption appeared to be reasonable since it amounted to supposing that if the sun were absent, the planets would not move as they did. On the other hand it created a difficulty, for it implied that Kepler's third law could not be correct. The fact is that if the sun were responsible for the forces exerted on the planets, Newton's mechanical law of action and reaction would require that each planet should exert an equal and opposite force on the sun. These forces of reaction would then communicate an accelerated motion to the sun, and this solar motion in turn would preclude the perfect accuracy of Kepler's third law. A conflict between empirical laws and a physical theory usually entails the rejection of the theory, but, for reasons which will be mentioned presently, the degree of inaccuracy which Newton's theory conferred on Kepler's laws was so slight that the discrepancy

---

* In the absence of the third law, it would be permissible to assume that if the various planets happened to be situated at the same distance from the sun, the intensity of the force acting on them might differ from one planet to another.

was attributed to the lack of refinement of Tycho Brahé's astronomical observations.

In the logical development of the theory the next step introduced the assumption that just as the force exerted by the sun on a planet was proportional to the mass of the planet, so also the equal and opposite force exerted by the planet on the sun was proportional to the mass of the sun. The assumption was plausible, for according to terrestrial mechanics the force of attraction pulling a body towards the earth was known to be proportional to the mass of the body. But we must remember that the forces holding between the planets and sun had not yet been identified with the force of gravitation, and it was therefore hazardous to generalize from one situation to the other. At all events, if this last assumption of Newton's is accepted, the force of attraction of the sun on a planet (and vice versa) is given by $k\dfrac{Mm}{r^2}$, where M and m are the masses of the sun and of the planet, r is the distance between the two bodies, and k a fixed constant (the gravitational constant). According to Newton's formula, if the mass of the sun were twice as great, the force of attraction on the planets would be doubled, and in particular the earthly year would be shortened. This conclusion in itself gives an indication of the speculative nature of the assumption, for Newton had only one solar system at his disposal and hence could not compare the actions of suns of different masses. Newton, on the basis of his formula, observed that in view of the enormous mass which it seemed permissible to ascribe to the gigantic sun, the discrepancy between his theory and Kepler's third law became so small as to be negligible in practice. For this reason the conflict between Newton's theory and Kepler's empirical laws was regarded as unimportant.

The Newtonian theory indicates that the mass of the sun and the masses of the planets generate fields of force the intensities of which are proportional to the masses. Newton therefore supposed that the mutual attractions between the sun on the one hand, and the planets on the other, should hold among the planets also. Let us observe that Newton was here guided solely by the inner consistency of his theory and not by Kepler's laws. Indeed his new assumption went counter to Kepler's laws, for it indicated that the planets should exert mutual perturbations on one another and hence that Kepler's laws should be even more inaccurate than they were already supposed to be. Calculation showed, however, that, in view of the relatively small masses of the planets, the mutual perturbations should be very small and would certainly pass unobserved unless the planetary positions were charted with great accuracy over protracted periods of time. For these reasons, the assumption of the mutual attraction of all the bodies in the solar system was seen to be compatible with the facts of observation. Strong support for the theory was found in a somewhat different direction. Thus the moon describes a quasi-circular orbit around the earth, and on the strength of the laws of mechanics this motion is possible only if the moon is attracted towards the earth. If then the earth attracts the moon, there is no obvious reason why it should not also exert attractions on the other planets. Furthermore, the weight of a body on the earth's surface indicates the existence of an attraction, and it would therefore seem plausible to suppose that the motion of the moon and the weight of a body are related phenomena. Newton made this assumption. Fortunately he was able to submit it to a quantitative test. Thus, if the assumption is correct, we must infer that the acceleration of the moon as it falls towards the earth and the acceleration of a falling stone near the earth's surface are manifestations of the same field of gravitational force generated by our planet. This conclusion may be tested provided we know the magnitude of the earth's radius and the distance between the moon and the earth. Newton, on the basis of the information available in his time, uncovered a serious discrepancy between his theoretical anticipations and the observed results. With the true spirit of scientific caution, he accordingly refused to publish his theory. Several years later more accurate measurements of the earth's radius and of the distance between the earth and moon were performed by Picard, and Newton availing himself of Picard's results found that the earlier discrepancies vanished. He then decided, in 1682, to publish his theory of gravitation. By induction he supposed that what was true for the masses of the solar system would also be true for all the masses of the universe. The theory of universal gravitation was the result.

Several important points are revealed by this analysis. We have listed a number of assumptions introduced in the theory, but of these the first alone, which assumes the planetary motions to be controlled by the laws of terrestrial mechanics, is truly speculative; the others are more or less inevitable. This first assumption, owing to its speculative nature, is typical of the assumptions encountered in the theories of physics.

Another feature brought to light by the foregoing analysis is the extreme fragility of a theory of mathematical physics. The slightest discrepancy between the anticipations of the theory and the facts observed may invalidate the theory. An instance was mentioned when we said that Newton did not publish his theory until a numerical discrepancy which he believed to exist was removed. Of course only when the discrepancies cannot be attributed to errors of observation is a theory in danger; the example of Newton's theory being incompatible with Kepler's laws illustrates a case where the discrepancies are attributed to faulty observations and do not demand the rejection of the theory. Many are the theories, however, which have had to be abandoned because of discrepancies that could not be ignored. The layman usually hears little of such theories, for they are soon forgotten and text-books do not mention them.

The fragility of theories of mathematical physics arises because these theories profess to anticipate accurate quantitative results, which is the feature that enables them to be tested. A theory which anticipates nothing definite or, at best, vague qualitative results is always more difficult to demolish, but this advantage is cancelled by its uselessness. These remarks are well illustrated in Descartes' vortex theory for the planetary motions. In the main all theories of mathematical physics present the aspects we have discussed, though minor differences distinguish one theory from another. In the kinetic theory of gases, the speculative assumption is that a gas is formed of perfectly elastic molecules colliding and rebounding according to the laws of mechanics. A second assumption is the so-called "ergodic hypothesis"; we shall explain it in the chapter on the kinetic theory. The first of these two assumptions is obviously speculative and is by no means one which every man would naturally make. In point of fact an alternative assumption was at one time contemplated; it was that gases were formed of molecules which repelled one another and thereby generated the expansive force. This latter theory was soon abandoned, for it did not appear to be compatible with the gas laws.

In Maxwell's electromagnetic theory the speculative assumption is that all electric currents are closed. Thus we know that if we join two conductors at different potentials by means of a wire, a current flows from one conductor to the other until the potential becomes the same on either conductor. Such a current is called an open one because the path it describes does not close round on itself. The originality of Maxwell's theory consists in the supposition that the circuit, though open in appearance, is in reality closed by the generation of another kind of current flowing from the second conductor to the first through the surrounding ether. This ethereal current, which Maxwell calls a displacement current, is assumed by him to produce the same magnetic effects as an ordinary current. Maxwell's assumption ensures the law of conservation for electricity. From the mathematical standpoint it entails the addition of a new term to the equations which, prior to Maxwell's time, were supposed to represent electromagnetic changes. But in whatever way Maxwell's assumption is expressed, it remains extremely speculative; and Maxwell's theory received serious attention only when Hertz, by direct experiment, corroborated one of its major anticipations (electromagnetic waves).

As a last illustration of assumptions at the basis of a theory, we may mention Lesage's assumption that gravitation is due to the bombardment of "ultramundane" corpuscles impinging on material bodies. Lesage himself was not a mathematician, and his theory did no more than state his assumption. But when the theory was subsequently worked out mathematically by G. H. Darwin and by Lorentz, it was found to entail consequences which were in disagreement with facts and was therefore rejected. Here we have one of the many examples of a theory which has been proved untenable.

In addition to the speculative assumptions that vary from one physical theory to another, there is one fundamental assumption which is implicitly contained in all these theories. It consists in the supposition that mathematical analysis is applicable to the physical world.* We shall refer to this supposition as "the mathematical assumption." The validity of the mathematical assumption hinges on the realization of several conditions in Nature, the most important of which is that Nature should manifest a deterministic scheme and hence be controlled by laws. But this first condition by itself would be insufficient; it is also necessary that Nature should display some measure of simplicity, uniformity, and unity. The necessity for this second set of conditions is well illustrated in the game of chess. As Poincaré observes, mathematics cannot apprise us of the strongest moves in chess because there are too many pieces whose moves are entirely different; simplicity, uniformity and unity are lacking.

It would be futile to decide by *a priori* arguments whether or not Nature exhibits the various conditions conducive to the mathematical assumption; all that we can do is be guided by experiment. The mathematical assumption is by no means a dogma which is accepted blindly; it is a mere tentative assumption, the validity of which is subsequently tested when the consequences of a theory are compared with the results

---

* We are referring here to the applicability of mathematics in its broadest sense, and are not concerned with the purely technical difficulties which the solution of the mathematical problems may present.

of experiment and observation. As such, it is an assumption similar to the one introduced by Newton when he assumed that the planets were controlled by the same mechanical laws as were ordinary terrestrial bodies. It may be that Newton himself started out with a firm conviction that Nature was "mathematical," but, whether this be so or not, the fact remains that the success of Newton's theory, not his personal beliefs, vindicated the mathematical assumption, at least in the restricted case of the planetary motions. In recent times discoveries in the quantum theory have led some of the foremost physicists to suppose that there are no mathematically exact laws in the physical world, and that the best we can attain are probabilities. If this be so, the mathematical assumption cannot have absolute validity, but even then it retains its usefulness to a high degree of approximation in many cases.

Thus far we have been concerned with the general requirements that must be satisfied if mathematics is to be applicable to physical processes. But we must realize that even if all these requirements are satisfied, we cannot be certain that any given physical problem can be solved in practice, even when the laws controlling the system are known. The fact is that the causes affecting the evolution of the phenomenon may be so numerous and varied that their transcription into mathematical form would yield a problem of insuperable difficulty. For example it would be idle for the mathematician to attempt long-range weather forecasting.

Let us then inquire into the further conditions that must be realized for mathematical solutions of physical problems to be obtainable in practice. A first condition is that, in the physical system of interest, the interacting parts which must be taken into consideration should not be too numerous. Now when we realize that our universe forms a unit in which each physical event influences every other physical event, we may well wonder how a sufficiently simple system can ever present itself for mathematical study. For example according to Newton's theory of gravitation, every mass in the universe attracts every other mass; we cannot therefore view the motion of a planet round the sun as due solely to the sun's attraction. A rigorous treatment would require that we take into consideration the actions of all the stars. Fortunately the attraction of the sun is so much greater in its effect than are any of the other influences that we can afford to disregard the other influences. Thanks to this fortunate circumstance the motion of the planet is furnished by the solution of a relatively simple mathematical problem. We conclude that one of the major reasons why mathematics can be applied successfully in practice is that, in many cases, only a very limited number of influences are predominant; the others may then be treated as non-existent. However, even when a simplification of this sort is permissible, the mathematical solution of a physical problem may still present insuperable difficulties. The relatively simple problem of N gravitating masses in celestial mechanics is an example.

The purely mathematical difficulties which may confront the theoretical physicist in his investigations are responsible for his dependence on the mathematical knowledge of the day and therefore on the discoveries of the pure mathematicians. In a general way the task of the theoretical physicist is to translate into mathematical language the physical problem of interest and the assumptions he has made. He then seeks to deduce the implications of his assumptions and hence of his theory. But to do so, he must solve a purely mathematical problem, and unless he can solve the problem, he is helpless.

To many, the mathematicization of physical phenomena is abhorrent. The contention is that mathematical analysis requires that we single out the quantitative features of a phenomenon, while the qualitative features are disregarded or else artificially reduced to quantitative differences. According to these objectors an undue stress is thereby placed on quantity, with the result that the outlook of the natural philosopher is biased.

The shallowness of such criticisms resides in a confusion between contemplation and investigation. Contemplation makes no demands on the powers of thought; it only affects our emotions, which by their very nature are imprecise and fleeting. In scientific investigation on the other hand, we seek to learn more than the immediately obvious, and since we are not demi-gods, we have to test our theories and expectations. But tests, to be of any value, must be precise. Vagueness necessarily covers a multitude of different possibilities. It is the desire to obtain accuracy which compels men to attach particular importance to quantity and to measurements. We may speak of an object as being light or heavy, but only when we have measured its precise weight in grams or in tons can we attribute any definite value to its lightness or its heaviness. If, for reasons of personal preference, we insist on describing Nature in terms of such vague qualitative attributes as heavy, soft, rough, green, or if we refuse to measure time under the plea that, by so doing, we are mutilating it, how shall we be able to establish any relations or laws that are sufficiently precise to predict anything definite? Our emancipation from quantity will thus have been purchased at the price of ignorance. Possibly in the dim future, the human species will have evolved to such an extent that man will gauge qualitative differences as accurately as he comprehends quantitative ones today. But until that time comes, we must be satisfied with the forms of knowledge at our disposal, and for this reason the stress must necessarily be placed on quantity in theoretical physics. The entire dispute, as we see it, arises from a conflict between emotion and reason. As a matter of fact, there is no justification for such a conflict, since both emotion and reason have their place in Nature and are not mutually exclusive; the important point, however, is not to confuse them.

CHAPTER III

THE SIGNIFICANCE OF THEORETICAL PHYSICS

PRIMARILY, the aim of a theory of mathematical physics is to coordinate and establish relations among observed phenomena. General laws are sought to unify isolated, disconnected facts. From the standpoint of the natural philosopher this aim in itself would be worthy of effort. But theories of mathematical physics, quite aside from their aesthetic and philosophic significance, have also a considerable practical interest. They often lead to the discovery of unsuspected phenomena, and on many occasions they throw light on matters which observation and experiment alone would be unable to elucidate.

Let us first give a few illustrations of new phenomena which were anticipated by various theories.

(1) Experiment shows that when a narrow beam of light is made to fall on a doubly refracting crystal, the beam is split into two separate beams which proceed in different directions. Hamilton, submitting this phenomenon of double refraction to mathematical investigation, proved that if the incident beam of light were to fall under a certain incidence, it should spread out into a cone on entering the crystal. This phenomenon, called "conical refraction," was unknown prior to Hamilton's calculations but was subsequently verified by experiment.

(2) Poisson deduced from Fresnel's undulatory theory of light that the shadow of a small disk might be represented by a bright patch surrounded by dark rings. The prediction was surprising but was soon verified.

(3) Maxwell predicted, as a result of his kinetic theory of gases, that the viscosity of a given gas at fixed temperature should be independent of the pressure. This fact, though unexpected, was subsequently proved correct.

(4) Lorentz predicted, on the basis of his electronic theory, that the light emitted by a monochromatic source and subjected to the action of a magnetic field should be split into three differently colored radiations. This phenomenon was verified by Zeeman; it is called the "Zeeman effect."

(5) De Broglie's theory suggested the possibility that a shower of electrons would be diffracted on passing through an aperture, much in the same way as waves of light. Diffraction phenomena of this type were detected later by Davisson and Germer.

These examples, though interesting, are less characteristic of the practical importance of theories of mathematical physics than are others we shall mention presently. It is true that they afford illustrations of cases where unsuspected phenomena are predicted and where experiments are suggested to test the correctness of the predictions. To this extent the theories concerned facilitate discovery. On the other hand, we may presume that the experiments suggested would eventually have been performed even in the absence of the theories. Someone, at some time or other, would accidently have made light fall on a crystal in the way requisite to yield the phenomenon of conical refraction predicted by Hamilton, so that this curious phenomenon would have been discovered in any case. Similarly for the Zeeman effect: Even the most casual observer could not fail to notice it by means of a modern spectroscope of high resolving power. Finally, de Broglie's theory had so little to do with the discovery of Davisson and Germer that these investigators made their important discovery by accident. We may add that the greater number of important experimental discoveries have been made by accident, or at least without the assistance of any mathematical theory. Radioactivity, X-rays, the diffraction of X-rays by crystals are phenomena which were discovered accidentally.

However, we may mention many illustrations in which the usefulness of physical theories is more apparent. For instance, before Newton constructed his theory, astronomers could have foretold the future position of a planet by applying Kepler's laws. But they could not have predicted the path and motion a stray comet would pursue on entering the solar system. Thanks to Newton's theory this information became available. Furthermore, as we mentioned on a previous page, the theory of gravitation leads to a correction of Kepler's laws and to a recognition of their approximate nature. Theory in this case remedies the imperfections of observation. It is also by relying on the theory of gravitation that we can compute the masses of the various bodies of the solar system. In the absence of the theory of gravitation, we should be helpless. Much of our knowledge in modern physics (for instance the mass of the electron) is obtained indirectly by taking into consideration some mathematical theory. Other illustrations follow:

Michelson, in one of his experiments, sought to measure the velocity of light passing through carbon disulphide. His experiment consisted in directing a beam of light on an aperture that could be opened and closed by a shutter. As the shutter was rapidly opened and closed, a jet, or "packet," of light passed through the hole and was transmitted through the carbon disulphide. Michelson assumed that the velocity with which this packet was propagated was the velocity of the waves of light through the medium. Great was his surprise, therefore, when he found that the velocity of the packet of light fell considerably below the velocity that light was expected to have in carbon disulphide. Michelson's result was so unexpected that physicists were at a loss to account for it.

The solution of the puzzle was given by a theoretical physicist, the late Lord Rayleigh. Rayleigh made a mathematical study of the transmission of wave motions and found that in some cases a packet of waves might advance with a velocity that differed considerably from that of the wave crests. Two different velocities were thus uncovered: the velocity of the packet, or the "group velocity," and the velocity of the individual wave crests, or the "wave velocity." In the case of waves of light transmitted through a highly dispersive medium, like carbon disulphide, Rayleigh's investigations showed that the velocity of a packet of light (the group velocity) should fall considerably below the velocity of advance of the individual waves (the wave velocity). Calculation then proved that the group velocity coincided precisely with the velocity measured by Michelson. The error was not in Michelson's measurements, but in his interpretation of his experiment. Unwittingly, he had measured the wrong velocity, namely, the group velocity, instead of the wave velocity (commonly called the velocity of light), which it was his intention to measure. At the same time, Rayleigh's investigations showed that the earlier measurements of the velocity of light, performed by Römer, Foucault, and Fizeau, suffered from the same confusion. In each case the group velocity, not the wave velocity, was measured. Thus, thanks to Rayleigh's theoretical investigations a new interpretation was placed on the significance of certain experiments, and the misconceptions of the experimenters were corrected. A more detailed analysis of Rayleigh's wave packets will be given in Chapter XIX.

A second example of the practical utility of physical theories concerns the principle of entropy. Common experience corroborates this principle. For instance, we have no reason to suppose that heat can leave a piece of ice and warm a neighboring fire. Yet, when Boltzmann developed the kinetic theory of gases, he showed that even if this theory were only approximately correct, the principle of entropy could not possibly have absolute validity. As may well be gathered, this discovery has considerable importance, and we note that it was arrived at by means of a mathematical theory and not by direct experiment. As a third example, we may mention the microscope. The construction of the microscope was revolutionized by Abbe, not by tentative modifications here and there, but as a result of an application of the undulatory theory of light.

CHAPTER IV

TWO KINDS OF SIMPLICITY

WHEN we say that one theory of mathematical physics is simpler than another, the epithet "simple" may be interpreted in two ways.

(a) *Simplicity in the Assumptions.* Initial assumptions are inevitable in any physical theory, for some assumptions must always be made when the physical facts are transcribed into mathematical language. But though initial assumptions are necessary, they must not be too numerous; and above all we must not introduce a new assumption to account for each new experimental fact. In more technical language, this last reservation is expressed by the statement that hypotheses *ad hoc* must be avoided. A physical theory which satisfies the foregoing restrictions is said to be "simple in its assumptions."

There are two reasons for requiring this form of simplicity. First let us recall that the aim of a physical theory is to deduce mathematically the implications of a given body of assumptions, or hypotheses. If, then, to the fundamental hypotheses others are added in the course of our deductions, we shall be unable to extrapolate the theory to any extent, for we shall never be assured that the introduction of new hypotheses is at an end. A second reason why theories that contain too large a number of hypotheses *ad hoc* must be rejected is that we can account for anything we wish if we allow ourselves the privilege of introducing corrective assumptions as and when needed. A theory that accounts for anything we choose is useless.

These considerations are illustrated in a theory tentatively advanced by Jeans at the Solvay Congress of 1911. The radiation which experiment proves to be present in a heated enclosure had been accounted for by Planck by means of his revolutionary quantum theory. To obviate the strange quantum theory, Jeans, supported by Lord Rayleigh, claimed that the radiation could be interpreted along classical lines if we assumed that the enclosure were to leak in an appropriate way. He pictured the situation by imagining an enclosure connected to pipes, and he assumed that these pipes leaked in such a manner as to account for the facts observed.

Commenting on Jeans's suggestion, Poincaré made the following observation:

"It is obvious that, by giving suitable dimensions to the connecting pipes between the enclosures and suitable magnitudes to the leaks, Jeans can account for any experimental result. But this is not the rôle of physical theories. Physical theories must not introduce as many arbitrary constants as there are phenomena to be accounted for; they must establish connections among the various experimental facts and, above all, must lead to predictions." *

Jeans's suggestion was not considered further, and Planck's theory was preferred in spite of its unfamiliar concepts.

Many are the physical theories that have been abandoned because of defects of the foregoing type. Usually, however, such defects do not appear in the early stages of a theory; only as the theory develops and is confronted with new facts does the necessity for introducing additional hypotheses *ad hoc* arise. The theory then becomes cumbersome and must be remodelled from its foundations. Bohr's theory of the atom is a case in point. We shall see nevertheless that the scrapping of a theory does not mean that the theory has been useless. Some of its more important results will remain, and usually it serves as a guide to the more accurate theory that supersedes it. The urge to attain simplicity in the assumptions is not characteristic solely of mathematical physics. In our commonplace attempts to coordinate facts, we also follow the same urge, and we usually prefer a simple, direct explanation to an involved one.

(b) *Mathematical Simplicity*. The kind of simplicity we have just discussed is an essential requirement in any theory of mathematical physics. But we must now mention another important kind of simplicity. It may be understood when we say that the theoretical physicist (and for that matter also the experimenter and the observer) operates as though he believed in the mathematical simplicity of Nature.

For example, Kepler concluded from Tycho Brahé's observations that the planets described those mathematically simple curves called ellipses. Yet Kepler must have realized that Tycho Brahé's observations could boast only of a limited measure of accuracy, and that orbits differing slightly from perfect ellipses would also be consistent with Brahé's findings. Nevertheless Kepler chose the mathematically simple curves. It is true that he was a firm believer in the simplicity of Nature, and that for this reason he would presumably have dismissed complicated orbits as impossible. But this is not the point at issue. Today we fully realize

---

* Solvay Congress of 1911.

that Nature is not simple, and yet any modern astronomer would follow the same course as Kepler. In any case if, in view of the inaccuracies of observation, the evolution of a phenomenon may be represented equally well by many different graph lines, one of which is mathematically simpler than the others, the simpler graph is preferred.

This same rule is observed in theoretical physics: The simplest mathematical solution is the one that is accepted. Newton started from Kepler's laws and found that, in accordance with his own mechanical assumptions, the sun must attract the planets with a force varying inversely to the square of the distance. There is no doubt that this is the only law that can be accepted if we assume Kepler's laws to be rigorous. But Newton on developing his theory was compelled subsequently to recognize that Kepler's laws could not be rigorous. As a result the necessity for retaining the law of the inverse square was no longer present, and Newton might just as well have selected any one of a vast number of other laws, differing slightly from the classical one and yet satisfying the requirements of the astronomical observations of the day. All other laws, however, would have been more complicated from the standpoint of the mathematician, and, besides, it would have been gratuitous to select one of them rather than the other. Newton accordingly retained the law of the inverse square—the simplest of them all. Only two centuries later, after more accurate observations of the planetary motions over long periods of time had been performed, was the classical law suspected of being incorrect. The more refined astronomical observations, in conjunction with the discoveries in electromagnetics, then gave rise to the general theory of relativity. Here, also, Einstein pursued the same method and chose the simplest law among the ones compatible with his assumptions and with the facts of observation. These considerations show that the modern scientist, who no longer believes in the inherent simplicity of Nature, can claim no theoretical justification for singling out the simplest mathematical law in any particular case. His urge is purely pragmatic and arises from the fact that the mathematical implications of a simple law are more easily obtained than are those of a complicated one.

The two types of simplicity we have discussed under the headings (a) and (b) have often proved confusing to the layman. There is some justification for this confusion, because we often find a competent author stating that Newton's law of gravitation is simpler than Einstein's and then a few pages later making the opposite statement. Both statements are correct, but they do not refer to the same type of simplicity. For instance, suppose we are called upon to determine, in accordance with

Newton's law, the motions of the two stars of a double-star system (Two-Body Problem). The solution is exceedingly simple; it was given by Newton himself in spite of the limited mathematical knowledge of his time. But suppose we wish to solve this same problem by utilizing Einstein's law of gravitation. The problem is so difficult that it has never been solved. The reason is that Newton's law is mathematically simpler than Einstein's; and hence from the standpoint (b) (of mathematical simplicity) Newton's law is the simpler.

On the other hand, when account is taken of certain astronomical phenomena unknown to Newton, we are justified in saying that from the standpoint (a) (of simplicity in the assumptions) Einstein's law is the simpler. Thus Einstein's law accounts for the peculiar motion of the planet Mercury, for the red shift in the lines of the solar spectrum, and for the precise bending of a ray of star-light that grazes the sun's limb. Newton's law accounts for none of these phenomena unless we supplement it with suitable assumptions devised for the occasion. But the complications we thereby introduce imply that the classical Newtonian theory, when confronted with the more recently discovered astronomical facts, is lacking in that form of simplicity which we called "simplicity in the assumptions." Obviously in this respect, Einstein's law of gravitation is the simpler.

CHAPTER V

SUCCESSIVE THEORIES OF MATHEMATICAL PHYSICS ARE PROGRESSIVE APPROXIMATIONS

CLOSELY allied with the ideas of simplicity developed in the foregoing chapter, is the fact that the successive theories of physics pertaining to the same field are progressive approximations. As we shall see, the implications of this situation have a heuristic value in the theory of relativity and in the quantum theory. Earlier experimenters gathered their information with crude instruments, and sometimes with no instruments at all. In addition, all the experiments they were able to make were performed under conditions that deviated only slightly from those with which we come in contact in our commonplace existence. Temperatures ranged from that of ice to that of fire; pressures were of the order of that of the atmosphere; velocities were low, far less than that of light; masses were of average magnitude, neither physically infinitesimal nor enormous. We may designate this range of experience as the commonplace level. Then, as knowledge increased and the technique of experimentation became more perfect, other levels were explored. Such levels refer, for instance, to temperatures in the neighborhood of the absolute zero, to exceedingly high vacua, to enormous pressures, to velocities nearing that of light, to masses that are physically infinitesimal. We shall call these levels the remoter levels. Even today there are levels which have never been explored. As yet we cannot experiment under the enormous pressures and high temperatures that are thought to reign in the interior of the stars.

The earlier experimenters were conversant solely with facts gathered from the commonplace level, and the aim of the earlier physical theories was to coordinate these commonplace facts. But a physical theory does not restrict itself to a coordination of what is already known; it also predicts other phenomena, many of which may pertain to the remoter levels of experience. Until such time therefore as these remoter levels are explored, we may be unable to submit the earlier theories to any decisive test. In practically every case the exploration of the remoter levels has necessitated a revision of the earlier theories, and the result has been that the history of theoretical physics is a story of successively discarded theories.

What we now propose to show is that although the successive theories of a given class may differ considerably in the concepts they introduce, yet from the mathematical point of view each theory is merely a refinement of its predecessor. The successive theories thus form a chain of progressive approximations.

Since the evolution of the law of gravitation is one of the simplest to follow, we shall devote some space to its study. Galileo's observations were restricted to regions in close proximity to the earth's surface; hence the experiments of his day showed no trace of any decrease in the weight of a body when its elevation was increased. The natural inference was that the gravitational field developed by the earth was the same at all elevations.

Newton shattered this belief when he proved that the intensity of the field decreases as the elevation is increased. But what is important for our present purpose is to examine the method that led Newton to his discovery. Newton could not have corrected Galileo had he merely confined his attention to the experiments of his great predecessor. Even today no experiment could detect a difference in the weight of a body when situated on the earth's surface or at a few feet in elevation. Only by seeking to coordinate a larger number of facts drawn from a remoter level, notably, by extending his observations to the moon and planets, was Newton able to progress.

Another point to be noted is that Newton's law of the inverse square passes over into a law of constant force when we restrict our attention to points in the immediate neighborhood of the earth's surface. The reason is that, whether a body be resting on the surface of the ground or be set at the top of a high tower, the difference in its distance to the earth's centre is negligible in comparison with the radius of our planet. Newton's law thus requires that the weight of a body should be practically the same in the two situations. Hence we are justified in saying that Galileo's belief expressed a first approximation to truth, an approximation which tended to become rigorous for small elevations. In short, Newton's law of the inverse square appears as a refinement of Galileo's idea.

For more than two centuries Newton's correction was deemed adequate. Then came the general theory of relativity. Newton's law was found in turn to be but an approximation, a far better one than that of Galileo, but an approximation nevertheless. The correction of Newton by Einstein came about in exactly the same way as that of Galileo by Newton. It was not by performing more refined measurements on the planetary motions or on the fall of bodies that Einstein corrected Newton's theory; it was by exploring new levels and extending thereby the range of the facts to be coordinated. The general theory of relativity is indeed but a plausible outgrowth of the special theory, and this latter theory is based on facts unknown to Newton; namely, on the results of highly refined electromagnetic experiments and on the behavior of beta-particles moving with speeds comparable to that of light. We are here in the remote level of extremely high velocities. As is well known to students of the mathematical theory of relativity, Einstein's law of gravitation differs less and less from Newton's as the intensity of the gravitational field and the velocities of the bodies concerned become smaller and smaller. In other words Newton's law tends to become correct under such limiting conditions. For this reason Einstein's law of gravitation may be regarded as a refinement of Newton's law, just as the latter is a refinement of Galileo's belief.

We have therefore to inquire whether Einstein's theory is the final revision or whether it is itself only a more refined approximation. Quantum mechanics proves conclusively that the theory of relativity cannot be the last word. Hence we might suppose that further investigation would confront us with an unending series of more and more accurate theories embracing an ever widening array of facts. Absolute truth and a final picture would be beyond our reach. A slight modification of this statement must be made if we accept the conclusions of the leading quantum theorists. According to Heisenberg, Bohr, Born, and Dirac, the scale of successive approximations will not extend indefinitely, for, owing to the Uncertainty Principle, the limit of accurate knowledge has already been reached in the quantum theory. Beyond, lies a fog of uncertainty due to the peculiarities of Nature herself. All that we can know are probabilities. It is unnecessary at this stage to venture an opinion on the merits of the new quantum philosophy, for in any case one conclusion remains unchanged: absolute truth is beyond our reach.

Let us now examine the implications that result from the progressive approximations of scientific theories. We have said that, when we pass from the remoter levels of the high velocities and enormous gravitational fields (in which the discrepancies between Newton and Einstein become perceptible) to the more commonplace level considered by Newton, the theory of relativity gradually merges into Newtonian science. Also, if we concentrate on the still more commonplace level near the earth's surface, investigated by Galileo, we find that both the relativity and the Newtonian theories merge into Galileo's beliefs.

Thus is illustrated an elementary rule of great importance in theoretical science. It may be condensed into the following statement: A more accurate theory passes over successively into the less accurate ones that have preceded it, as we step down from the remoter levels to the more commonplace ones. This statement is general, and its justification may be understood when we recall that the more accurate theories are mere refinements of the preceding ones and are arrived at by considering wider and wider arrays of facts. Numerous illustrations of this general rule may be found in the history of science. Let us mention some of them.

In the theory of relativity, gravitational forces, when compounded, do not obey the elementary rule of the parallelogram, as Newton believed. This is so because Einstein's gravitational equations are not linear. Yet, if we assume the forces to become weaker and weaker, the rule of the parallelogram holds in the limiting case. Similarly if, instead of considering a fast-moving planet like Mercury, we are dealing with one moving more leisurely with respect to the sun, e.g., Neptune, Einstein's law of gravitation approximates more and more to Newton's law of the inverse square.

The curious rule for the addition of velocities in the special theory of relativity passes over into the classical rule when the relative velocities involved become smaller and smaller. Simultaneity tends to become absolute, and 4-dimensional space-time passes over into separate space and time.

In the new statistics of Bose-Einstein and of Fermi-Dirac we also find that the formulae pass over into the classical formulae of Boltzmann's kinetic theory when the temperature and pressure are assumed to approximate to their more habitual commonplace values.

These examples also show why textbooks continue to utilize Newtonian mechanics even though their authors know that relativity mechanics is a better approximation. The reason is that, for the small velocities with which we are usually concerned, the two mechanics yield results so nearly alike that in practice no experiment would be sufficiently refined to detect their difference. And since Newtonian mechanics is mathematically the simpler, as explained in the previous chapter, there is every advantage in retaining it.

Heuristic Considerations—Of invaluable help in discovering the laws of the newer theories is the realization that the successive theories must become indistinguishable when the commonplace level is neared. In the special theory of relativity for instance, one of the fundamental assumptions is that all natural laws should satisfy a certain condition of invariance.* This requirement rules out the classical mechanical laws

---

* Invariance under the Lorentz transformation.

of Galileo and Newton.   Unfortunately it leaves open a wide choice of conceivable mechanical laws and does not restrict us to any specific law. We might, of course, avail ourselves of the experimental measurements performed on swiftly-moving beta-particles, and from these experiments derive the amended mechanical laws.   But the experiments are so difficult and so scant that their results could not furnish a sufficient basis for the formulation of exact laws.   It is here that the heuristic considerations come into play.   Thus we know that the mechanical laws of the special theory of relativity must tend to merge into the classical ones when we are dealing with situations on the commonplace level (low velocities). This clue, taken in conjunction with the general requirements of invariance of the theory of relativity, suffices to establish the new mechanical laws.

Precisely the same situation occurs in the general theory of relativity. Here the requirements of invariance show that Newton's law of gravitation cannot be correct, and, as before, these requirements are not sufficiently stringent to impose any specific amended law.   But when we remember that for low velocities and weak gravitational fields the new law must pass over into the classical law of the inverse square, we are furnished with a clue that enables us to obtain the amended law of gravitation.

More so than the theory of relativity, does the quantum theory offer examples of the application of the method just outlined.   In the quantum theory we are no longer guided by the general requirements of invariance, which play so large a part in the theory of relativity; and so more groping in the dark must precede our formulation of the quantum laws. Before showing how the quantum theorists proceeded, we must mention that the refinements of the quantum theory become noticeable only when we pass from the commonplace level to the level of very small masses, the level of the electron for example.   More precisely the level we refer to is the one in which the "action" * of the mechanical system of interest is exceedingly small.   If, then, we follow the inverse course, passing from the remoter level to the level of common experience, we must expect the quantum laws to pass over into the classical ones.   Thus, the same heuristic clue, which we found so effective in the theory of relativity, is also available in the quantum theory.   But whereas in Einstein's theory no particular name was given to this clue, in the quantum theory a name has been coined.   The clue is called "Bohr's Correspondence Principle,"

---

* Action is a technical concept that it has been found advantageous to introduce into dynamics.

and it is of considerable importance in the formulation of the quantum laws.

A Comparison of the Levels of the Theory of Relativity and of the Quantum Theory—Both the relativity and the quantum theories have issued from the necessity of amending the classical laws when remoter levels were considered. This prompts us to inquire: Which of the two levels, that of relativity or of the quantum theory, is the more remote from the commonplace level? If we can answer this question, we shall have determined which of the two theories is the more accurate. Unfortunately no straightforward answer can be given because the levels involved are complementary rather than successive. The level of high velocities and powerful gravitational fields (for which the refinements of the theory of relativity are needed) is totally different from the level of microscopic masses and microscopic quantities of action characteristic of the quantum theory. And so, if we have average masses moving at enormous speeds, we must apply the relativity theory but may disregard the quantum theory. Conversely, if our interest lies with microscopic masses moving at low velocities, we must have recourse to the quantum theory and may disregard the relativity theory. It is obvious, however, that a theory, more rigorous than either of the two theories individually, would result from their fusion. Such a theory would correspond to the level relating simultaneously to high velocities, to microscopic masses, and to microscopic quantities of action.

The first attempt to combine the two theories was made by Sommerfeld when he applied relativity mechanics to Bohr's atom. But Sommerfeld's results have lost much of their interest, because Bohr's atom has since been shown to be only an approximation to the more general concepts developed in the quantum mechanics of Heisenberg, Dirac, and Schrödinger. An amalgamation of the more advanced quantum theories with the theory of relativity was effected subsequently by Dirac in his celebrated paper on the nature of the electron. The two theories here impose mutual refinements one on the other. Dirac's theory has since inspired other investigators.

CHAPTER VI

CONCEPTUAL CHANGES IN PHYSICAL THEORIES

IN THE previous chapter we have seen that the successive physical theories proceed by progressive mathematical approximations. But we must be careful to understand that this statement is intended to apply more particularly to the mathematical equations which form, so to speak, the scaffolding of the theory. The revision of concepts, when we pass from one theory and level to another, is usually much more drastic. For instance, according to Newtonian science, the planets are subjected to forces pulling them towards the sun—an example of action at a distance. The general theory of relativity, though but a mathematical refinement of the Newtonian theory, rules out completely any such action at a distance. We must now say that the planets follow the geodesics of 4-dimensional space-time which is warped in the neighborhood of the sun. It is this warped structure of space-time, in the immediate vicinity of the point occupied by the planet, that guides the planet along a curve from one instant to another; and Einstein's law of gravitation expresses the general nature of the space-time curvature, or warping. As may be gathered, the conceptual departure from the Newtonian scheme is very great.

In a number of cases concepts which appear natural on the commonplace level lose all meaning on the remoter levels. One of the simplest examples is afforded by the concept of temperature. On the commonplace level, warmth and cold appeal directly to our senses, and the concept of temperature is most useful in a science like thermodynamics, which deals largely with commonplace phenomena. But when we pass to the level of molecular dimensions, the entire concept of temperature becomes meaningless. There is no sense in speaking of the temperature of a single molecule, for a large number of molecules in motion is necessary to give meaning to temperature. And just as some concepts may become meaningless when we pass to remoter levels, so also new concepts may arise. We have only to mention the 4-dimensional space-time of relativity. Here we have a blending of space and time into a 4-dimensional continuum, whereas on the commonplace level space and time appear to be entirely separate. We cannot usually give expression in words to new concepts that differ widely from those of the commonplace level. Only by perseverance and prolonged meditation can an idea of what is implied force itself upon us. To impediments of this sort most of the leading theoretical physicists ascribe the present conceptual difficulties in the quantum theory. As Bohr writes:

"In the adaptation of the relativity requirement to the quantum postulate, we must therefore be prepared to meet with a renunciation as to visualization in the ordinary sense going still further than in the formulation of the quantum laws considered here. Indeed, we find ourselves here on the very path taken by Einstein of adapting our modes of perception borrowed from the sensations to the gradually deepening knowledge of the laws of Nature. The hindrances met with on this path originate above all in the fact that, so to say, every word in the language refers to our ordinary perception." *

Even mathematics, which had long been thought immune from conceptual changes of this sort, appears today to be entering the same troubled region. According to Brouwer, the logical principle of excluded middle, which appears so inevitable in our everyday life, must be rejected where infinite aggregates are concerned. The exalted laws of logic, long assumed to be the laws of the human mind, thus appear to be mere empirical laws, subject to revision as the field of our mathematical experience is widened.

Modern physics owes a large part of its difficulty to the apparent inappropriateness of many of our former concepts; and to enlarge and modify these concepts much of the labor of modern theoretical physicists is directed. The necessity for enlarging our ideas in natural philosophy is not surprising. To take a simple illustration, we have but to imagine a man who, after having lived for years in his home town out of contact with the world at large, decided to travel. He would discover strange people, climates, trees, and animals, of which he had never dreamed, and he would return with a less dogmatic viewpoint. His views would conflict with those of his fellow citizens who had remained at home, believing the entire world to be like the little world of their limited experience—a belief they would doubtless back up with arguments they would call "logical." It is the same in natural philosophy. Cooped as we are in the world of common experience, we must endeavor to break away and explore the remoter levels, and construct theories befitting not solely the commonplace levels, but also all those we have explored. Only thus can knowledge (limited at that) be acquired. Those who have failed to explore will naturally construct theories compatible with their restricted

---

* Atomic Theory and the Description of Nature; Cambridge, 1934; p. 90.

experience but entirely out of touch with the wider fields which they ignore.

The Greeks have given an illustration of the stagnation of knowledge that results from a restricted viewpoint, and it will always be to the immortal glory of Galileo and Newton that they founded the methods of investigation which initiated intellectual emancipation. At the same time, the restricted viewpoint accounts for the feeling of diffidence which the layman experiences when informed of many of the conclusions of the modern theories, e.g., the paradoxes of the theory of relativity. He feels that these conclusions conflict too strongly with the notions he has formed from his long association with the commonplace level, and, like the man who has never left his home town, he believes that the whole unexplored universe should conform to the ideas he happens to cherish. His arguments are vain, for we have seen that there is no real conflict between the disclosures of the more advanced theories and those drawn from our daily activities: the former tend to merge into the latter when we pass from the remoter levels to the level of common experience. An apparent conflict arises only when we pursue the unjustifiable course of assuming that all the levels are mere repetitions of the one we know best. The greatest revolutions in modern thought have issued from a study of the remoter levels, and they would never have arisen had these levels been ignored.

CHAPTER VII

CAUSALITY

COMMON observation suggests the presence of causal connections in inorganic Nature; even the savage or child acts on the assumption that the same situations produce the same effects. In Greek antiquity there arose the doctrine of fate and destiny. We find the doctrine expressed in the tragedy of "Oedipus Rex," and it survives to the present day in such expressions as "being born under a lucky star." But this form of predetermination is vague and fails to specify exactly when and how the fortunate or unfortunate events will occur. Furthermore no unbroken causal chain extending from the initial event to its consequences is indicated.

The first rigorous formulation of the doctrine of causality and its mathematical expression by means of a differential equation are due to Newton. Until recent times Newton's presentation has satisfied the requirements of natural philosophy. Whether the doctrine can still be entertained or whether it must be rejected entirely is a moot point, the discussion of which will be taken up in later chapters.

The popular understanding of the doctrine of causality may be condensed into the statement: "The same causes generate the same effects." So vague a statement can be of little use in a rigorous scientific discussion. For instance, a stone is released and falls to the ground. What constitutes the cause of the fall? Is it the gravitational pull of the earth, or are we to regard the gravitational pull as given, and identify the cause with the act of releasing the stone? We might also suppose that the release of the stone and the gravitational pull are to be regarded jointly as representing the cause. Further definitions are obviously needed before we can agree on what constitutes the cause in any particular case.

Nevertheless the popular idea of causality may serve as an introduction to the scientific formulation of the doctrine. The doctrine necessarily embodies a belief in the presence of order and regularity in Nature, for without these characteristics the same causes would not always give rise to the same effects. In the scientific approach, regularity in Nature is connected with the existence of natural laws, which govern the different classes of phenomena (*e.g.*, the law of gravitation). The natural laws,

45

however, owing to their generality, impose only a limitation on possibilities and do not map out any definite course for future events. For this reason, even if we are acquainted with the law which controls the workings of a system, we cannot predict the evolution of the system. To determine the latter, we must supplement the general law with additional information. We shall illustrate these features by discussing a mechanical system.

Suppose our system is formed of point-masses moving freely under their mutual gravitational attractions. The general laws controlling the evolution of the system are known: they are the laws of mechanics and of gravitation. We wish now to determine how the configuration of the system will change in the course of time. The laws alone cannot give this information, for the future configurations of the system are necessarily affected by the initial configuration from which we start. Additional information is thus required. The additional information which suffices to make the problem determinate may be expressed in various ways, but there is one mode of expression which harmonizes with the requirements of the doctrine of causality. It consists in the specification of the positions and of the velocities (in magnitude and in direction, as referred to a Galilean frame) of the several point-masses at some initial instant. The additional information, when expressed in this way, is usually said to constitute the *initial conditions* of the system. When the initial conditions are specified, the application of the general laws permits us (in theory) to predict the configuration of the system at any future instant, and also to determine its past configurations.

The contingency of the initial conditions is in marked contrast with the permanency of the laws. Thus we may assign any initial positions and velocities we choose to the point-masses at the initial instant, and, according to the initial conditions selected, the evolution of the system will proceed along one line or another. But the laws governing the evolution will always be the same. The instant of time which we have referred to as the initial instant may be chosen arbitrarily, and we may assume it to be the present instant or some past instant or even some instant in the future. This last choice might, however, lead to confusion, and we shall not consider it for the present. It is often convenient to say that the initial conditions define the initial *state* of the system. The initial state of a mechanical system of point-masses is therefore represented by the positions and velocities of all the point-masses at the initial instant. More generally the state of the mechanical system at any instant—past, present, or future—is defined by the positions and velocities at the instant of interest. Generalizing from mechanics to all physical systems, we may formulate the doctrine of causality as follows:

The evolution of every physical system is controlled by rigorous laws. These, taken in conjunction with the initial state of the system (assumed to be isolated), determine without ambiguity all future states and also all past ones. The entire history of the system throughout time is thus determined by the laws and by the initial state.

The scientific presentation of the doctrine of causality removes much of the vagueness which surrounds the commonplace concept of "cause." Let us revert to the example of the falling stone. The cause of the fall might popularly be identified with the act of releasing the stone, or with the law of gravitation, or with both. Thus, the cause might be the initial state, or the law controlling the evolution of the system, or both. Usually, however, in the ordinary vernacular, a "cause" refers to an instantaneous event; and if we accept this restriction, the law, which acts over a protracted period of time, cannot be identified with the cause. The sole candidate for this title is then the initial state of the system. We shall therefore identify the cause with the initial state. As a result of this identification, the law controlling the evolution of a phenomenon is seen to act as the connecting link between the cause and its effects: it represents the causal chain.

Several points in our statement of the scientific doctrine of causality require elucidation.

1. Let us assume that the initial instant is the present one. The scientific doctrine of causality then implies that the present state determines not only the future states, but also the past ones. However, we are violating the customary use of words when we speak of the present as determining the past; for this mode of expression would imply that a present condition is the cause of past events, whereas in the commonplace understanding of causality, the passage of time plays an important part, and it is customary to view the cause as preceding the effect. We may, however, attenuate the conflict between the scientific doctrine and the more familiar conception by noting that the statement "the present determines the past" is equivalently expressed by saying: "If the present state is such and such, the past state must necessarily have been so and so."

It is easy enough to give any number of illustrations which seem to refute the scientific law of causality in its reference to the past. For instance a small piece of ice is thrown into hot water. We know that the ice will melt and disappear and the water attain a lower temperature. There is no difficulty in drawing conclusions in this case because the conclusions concern the future. But suppose now we wish to reverse the order of our deductions and discover the past from the present state of the water. This appears impossible, for how can we possibly tell whether there ever was any ice in the water? The same present state of the water might have been attained in many different ways. These conclusions, however, are too hasty. To investigate the situation properly, we must take into consideration the microscopic constitution of the water. Thus, if we view the liquid as a mechanical system formed of myriads of water molecules, its present state is defined not by the uniform temperature of the mass of water, but by the positions and velocities of all the individual molecules; and we may perfectly well suppose that the present state, when defined in this way, determines the past history of the water. At all events, none of the commonplace objections can refute this view.

2. In ordinary conversation we say that the present is the cause of the future. But the present to which we refer is not necessarily assumed to consist in a mathematical instant of time. Usually, by the word *present* we wish to imply a more or less protracted period. For instance we receive a telegram and act accordingly. The reception of the telegram is the cause of our subsequent actions. Yet it is not the actual reception of the telegram that dictates our future course; it is the reading and the comprehending of the message—and this reading is not instantaneous. In the scientific formulation of the law of causality, the situation is different, for the cause (such as we have defined it) refers to the initial state and is thus an instantaneous event. We might, if we so chose, extend the scientific doctrine so as to cover cases where the initial conditions would no longer be instantaneous but would be spread over some interval of time. This course has not been followed, however.

3. We have said that the initial state determines all other states. But the fact that the initial state determines all others does not necessarily mean that a knowledge of the initial state automatically confers a knowledge of the evolution of the system. No information can be derived unless we are acquainted with the general laws which control the workings of the system. In practice, even a knowledge of the laws might prove useless, for, in order to make use of this knowledge in any given situation, we should have to solve a mathematical problem whose difficulty might defeat all efforts.

4. Our allusion to an isolated system in the statement of the law of causality is readily explained. If our system is not isolated, it is subjected to the influences of the outside world, and we are no longer justified in saying that the history of the system is determined solely by the initial state and the internal laws. To overcome this difficulty, we must enlarge the system so that it will comprise the external influences alluded to.

Proceeding in this way, we should eventually have to consider the entire universe, for no limited portion of the universe can be regarded as strictly isolated. We conclude that the causal doctrine (if it be valid at all) can apply only to the entire universe. Laplace, in his presentation of the doctrine, stressed this last feature. In practice, however, some limited systems, such as the solar system, may be viewed as very approximately isolated, and it is with such systems that we shall be concerned.

5. We have illustrated the principle of causality in connection with mechanical systems formed of point-masses, and have seen that in such systems a state is defined by the positions and velocities of all the masses at the instant of interest. This definition of a state is necessarily peculiar to mechanical systems, and we must now give a more general definition which will apply to physical systems of all kinds. The general definition of a state is the following: The state of a physical system is specified by the aggregate of those items of information which pertain to the system at any given instant and which, when taken in conjunction with the laws controlling the system, suffice to determine its history.

Let us now examine the procedure which will furnish the items of information necessary to determine a state in any specific physical system. The first step will be to establish the laws which control the evolution of the system and to express these laws in the form of differential equations. Now a characteristic of differential equations is to define restrictions which are not sufficiently comprehensive to determine any precise result. (This feature is indeed one of the reasons which make differential equations so suitable for the expression of natural laws.) To render determinate the implications of a differential equation, we must furnish complementary specifications. A study of the differential equation shows what forms these specifications must assume. Suppose, then, that we have translated the laws controlling the system of interest into the mathematical form of differential equations. If we can establish the specifications which render these equations determinate and which refer to conditions holding at an instant of time, we shall automatically have determined the general items of information which define a "state" of our system. The initial state of the system will then be expressed, as usual, by the aggregate of these items of information at an initial instant. It was by following this method that the conditions defining the initial state of a mechanical system were discovered by the founders of mechanics.

The definition of the initial conditions, or initial state, varies with the physical system considered. For example, let us suppose we are dealing with the flow of heat by conduction through an infinitely extended solid. Fourier established the law of heat conduction and expressed it by means of a differential equation (Fourier's equation). A study of this equation then shows that the initial state of our system is defined by the temperature distribution throughout the solid at the initial instant. In the case of electromagnetic phenomena occurring *in vacuo,* the laws, or equations, were discovered by Maxwell. From the form of these equations it can be proved that the initial distribution of the electric and of the magnetic vectors throughout infinite space defines the initial conditions, or initial state.

But the situation is not always so simple. If electrons are present, Lorentz's electromagnetic equations show that for the future to be determined, not only must we specify the initial values of the electric and magnetic vectors, but that we must also specify the motions of the electrons throughout a certain interval of time. The possibility of defining an initial state thus eludes us, and we may no longer say that the evolution of the system is determined by any initial conditions. This situation seems strange, for it implies that the causal doctrine is no longer valid. More attentive consideration shows, however, that the causal doctrine is not endangered. The fact is that, when electrons are present, Lorentz's electromagnetic equations are incomplete because there are less equations than variables. In more familiar language, the laws are incompletely stated. If these laws were rendered complete, an initial state could be defined and the evolution of the system could be determined without our having to specify the motions of the electrons throughout a certain time. Thus, the requirements of the causal doctrine would be satisfied. Incidentally, we may mention that various attempts were made to complete Lorentz's laws. Such attempts have been abandoned, however, on account of their speculative nature.

A somewhat similar illustration is afforded by the theory of elasticity. In some cases the equations, or laws, require that the entire past history of the elastic body be specified before the future evolution can be gauged. A situation of this sort seems to conflict with the simple conception of causality that we have been discussing, since one of the characteristics of this conception is to assume that the initial state which determines the evolution is an instantaneous one. The difficulty is overcome when we find that the past history of the elastic body may equivalently be expressed by a number of separate items of information at an initial instant.

Tests of the Doctrine of Causality—Our analysis shows that even if our measurements were perfect, an accurate test of the doctrine of causality would require that we experiment on a truly isolated system. Since, however, the only truly isolated system of which we can conceive is the entire universe and since it would be impossible to experiment on that system, we conclude that in any case an accurate test of the doctrine of causality cannot be performed.

We may, however, be satisfied with experiments on a system of manageable proportions which is approximately isolated. For instance, experiments performed in the laboratory appear unaffected by the changing conditions of the universe at large. But here another difficulty confronts us. In any observation we necessarily introduce extraneous factors into the system (*e.g.*, our measuring instruments), and these disturb the system and destroy its isolation. Thus we may determine the temperature of a liquid by placing a thermometer in it. But unless the thermometer happens to be at the same temperature as the liquid, it will heat or cool the latter, with the result that the temperature finally read will not be the original temperature of the liquid. Furthermore, we know that human measurements are necessarily imperfect, so that even if they did not disturb the magnitudes to be measured, they would still yield only approximate results.

In view of the practical impossibility of testing the doctrine of strict causality by experiment, we must regard this doctrine as a dogma which may be accepted by some and rejected by others. Classical science was well aware of these elementary difficulties, but the contention was that although rigorous causality could not be tested, approximate causality had proved itself to be valid in experiments conducted on approximately isolated systems. The objection that all observations necessarily disturb the magnitudes to be measured and thereby render the readings uncertain was dismissed on the plea that, by exerting sufficient care, we could minimize the effects of these disturbances and, in theory at least, reduce them to a vanishing limit. Thus in the illustration we gave of the liquid and the thermometer, we may minimize the disturbance by taking a large mass of liquid and a thermometer of small heat capacity. In short, in the opinion of classical scientists the impossibility of testing the law of causality was merely of a practical nature; it was due to casual limitations, and no theoretical impossibility was involved. This distinction between a practical and a theoretical impossibility is fundamental in the recent controversies to which the law of causality has given rise.

During the classical period many arguments in support of, and in opposition to, the doctrine of causality were advanced by metaphysicians. The majority of these arguments do not deal with the precise scientific doctrine but with the vague understanding of causality entertained in metaphysics and in our daily activities. According to some, causal connections are the mere expression of regular repetitions. Thus, if *B* always follows A, then A is to be regarded as the cause of B. This contention was countered by inquiring whether "day" should be regarded as the cause of "night." But all such arguments and counterarguments have no bearing on the scientific doctrine of causality, for in them the word "cause" is used too loosely. The mere fact that B follows A does not entitle us to view A as the cause of B. To be justified in saying that the same cause generates the same effects, we must include under the appellation "cause" all the items of information that are required to render the future sequence of events determinate. The sophisticated notion of "initial state" was devised for this express purpose; and we have seen that the definition of a state cannot be obtained until the laws controlling the system have been established. It is idle therefore to speak of cause and effect in any given situation until we have agreed on the laws. In the illustration mentioned, "day" and "night" do not define states and for this reason cannot be treated as cause and effect. If we wish to introduce the notion of state, and thereby of cause, we may do so, but we must understand that the state of the earth in relation to the sun is defined by the specification of that half of the earth which faces the sun (this corresponds to the specification of day or night), and also by the specification of the rate of rotation of our planet. This last item of information is essential in the definition of the state. Its importance is readily understood when we note that, if the earth were to rotate more slowly and to turn always the same half towards the sun, day or night would be eternal and would not follow one another.

Newton's Mathematical Expression of Causal Connections— One of Newton's greatest achievements is his discovery of the means of representing a physical law, and hence a causal chain, by a mathematical scheme. A more detailed explanation of Newton's method is given in Chapter XV. For the present we shall be concerned only with the less technical aspects.

In our presentation of the doctrine of causality we have mentioned that the initial conditions, or state, taken in conjunction with the law controlling a system, determine the history of the system. For the purpose of discussion we may suppose that the continuous evolution of the system is represented by a continuous curve. The various points of the curve represent the various states of the system at successive instants of time; the point of departure of the curve denotes the initial state. According to this picture, if we move continuously along the curve, the continuous succession of states is described. Let us select a succession of points A, B, C, D along the curve; the point A being situated at the point of departure, represents the initial state. We assume these points packed very closely, so that the states they represent follow one another in rapid succession during the evolution of the system.

To this picture we apply the doctrine of causality, which asserts that the initial point, or state, A determines all the others. But, as Newton observed, we may also say that the initial state A determines the contiguous state B; and that this state B, viewed as a new initial state, determines the next state C; and so on. The fact that the point, or state, A determines the state B implies a connection, or a relation, between the states A and B; the same holds for the other couples of states, B and C, and C and D, and so on. Now the relation retains exactly the same mathematical form whether it refer to the first two states, A and B, or to any other of the couples of consecutive states. In other words, the permanent relation does not change from point to point or from instant to instant during the evolution of the system. It is this permanent relation which represents the law controlling the system in Newton's form of representation.*

Suppose we change the initial conditions while retaining the same system. In our simile, the path of the curve which represents the evolution of the system will be changed, but the permanent relation between consecutive points, or states, will not change. We have here the expression of the permanence of the law as contrasted with the contingency of the initial conditions. The law controlling the evolution of a system is thus seen to express a permanent relation between any two successive states that follow each other at an infinitesimal separation in time; it represents a continuous causal chain forming a link between each ante-

---

* A more precise presentation would require us to add that the relation which constitutes the law is arrived at by a limiting process, *i.e.*, by supposing that the interval of time between the occurrence of the two successive states contemplated becomes infinitesimal.

For example, the states of a mechanical system, at consecutive instants of time $t_0, t_1, t_2 \ldots t_n \ldots$, are defined by the positions and velocities at these instants. Suppose, then, we are dealing with a particle of mass m acted on by a permanent force along a straight line. If we set x and v to measure the position and velocity of the particle at any instant t, the consecutive states are defined by $(x_0, v_0), (x_1, v_1) \ldots (x_n, v_n) \ldots$. The permanent relation to which we have referred in the text is then of the general form

$$\lim_{t_{n+1} \to t_n} m \frac{v_{n+1} - v_n}{t_{n+1} - t_n} = f(x_n),$$

where the function $f(\ )$, which defines the force, is the same for all values of n, and hence is the same at all instants of time.

cedent state and its consequent. The mathematical expression of the relation, and hence of the law, is given by a differential equation. Thus, natural laws (in the form given by Newton) are expressed by differential equations. For this reason, differential equations are the mathematical instruments which express Newton's presentation of the causal doctrine.

Einstein, commenting on Newton's formulation of the causal principle in mechanics, writes:

"In order to give his system mathematical form, Newton had first to discover the concept of the differential coefficient, and to enunciate the Laws of Motion in the form of differential equations—perhaps the greatest intellectual stride that it has ever been granted to any man to make." *

When the law governing the evolution of a system is known and its mathematical transcription, the corresponding differential equation, is obtained, we are in a position to derive a knowledge of the evolution of a system from any given initial state. Thus, if A represents the initial state, the differential equation, by expressing the relationship between A and the next state B, enables us to obtain a knowledge of B. Since we know the state B, a second application of the differential equation yields the state C, and so on. We must remember, however, that these states occur at instants of time that are separated only by infinitesimal intervals. Consequently, if we wish to obtain a knowledge of the states that the system will assume after some finite interval of time, we shall have to repeat the foregoing procedure an infinite number of times in succession. Of course this would be impracticable. The difficulty is overcome by Newton's discovery of the method of integration. Thanks to this powerful method it is always possible, in theory at least, to follow the causal chain over finite intervals. In practice, however, rigorous integration may present insuperable mathematical difficulties, and in such cases methods of approximation must be resorted to.

The Indeterminism of Quantum Mechanics —Today, less than three centuries after Newton made his momentous discovery, doubts are being cast on the validity of the rigorous causal connections of classical science. The attack is due not to the difficulties previously noted (namely, to the impossibility of testing the doctrine in practice), but to totally new discoveries in the subatomic world, where the mysterious quantum phenomena become noticeable.

---

* James Clerk Maxwell; Cambridge, 1931; p. 69.

We mentioned that the practical impossibility of testing the doctrine of rigorous causality arises from the impossibility of operating on perfectly isolated systems; first of all, because no such systems exist; and secondly, because, even if we grant the existence of such systems, we cannot observe their internal workings without disturbing them and thereby destroying their isolation. In addition, human measurements are necessarily imprecise. These difficulties were not regarded as fatal to the doctrine, because approximately isolated systems could be found, and our measurements could be so refined that they would not perceptibly disturb the magnitudes to be measured. The situation was expressed by the statement that, in practice, rigorous causality could not be tested, but that no theoretical obstacle barred the way to a rigorous test.

The novelty resulting from the discoveries of the quantum theory is that we now have reasons to suspect a definite theoretical impossibility which would render illusory any attempt to test rigorous causal connections. The quantum theorists, under the lead of Born, Heisenberg, Bohr, and Dirac, agree with the classical scientists in recognizing that the practical difficulties of testing rigorous causal connections may be disregarded; but they are adamant in their claim that the recently discovered theoretical impossibility cannot be dismissed so lightly. As may be gathered from this synopsis, the philosophy upheld by the quantum theorists has not arisen from a revised interpretation of classical facts but from the necessity of coordinating a new body of facts unsuspected in former days.

The various aspects of the new philosophy will be considered in greater detail in the latter part of this book. For the present we shall restrict ourselves to the more salient points. The developments of the quantum theory indicate that the "uncertainty relations," discovered by Heisenberg, prohibit us *in principle* from effecting simultaneous accurate measurements of so-called conjugate magnitudes. This very general rule applies in particular to position and momentum, or velocity, to time and energy, to the electric and the magnetic intensities at the same point in an electromagnetic field. Thus, if the position of the centre of a particle is measured with accuracy, the unpredictable disturbance, entailed by the measurement itself, causes the particle's momentum to become vague. The classical contention that, by exercising sufficient care, we might reduce the disturbance indefinitely is here no longer valid, for the essence of the uncertainty relations is that the limit we might hope to attain is not vanishing but is finite. Obviously, if this principle is accepted, the state of a mechanical system, involving as it does a simultaneous knowledge of the positions and momenta of the various masses, cannot be known with accuracy. Consequently, a test of rigorous causality is impossible in mechanics. The same conclusion may be extended to all departments of physics.

It is important to understand that the limitations imposed by the uncertainty relation are theoretical, and not practical. Thus the uncertainty relations do not interfere with our measuring, as accurately as we choose, the position alone or the momentum alone of a particle; they only prevent us from executing simultaneous measurements with accuracy. Practical difficulties, on the other hand, would render illusory the perfect accuracy of either kind of measurements. According to the quantum theorists, however, practical difficulties may be waived aside exactly as they were in the classical critique of the causal doctrine. We must therefore conclude that accurate simultaneous measurements are ruled out by the uncertainty relations, not because these measurements are *practically* impossible, but because they are *impossible in principle*. The source of this theoretical impossibility, unsuspected by classical science, must be sought in the very nature of things. It is intimately connected with Planck's constant h. The theoretical impossibility would vanish if this constant, the value of which is finite, were to be infinitesimal, as classical science had implicitly assumed.

These summary remarks may help to clarify the points at issue. The problem is quite general and does not affect solely the doctrine of causality. We are called upon to decide whether a concept which cannot be tested or a magnitude which cannot be measured, *in principle* (in contradistinction to *in practice*), should forthwith be classed as meaningless and cease to play any part in a theoretical discussion.* The dilemma is not entirely new, for it arose in the theory of relativity. Thus according to the special theory of relativity, we must assume that, for reasons which have nothing in common with practical experimental difficulties, a velocity through the stagnant ether cannot be detected. Einstein accordingly claimed a velocity of this type to be meaningless, and all theoretical physicists today endorse his stand. Absolute velocity is thereby dismissed because it cannot be observed in principle. On the other hand, only a practical difficulty prevents us from detecting the possible existence of small planets beyond the orbit of Pluto. For this reason we should not class as meaningless the assertion that such planets may exist; at best, we should claim it to be unproved.

In view of Einstein's attitude towards absolute velocity, we might expect him to adopt a similar one with respect to rigorous causality.

---

* This brief statement does not do full justice to the complexity of the question. The problem will be examined in greater detail in the last chapter.

Yet this he does not do. Einstein and Planck both retain a belief in rigorous causality, extending it even to living matter. But those who have contributed most to the development of the new quantum mechanics resist Einstein's views and insist that rigorous causality is a myth.

Despite appearances to the contrary, the contentions of the new school are not in obvious conflict with the very high measure of approximation with which rigorous causal connections seem to be realized on the commonplace level of experience, e.g., in ordinary mechanics. Analysis shows that the theoretical uncertainty, which prohibits a simultaneous accurate measurement of position and of velocity, is noticeable only when we are dealing with the very minute masses of the subatomic world. With ordinary masses the theoretical uncertainty, though still existing, falls below the practical uncertainties, which are due to the imperfection of human observations, and is completely submerged by the latter. This gradual obliteration of the quantum uncertainties, as we pass to the commonplace level of average masses, is associated with the gradual passage of the quantum theory into classical science. We have here an illustration of Bohr's correspondence principle, to which reference has already been made.

CHAPTER VIII

CONSERVATION

COMMONPLACE observation teaches us that, despite the presence of change in Nature, some properties seem to remain unaffected. For instance, when water is poured from a jar into a glass, the space filled by the water changes in shape but the volume of the water is unchanged. Thus, the volume of the water appears to exhibit a conservative property. From homely examples of this sort may have arisen the belief in the existence of a fundamental space-filling substance which could neither be created nor destroyed, and which would endure unchanged throughout time. In more technical language a substance of this sort is said to be conserved, or to satisfy a law of conservation.

Modern science does not lend support to this primitive belief in conservative substances. In the first place the existence of rigorous laws, and of laws of conservation in particular, is possible only if Nature exhibits a deterministic scheme. But the trend of modern science is to deny strict determinism, so that on this score alone rigorous conservation can have no place in natural philosophy. The suspicion that Nature betrays no deterministic scheme is, however, recent; and for argument's sake we may suppose that the modern view is erroneous and that rigorous laws are valid. If rigorous laws are assumed, it will always be possible to construct by mathematical means physical magnitudes having conservative properties, but these magnitudes may bear little or no resemblance to the intuitive notion of a substance. Thus, in any case there is no justification for asserting in a dogmatic way that a fundamental conservative substance is a necessity. As an illustration of the procedure whereby conservative magnitudes can be constructed when a rigorous law is known, we select the following:

The law of perfect gases for any given mass of gas is

$$pv = AT,$$

where p, v, and T denote the pressure, volume, and absolute temperature respectively, and where A is a constant characteristic of the gas and of the mass of gas considered. By dividing this equation through by T, we see that the product of the pressure and volume, divided by the absolute temperature, always retains the same value. The artificial magnitude thus built up has the conservative property; and we note that this magnitude has nothing in common with the naive conception of a substance. Our example brings out a further point. The magnitude we have constructed is rigorously conservative only insofar as the law of perfect gases is itself rigorously accurate. And since we can never boast of the perfect accuracy of any empirical law, we can never be certain that a magnitude is strictly conservative.

Usually, no particular names are coined for the conservative magnitudes that can be constructed more or less artificially. In other cases names are invented. The "total energy" of a so-called conservative mechanical system is an illustration of a conservative magnitude which is built up artificially and which has received a name. The total energy is the sum of the kinetic and potential energies, which individually are not conserved. "Magnetic induction" is another illustration of an artificially constructed conservative entity. A general class of manufactured magnitudes which are conserved is encountered in classical mechanics. Such quantities are known as the "first integrals" of the motion. The total mechanical energy of a conservative dynamical system is precisely one of these first integrals; and so are the total momentum and total angular momentum of an isolated mechanical system.

The illustrations we have given are of a sophisticated kind in that they assume the prior discovery of natural laws. But we may also examine the concept of conservation more directly. As an example, let us imagine a hypothetical investigator bent on discovering by direct measurement some magnitude which is conserved. His first idea might be to view the volume of a liquid as unalterable. He would soon find, however, that a liquid was not rigorously incompressible and that its volume could be decreased by pressure. "Volume," therefore, could not be regarded as a conservative property. Subsequently he might notice that, insofar as his measurements could detect, the weight of the liquid remained unchanged whether the liquid were compressed or not. He would then claim that a conservative magnitude was manifested in "weight." But Newton's theory of gravitation would dispel this belief, for it would require that the weight should decrease considerably if the volume of liquid were transported to the moon's surface. According to Newtonian mechanics, the correct conservative magnitude is "mass." Mass is a more abstract concept than weight since it is the quotient of weight by acceleration. Then comes the special theory of relativity according to which mass cannot be conserved, for the total mass of an electron and proton, when these two corpuscles are packed closely, is less than the sum of the two separate masses. The theory of relativity furnishes, however, a conservative entity—energy. Our hypothetical investigator would thus conclude that energy is the entity which is rigorously conserved. Incidentally, in spite of its wide use in everyday discussion, the word energy in science refers to a concept still more abstract than mass; and if we except its conservative properties, it bears no resemblance to a space-filling substance.

We may now inquire whether the conservation of energy is rigorous or whether energy, like mass, weight, and volume, may not have to be rejected in our search for a truly conservative magnitude. As things stand today, it is difficult to answer the question with assurance. Bohr at one time rejected the rigorous law of the conservation of energy, viewing conservation as a statistical manifestation. The experiments of Bothe in connection with the Compton effect led Bohr subsequently to revise his views. His new attitude, which is widely accepted, is embodied in his "Principle of Complementarity," a principle that is intimately connected with Heisenberg's "uncertainty relations." Bohr's principle does not deny the conservation of energy (and of momentum) in certain cases, but in such cases it deprives conservation of its intuitive appeal, for it renders impossible the space-localization of the energy which is conserved. On the other hand, if the energy and momentum are located with any degree of accuracy, the principles of conservation cease to be rigorous. Besides, as we mentioned at the beginning of the chapter, if rigorous causal relations do not exist in Nature (and this is the claim made by the quantum theorists), the strict conservation of any magnitude, be it energy or anything else, appears impossible.

CHAPTER IX

MECHANISTIC THEORIES

MECHANICAL phenomena were the first physical phenomena in which supposedly rigorous laws were detected. Mechanics thus appeared as the stronghold of rigorous determinism, and it stood in marked contrast to teleological doctrines and to those in which free will seemed to play an important part. Owing to these peculiarities a mechanistic interpretation of a phenomenon was construed to mean one in which rigorous determinism was assured. Indeed, with some writers, the words "mechanism" and "determinism" were often treated as practically synonymous.

But in view of the subsequent development of science, this confusion of words is apt to be misleading. In the first place, rigorous determinism is no more characteristic of classical mechanics than of many other departments of physical science, e.g., field theories or thermodynamics. Furthermore, the laws of mechanics may, if we choose, be expressed in a form which smacks of teleology.* Finally, recent discoveries in quantum mechanics have prompted some of the leading physicists to question whether the true mechanics of Nature is in any sense deterministic. Our first concern will therefore be to define the meanings of the words we propose to use.

By a mechanistic theory, we shall mean one in which the only physical magnitudes considered are those which are characteristic of mechanics, e.g., force, momentum, mass, strains, and stresses. A theory which is not restricted exclusively to magnitudes of this sort will be called non-mechanistic. The non-mechanistic theories comprise the field theories and phenomenological theories, such as thermodynamics. These other kinds of theories will be discussed in Chapters X and XI. For the present we shall be concerned with mechanistic theories.

Mechanistic theories may be classed into two main groups according to whether they deal with discrete or with continuous media. In the first group we have the atomistic theories (e.g., the kinetic theory of gases). In the second group we may mention the theory of elasticity, and Fresnel's theory of the continuous and elastic ether. A noteworthy fact is

---

* See Chapter XVIII.

that mechanistic theories preceded field theories and that the mechanistic theories of the corpuscular type preceded the continuous theories. As we shall see, this historical order may be attributed to a number of causes, some of them physical and others mathematical.

Greek Atomism—Even in the remotest civilizations men were presumably impressed by the difference between substances like water, which appear continuous, and others such as sand, which are discrete. Presumably also, some early thinker may have wondered whether, in spite of its continuous appearance, water was not in reality formed of microscopic corpuscles. Fragments from the earlier Greek writers indicate that problems of this sort formed the starting point of the doctrine of atomism. Democritus is usually credited with the invention of this doctrine; more probably, however, the atomic hypothesis arose in remoter days among the Chaldeans and the Egyptians; it is not sufficiently original to be credited to any one man in particular. Inasmuch as Democritus's writings have been lost, we must turn to the writings of Lucretius for the most exhaustive presentation of the doctrine.

The major aim of atomism is to interpret the qualitative differences, which impress our senses, by means of hidden occurrences of a more or less uniform nature. Atomism thus presupposes that the real world is not the world of our sensations, but is a hidden world which we perceive only indirectly through our senses. In this respect atomism is in agreement with modern science, for science also assumes that a sensation such as "green" is due to a certain number of vibrations taking place every second in the electromagnetic field. In Greek atomism the hidden occurrences postulated are the motions of atoms through the void. The atoms are impenetrable, unchangeable, and eternal; they have different shapes and sizes, the tiniest atoms of all being the atoms of fire. The motions of the atoms produce the subjective sensations of warmth, light, smell, and they are the cause of all the manifestations of change in the universe. Solid and semi-permanent bodies, such as stones, are mere agglomerations of atoms. Beyond the atoms there is no further mystery, for by hypothesis the atoms are the ultimate building blocks of the universe; and it is therefore useless to inquire into the secrets of their inner parts. Thus, all further mystery is banished *a priori*.

But this simple scheme affords no means of explaining the existence of semi-permanent bodies. For, if the atoms be moving about at random, how can we ever expect them to hold together? The difficulty was removed by assuming the atoms to have hooks and eyes, thanks to which they could interlock. Thus, the coherence of matter was accounted for.

Like all primitive people, the Greeks experienced an uncontrollable desire to compress all knowledge into a single formula. Whereas today a man who would claim omniscience would be regarded as a charlatan, the wise men among the ancients boasted of their ability to answer all questions. Small wonder, then, that the simplified world picture presented in the hypothesis of atomism should have appealed so strongly to many of the Greeks. In our opinion the sole merit of atomism is that it interprets quality by means of quantity, and by so doing enables the human mind to reason with greater precision.

As usually occurs when a doctrine becomes popular, the inevitable metaphysician appeared upon the scene. He proceeded to explain that the entire doctrine was a necessary consequence of some of his tenets. Atomism in particular was, according to him, a necessary consequence of the presence of change and of permanency in Nature. Permanency was illustrated by the indestructible eternal atoms, and change by the varying configurations of the atomic aggregate. We need scarcely point out that any conservative property postulated for any changing configuration would be just as suitable in accounting for permanency and change as is the atomic hypothesis; but perhaps one should not be too critical of the naivete of these early speculations, for they were advanced practically at the dawn of thought. There is less excuse, however, for those who defend them today.

If we purge Greek atomism of its unnecessary metaphysical trappings, it has more to commend it; but even so, it exhibits only a superficial similarity with the corpuscular theories of science. In the first place, scientific atomism does not profess to furnish a universal interpretation of all things. Originally, it was put forward as a hypothesis so as to account for certain restricted empirical facts, notably for some of the properties of matter. Other existents were not regarded as atomistically constituted. Electricity and heat, for instance, were at one time supposed to be continuous fluids. At a later date the continuous electromagnetic field was introduced. At no time was the force of gravitation interpreted in terms of chains of hooked atoms extending from the sun to a planet. Instead, the force of gravitation was postulated as an independent hypothesis.

A further difference between ancient and modern atomism becomes apparent when we recall that the unalterable, eternal existence of the atoms, which the Greeks assumed essential to their ideas, is precisely one of the beliefs that is rejected by modern science. The closest approach to the Greek atom is illustrated by the electron or the proton. Yet these two existents are so far from being deemed unalterable and eternal that physicists have assumed they may coalesce, give rise to radiation and hence to something totally different. With the discovery of wave mechanics, it is doubtful whether even matter can be interpreted solely in terms of corpuscles and of electromagnetic fields; wave magnitudes of a mysterious nature also appear to be required.

Having noted the differences between the two varieties of atomism, we may examine the reasons which gave rise to atomism in science. Matter was the first existent to suggest itself as being formed of corpuscles. Thus, if we mix a pint of water with a pint of alcohol, we obtain less than two pints of the mixture. The phenomenon is easily accounted for if we assume the water and the alcohol to be granular; the finer water corpuscles may then squeeze in between the coarser corpuscles of the alcohol. Similarly, when subjected to sufficient pressure, water can be made to ooze through a sheet of gold; the porousness thereby revealed by the gold suggests a granular structure. Finally, the compressibility of gases and the basic laws of chemistry furnished additional evidence of the atomicity of matter.

The first physical theory of an atomistic nature was not, however, a theory of matter, but a theory of light. Newton believed that the phenomenon of double refraction could more easily be accounted for by supposing that light was formed of corpuscles instead of being a wave manifestation, as was suggested by Huyghens. But Newton's theory was more in the nature of a sketch than a well-developed mathematical theory; and when the theory of light was subsequently constructed in the nineteenth century, Huyghens's assumption and not Newton's was accepted.

Mechanistic theories of a corpuscular type attained a considerable degree of development when phenomena dealing with ordinary matter were studied. We have mentioned the empirical evidence that suggested the corpuscular structure of matter, but, as we shall now see, other reasons relating to Newton's theory of the solar system conspired to give priority to corpuscular theories.

The study of moving solids was one of the most immediate which forced itself on the attention of men, and for this reason the laws of dynamics were formulated before those of electromagnetics and thermodynamics, for example. Newtonian dynamics was thus the first physical doctrine to be established on a secure mathematical basis. When Newton's attention was directed to the planetary motions, he assumed the planets and sun to be so many material bodies moving in accordance with the same mechanical laws that hold on the earth's surface. Even if we grant that this assumption was not inevitable, it was at all events plausible, and it has never been challenged by subsequent scientific developments. If this assumption is made, the interpretation of the planetary motions necessarily becomes a mechanistic one. Furthermore, the planets and sun are spherical bodies which we may suppose to be approximately homogeneous; and Newton proved that, from the standpoint of gravitational action, a spherical homogeneous mass is equivalent to a point-mass having the same total mass. Hence, in Newton's theoretical treatment the sun and the planets may be replaced by so many point-masses. Thus, the first grand theory of mathematical physics was, to all intents and purposes, a mechanistic theory involving the motions of point-masses. Let us observe that the mechanistic nature of Newton's theory was not prompted by any particular preference for mechanistic interpretations. As for its association with point-masses, it was obviously imposed by the very configuration of the solar system and was quite foreign to any atomic hypothesis.

The following century (the eighteenth) witnessed an intensive development of pure mathematics, thanks to which a systematic attack was made on many of the problems of celestial mechanics left unsolved by Newton. Little was achieved in the other departments of theoretical physics, chiefly for the reason that very few new experimental facts were established. Certain properties of matter, such as adhesion, cohesion, and the phenomenon of capillarity, were, however, investigated by the various experimenters, and, on the basis of their findings, attempts were made to construct mathematical theories. Owing to the empirical evidence, a corpuscular structure was credited to matter in these theories. Newton's treatment of the planetary motions served as a model for the new theories, the molecules of matter being likened to point-masses controlled by the accepted laws of mechanics. It is true that conditions were considerably more complicated than they were in the solar system. In particular, ordinary gravitational forces between molecules proved insufficient to account for the cohesion of matter, so that other laws of attraction differing from that of the inverse square were introduced. But from our present standpoint such changes are of minor importance, because in any case the new theories were necessarily mechanistic and corpuscular.

Among the most celebrated of these mechanistic theories of the eighteenth century is the one elaborated by Laplace in connection with the phenomenon of capillarity. Also worthy of mention is Bernoulli's suggestion of the kinetic theory of gases—a theory according to which a gas is formed of perfectly elastic molecules in rapid motion, colliding and rebounding in conformity with the laws of mechanics. The kinetic theory was developed in the following century by Maxwell and Boltzmann; it was also extended to liquids, but with less success. In connection with these kinetic theories we may note that the corpuscles of matter, which constitute the ultimate elements on which the theories are based, are not assumed to be ultimate in any absolute sense. The molecules of the gas are known to be divisible into atoms, and these atoms in turn may be disrupted into smaller components. Nevertheless, since the molecules behave as though they were permanent units in the phenomena contemplated by the kinetic theory, they may be treated as such in the elaboration of the theory. Atomism, from this standpoint, implies merely a high degree of stability; nothing further is intended.

The next theory of corpuscular physics to appear was formulated so as to coordinate certain electrical phenomena. In the early years of the nineteenth century, the electric current that flows in a wire was thought to be a continuous fluid in motion, and not until Faraday established the laws of electrolysis was the corpuscular nature of electricity suspected. This suspicion received added support when cathode rays were discovered (these are produced by causing an electric current to pass through a highly rarefied gas). Hertz, in his last experiments, proposed to determine whether the cathode rays were streams of swiftly moving particles or wave manifestations. Subsequently the corpuscular view triumphed, the cathode particles being called beta-particles, or electrons. Since then, numerous experiments have established the corpuscular nature of electricity, though we may add that in view of the latest developments in the quantum theory, the concept of corpuscles is far from being as clear cut as was formerly believed.

The first mathematical theory of electric phenomena viewed from a corpuscular standpoint is Lorentz's theory of electrons. When we say that this theory is of the corpuscular variety, our statement is not quite correct, for Lorentz retains the continuous electromagnetic field side by side with his corpuscular electrons. The motives which led Lorentz to introduce electrons arose from the failure of Maxwell's continuous theory to account for several well-known physical phenomena, notably for dispersion. These considerations are clarified by the following short digression on Maxwell's theory.

Maxwell's equations determine the possible distributions and variations of the electromagnetic field. If the region of space in which the field is situated is a vacuum, Maxwell's equations which control the field are called "the equations of electromagnetics *in vacuo*." Maxwell also considered the case where the field was present in the interior of matter.

He supposed that, in a transparent medium such as glass (assumed homogeneous and isotropic), the field would be modified, and that the modification could be expressed by supposing that the space was characterized by an appropriate numerical constant K the value of which depended on the nature of the matter. Space empty of matter was characterized by the value unity attributed to the constant K, but in the interior of matter K exceeded unity. Maxwell's theory leads to the identification of waves of light with electromagnetic waves. According to his theory, the velocity of propagation of electromagnetic waves (and hence of light waves) in a homogeneous medium has the value $\dfrac{c}{\sqrt{K}}$ where c is the velocity of light *in vacuo* and K characterizes the medium. Since the value of K exceeds unity inside matter, we conclude that waves of light are propagated with reduced speed in the interior of matter. The velocity of light will consequently suffer a drop when the light passes from a vacuum (or air) into glass. Now the wave theory shows that this decrease in velocity will cause a ray of light to be bent when it penetrates from air into glass under a slanting incidence. In more technical terms, the ray of light will be refracted. Furthermore, Maxwell's theory suggests various experimental means of measuring the value of the constant K for glass, and thereby enables us to predict the precise degree of bending a ray of light will sustain. Suffice it to say, the predictions drawn from Maxwell's theory are in general agreement with the well-known phenomenon of refraction.

Though Maxwell's theory accounts for refraction, it is unable to predict the important phenomenon of dispersion, according to which rays of light of different color are not bent in the same degree (violet light being refracted more strongly than red light). It is to dispersion that the appearance of the rainbow is due. The reason Maxwell's theory fails to account for dispersion is easily understood. Thus we have seen that the bending of a ray of light falling on a slab of glass under a given incidence is governed by the value of the constant K for glass. Inasmuch as, in Maxwell's theory, this constant is determined solely by the glass and not by the color of the light, all rays of light should be bent in exactly the same degree, and dispersion should not arise. To account for dispersion, we must assume that K is characteristic not only of the matter, *e.g.*, the glass, but that it must also vary according to the color of the light. In short, Maxwell's theory in the presence of matter is seen to be but an approximation. This realization prompted Lorentz's refinement.

At the time Lorentz was initiating his theory, the empirical evidence suggested the existence of corpuscles of electricity, and so Lorentz availed himself of this discovery in order to remedy the difficulty connected with the constant K. He sought to give a physical interpretation of the dependence of K on the color of the incident light by supposing that, in the interior of transparent matter, electrons were distributed and were free to vibrate about fixed positions. Under the exciting influence of the incident light, these electrons would be set into vibration with the same frequency as that of the incident light. In this way the required dependence was established mathematically between the characteristics of the glass and the frequency of the light. Dispersion was thus explained by refining Maxwell's oversimplified treatment of matter. Dispersion, however, was not the only phenomenon that Maxwell's theory had failed to interpret; magnetic rotary polarization was another. But here, also, the electron theory cleared up the major difficulties.

Lorentz's theory readily accounts for the difference between conductors and non-conductors. Thus in dielectrics (*i.e.*, non-conductors), such as glass, the electrons cannot move appreciably from their fixed positions, but in conductors, such as metals, the electrons are free to move throughout the substance. If, then, we create a difference in potential in a metal, all the electrons will swarm in the same direction, and an electric current will result.

Mechanistic Theories Based on Continuity—We have seen why mechanistic theories were the first to be considered, and we have also seen that, in view of the corpuscular structure credited to matter, these theories were of the atomistic type. In the course of the nineteenth century, mechanistic theories of a different kind were constructed. In these, matter was treated as though it were continuous. The reasons that led to the formulation of the continuous theories may be understood from the following considerations.

If we accept the corpuscular theory of matter, we must assume that the water in a glass is formed by the agglomeration of myriads of tiny molecules acting on one another in accordance with more or less complicated laws. But in practice, whether the corpuscular theory be valid or not, water in bulk behaves like a continuous fluid and we may treat it accordingly. In the early theories of mathematical physics, where the corpuscular theory was adhered to, the rigorous treatment of problems dealing with matter required that we consider the motion of each individual molecule. As may well be imagined, the rigorous procedure was impossible in practice, and recourse was had to average effects and to approximations. But then, since the corpuscular theory did not hold its promise of leading to rigorous predictions and since it had the disadvantage of involving complicated mathematics, it seemed simpler to disregard the corpuscular theory entirely and to proceed as we should in ordinary life by viewing matter as continuous. Considerations of this sort led to the development of mechanistic theories of the continuous type.

A noteworthy fact is that the mechanistic theories of the continuous type began to appear only in the first half of the nineteenth century. In view of the plausible reasons that prompted their construction, we might be surprised at the tardiness of their arrival. The reason must be sought in the mathematical limitations of the day. The mathematical treatment of corpuscular media involves the application of differential equations of the simpler type studied by Newton and by the mathematicians of the eighteenth century. On the other hand, the treatment of continuous deformable media requires the application of a more complicated kind of differential equation, the so-called "partial differential equation." The mathematicians of the eighteenth century, notably Euler and Lagrange, had made some investigations on partial differential equations, but only in the following century, thanks to Cauchy, Poisson, Jacobi, and others, were the more important properties of these equations established. It then became possible to study mathematically the statics and dynamics of continuous deformable media, and the mathematical theory of elasticity was founded. In this way the first mechanistic theory of the continuous type arose.

Contemporaneously with the mathematical study of elastic continuous media, the early nineteenth century witnessed the rise of the undulatory theory of light. Experiments performed by Young and by Fresnel suggested that light consisted of waves propagated through space, much as elastic waves are propagated through an elastic medium. In many respects the analogy appeared so striking that Fresnel postulated a space-filling elastic medium, the ether, the deformations of which were controlled by mechanical laws and in which waves could be propagated. Fresnel's assumption constitutes the hypothesis of the elastic ether. Physicists realized that certain properties of Fresnel's elastic ether would have to differ considerably from those of material bodies, but since it was unnecessary to suppose that the ether was matter in the ordinary sense of the word, the features of dissimilarity did not appear alarming. It seemed gratuitous and unnecessary to make the assumption that this postulated ether was corpuscular in structure; the ether was therefore regarded as a continuum. The mathematics required by Fresnel's theory was at the disposal of investigators thanks to the prior study of partial differential equations and of elastic media. Fresnel's theory was thus mechanistic but based on an underlying continuous medium.

Maxwell's theory, to which reference has already been made, seemed at first to be of the same general kind as Fresnel's. The electromagnetic phenomena studied by Maxwell were regarded as manifestations of mechanical changes occurring in the continuous ether. But, for reasons which will be understood in the next chapter, mechanical attributes were eventually removed from the ether, and mechanical interpretations of electromagnetic phenomena abandoned. It was the failure of mechanical interpretations to account for electromagnetic phenomena that led to the development of the field theories. With these we shall be concerned in the next chapter.

CHAPTER X

FIELD THEORIES

In physics the world *field* is given to the continuous distribution of some "condition" prevailing throughout a continuum. We are using the vague word "condition" purposely, because the precise nature of the magnitude, the distribution of which constitutes the field, varies according to the problem at issue.

When the condition is adequately described at each point of space by a simple number (*i.e.*, a scalar), we have what is known as a scalar field. Temperature is such a condition, and so the temperature distribution throughout a volume is a physical illustration of a scalar field. In many cases the condition at each point of space has a direction as well as a magnitude; in this case an arrow, or vector, is required to represent it. The field is then called a vector field. A field of force or the field defined by the instantaneous velocities of the various points of a fluid in motion are illustrations of vector fields. Finally, if the condition is described by a tensor at each point, we have a tensor field. The distribution of the stresses in an elastic medium is of this type; and so also is the field of gravitation in the 4-dimensional space-time of the general theory of relativity. Whatever its nature, the field is called "permanent" or "variable" according to whether the distribution of the condition remains unchanged or varies with time. We might also consider a discrete field, for instance the field defined by the velocity vectors of the various particles of sand in a sandstorm. In a field of this sort each point at which a particle of sand is situated, at the instant considered, is associated with a velocity vector defining the velocity of the particle; the intervening points at which no sand particles are present have no vectors attached to them. It must be emphasized, however, that discrete fields are seldom considered. Usually, when we speak of a field, a continuous one is understood; and the fields we shall consider in the present chapter will always be continuous.

The physical condition, whose continuous distribution constitutes the field, differs from one situation to another. In some cases it may be interpreted by means of mechanical categories, such as force, velocity, acceleration, displacement, strain or stress. The field is then of a mechanistic nature. Sometimes, however, mechanical categories cannot account for the condition. In this event, the field is non-mechanical.

We must now examine what is meant by a field theory. A field theory is a theory whose aim is to study the peculiar condition, or field, which is thought to pervade the ether of space in certain cases. Formerly, the ether was viewed as an elastic medium, having many of the properties of matter, and so a field theory did not appear to differ essentially from a mechanistic theory of the continuous type, such as the theory of elasticity. But according to modern views, the two kinds of theory are totally different, for the ether is now assumed to have no mechanical properties, and hence its field is sharply distinguished from the field of mechanical stresses which pervades an elastic body.

The withdrawal of all mechanical properties from the ether has caused some physicists to regard the word "ether" as merely another name for empty space. According to this conception the field of the ether is a field which is due to the properties of space itself (or space-time) ; it is thus the field of space, and so space must be regarded as an active agent instead of a passive void.

The point to be kept in mind for the present is that field theories differ from mechanistic theories of the continuous type since they do not involve the mechanical categories; and that they differ still more from the mechanistic theories of the discrete type (such as the kinetic theory of gases) because of the additional contrast between the continuity of the field and the discreteness of matter viewed atomistically.

Several considerations account for the relatively recent appearance of field theories. First of all, fields, since they are continuous, require for their mathematical formulation the introduction of partial differential equations. Accordingly, field theories (as also mechanistic theories of the continuous type) could not have been constructed until the early part of the nineteenth century, when the properties of partial differential equations began to be understood. But even after the purely mathematical difficulties were overcome and continuous distributions of a physical condition could be studied mathematically, field theories did not suggest themselves immediately. The fact is that the first kind of continuous distribution which attracted the attention of physicists was the continuous distribution of stresses in an elastic medium; and these stresses are obviously mechanical magnitudes. Only when the electromagnetic field was discovered by Maxwell, did physicists suspect that a new, non-mechanical physical existent had been revealed. Maxwell's discoveries could not have arisen at a much earlier date, for Maxwell was guided by many of the experimental and theoretical results established by the pioneers in electrical research. Thus, considerations of an experimental and of a mathematical nature both conspired to prevent the development of field theories in the seventeenth and eighteenth centuries.

But we might be tempted to inquire why a field theory was not developed in connection with the gravitational field. The very appellation "gravitational field" would seem to imply a belief in the existence of a field. In point of fact, however, prior to Einstein, a gravitational field was not meant to suggest a condition of space or of the ether. When physicists spoke of a gravitational field surrounding matter, they meant that a mechanical field of force surrounded the matter; the space itself around the matter was assumed to be inert and to have no direct connection with the field. For this reason Newton's theory of gravitation was not a field theory. These considerations will be better understood when we examine some of the peculiarities of field theories.

Let us imagine that a massive body, such as the sun, suddenly comes into existence. A planet situated in its vicinity is then subjected to a gravitational pull. We wish to understand how the gravitational action of the sun may have been transmitted to the planet. One way of accounting for the phenomenon is suggested by commonplace observation. Thus let two pieces of cork be floating on the tranquil surface of a pond. We move one of the corks up and down; a train of expanding circular waves is generated; and when these waves reach the second cork, it is set into motion. There is no direct influence of one cork on the other; the transmission of the action occurs from place to place through the intervening water, and a certain period of time necessarily elapses before the action is communicated from one cork to the other. Any transmission of the type just discussed is called a transmission by "contact action."

We now attempt to interpret gravitational action along the same lines. We must suppose that in our hypothetical illustration the sudden birth of the sun disturbs the condition of surrounding space (or of the medium filling space). The disturbance is then propagated in expanding spherical waves by contact action until it reaches the planet. At this instant the attractive force manifests itself on the planet. In our present analysis we need not be concerned with the precise nature of the disturbance sustained by the intervening medium, whether it be a mechanical compression or anything else; we may therefore refer to the disturbance by the vague name "gravitational condition." The intensity of this condition will vary from place to place in the space surrounding the sun, and the distribution of this condition defines what we have called a field. The attitude here tentatively adopted leads to a field interpretation of gravitation: for the sun does not act directly on the planet; it merely generates a change in the condition of the ether, and it is this condition of the ether which betrays itself in a gravitational pull. The ether, or space, surrounding the sun thus plays an active part in the phenomenon. We also note that, according to the foregoing field interpretation, a certain period of time must elapse between the instant at which the sun comes into existence and the instant at which the gravitational pull is experienced by the planet. Attempts were once made to view gravitation as a field manifestation; and experiments were devised to illustrate the situation. We refer to the experiment of Bjerkness's pulsating spheres. Two rubber spheres which contract and expand rhythmically are placed in water. Theory and experiment show that the spheres act as though attracted to each other. Such mechanistic interpretations of gravitation were soon abandoned, however.

The problem of gravitation may be approached in a different way. We may suppose that the action of the sun on the planet is direct and that the intervening medium remains passive, playing no part in the phenomenon. This interpretation compels us to assume that the transmission of the action is instantaneous, so that the planet is submitted to the gravitational pull at the very instant the sun comes into existence. Herein consists the hypothesis of "action at a distance." It is incompatible with a field theory.

A third interpretation, which differs essentially from the former two, is the one advanced by Lesage. According to Lesage, gravitational attraction is due to the incessant bombardment of particles moving through space in all directions. Two bodies which would normally be said to be attracting each other are in reality inert, but as they exert a mutual screening, the resultant effect of the incessant bombardment is to draw them together. Lesage's hypothesis was subsequently investigated and found to be untenable. We mention it because it illustrates the possibility of avoiding "action at a distance" without invoking the activity of the intervening medium—the latter remains passive.

A decision between the hypotheses of action at a distance and of contact action is not an easy one if we are to be guided solely by experimental tests. We cannot, of course, bring a large mass into existence suddenly (or annihilate it suddenly) and then measure the time taken by the gravitational force to manifest itself (or to cease) at a distant point. Newton himself was averse to action at a distance, and at one time invoked the pressure of the surrounding medium as a vehicle for the transmission of the gravitational action. But in any case the mathematical difficulties alone would have prevented him from developing a field theory, and in addition the hypothetical nature of the assumptions which he would have been compelled to make did not satisfy his cautious nature. His celebrated dictum "Hypotheses non fingo" may be interpreted, in our opinion, by the statement: If we cannot furnish an interpretation which is in agreement with a large number of facts, it is better to give none at all. The net result was that Newton's theory of gravitation, in its actual application, assumed action at a distance and was not a field theory.

We said that a direct test of the opposing interpretations is exceedingly difficult. Calculation does, however, furnish some information. Thus in the solar system, a given planet, say the earth, is submitted to the gravitational actions of the sun and of all the other planets. If the gravitational action is instantaneous, the force exerted by the planet Jupiter on the earth at a given instant is directed toward the *present* position of Jupiter. If, on the other hand, a certain time, say half an hour, is required for the gravitational attraction to be propagated, the earth will be attracted towards the position which Jupiter occupied half an hour earlier. The motion of the earth cannot therefore be quite the same in the two cases.

Laplace submitted the problem to a test by assuming a delay in the transmission of the gravitational force. His aim was to determine how great a delay was compatible with the observed planetary motions. As in all problems of celestial mechanics where more than two bodies are involved, exact calculations are hopelessly difficult, and so Laplace resorted to approximations. These he obtained by omitting to consider influences which he assumed to be of relatively small importance. Unfortunately, the effects he neglected were subsequently shown by Poincaré to be as important as those he had retained, and so Laplace came to the erroneous conclusion that the force of gravitation was transmitted instantaneously or, at any rate, with a velocity far greater than that of light. Laplace's results were accepted in his day and were regarded as affording an important confirmation of "action at a distance." Any attempt to develop a field theory for gravitation was therefore deemed useless.

In the study of electromagnetic phenomena, the problem of deciding whether to adopt a theory of action at a distance or a field theory came up for consideration as it did in the case of gravitation. The first empirical law established was Coulomb's law of electrostatic actions, which states that opposite electric charges attract each other in accordance with the Newtonian law of the inverse square, and that charges of the same sign repel according to the same law. The law of the inverse square also holds for magnetic poles. Possibly by reason of the similarity between these laws and the law of gravitation, but more probably because as yet no field theories had been constructed, the hypothesis of action at a distance was retained. Likewise, the action of currents on currents and on magnets was viewed as an instance of action at a distance.

Faraday and Maxwell were the first to suspect that the seat of electromagnetic phenomena should be sought in the surrounding medium. The next step was therefore to construct a field theory of electromagnetism. But first let us recall what consequences a field theory will entail.

Suppose that a current is made to pass through a straight wire. We know that a magnetic needle placed in the neighborhood of the current is acted upon by a force and sets itself crosswise to the direction of the current. If we accept a field theory, we must assume that the needle is subjected to the deflecting force, not instantaneously, but only some time after the current has started. Similarly, if the current is stopped suddenly, the needle continues to be deflected until the disturbance has passed it; and this will necessarily be some time after the current itself has ceased. The disturbance, or field, then spreads out into space, decreasing in intensity as it proceeds. A successful field theory of electromagnetism must be in agreement with the foregoing conclusions. Maxwell took into consideration the empirical laws already known, and, by combining them with an assumption of his own, constructed his celebrated field equations of electromagnetics.

In Maxwell's theory, the condition of the ether is determined at each point by two magnitudes: the electric intensity (a vector), and the magnetic intensity (a tensor). The distribution of these two magnitudes at an instant determines the electric and the magnetic fields respectively. The two fields are not independent, for any variation in the one entails a change in the other. Maxwell's equations express the interconnection between the distributions and variations of the two fields. The intimate connection between the electric and the magnetic fields suggests that we view them as forming one single field, the electromagnetic field. Maxwell's equations may therefore be said to express the mathematical laws controlling the changes of the electromagnetic field. As is well known, a direct consequence of Maxwell's theory is that an electromagnetic disturbance should be propagated *in vacuo* with the velocity of light. Hertz, by submitting this important prediction to a successful experimental test, vindicated the theory and secured its acceptance.

There is considerable similarity between the condition of the ether when an electromagnetic field is present and that holding in an elastic solid submitted to deformations. The similarity is still more striking when the evolution of the field assumes the form of waves—a situation which is realized when we cause a current to oscillate back and forth along a straight wire, as in a radio antenna. But the similarity is superficial, for on closer analysis the mechanical interpretations prove cumbersome and often contradictory. Maxwell therefore abandoned his initial attempts at giving a mechanical interpretation of the field. His theory thus assumed the aspect of a field theory in the modern sense. The condition of the ether betrayed by the field was a physical condition, but it could not be interpreted in terms of the mechanical categories; and so physicists came to regard the electromagnetic field as a new type of physical reality.

In Lorentz's theory of electromagnetics, this non-mechanical attitude was stressed, and all mechanical interpretations of the field were ruled out on principle. The Lorentzian ether was rigid, indeformable, and so it was taken to define the reference frame of absolute rest. Lorentz's ether is thus but another name for Newton's absolute space. These attributes of the ether automatically exclude the possibility of our viewing the field as due to displacements or to stresses or indeed to modifications of any type that is familiar. So long as we are dealing with a region of space free from matter, the field is the only physical existent to be considered, but in many cases matter and ordinary electric currents may also be present. For the reasons mentioned in the preceding chapter, Lorentz assumed that electricity was corpuscular and that matter was built up from the electrons and protons. The corpuscular electrons are, to all intents and purposes, matter, for they have mass and their motions are controlled by the ordinary laws of mechanics. The word "electricity" refers to the electrons rather than to the field; and the electric current, which is due to the rushing of electrons along a wire, is regarded as something entirely distinct from the field and its changes. Lorentz's theory thus endorses a dualistic conception. On the one hand, we have the continuous non-mechanical field, represented by physical magnitudes existing on their own account. On the other, we have matter, represented by the corpuscular electrons and protons, the motions of which are controlled by the laws of mechanics. The field and the electrons are not altogether independent, for the presence and the motions of the electrons affect the field; and conversely, the field, even though it is non-mechanical, develops forces on the electrons. The mechanical force exerted by the field is called by Lorentz the "ponderomotive force"; its mathematical expression is obtained by combining in an appropriate way the electric and magnetic magnitudes of the field. The ponderomotive force therefore acts as a link between the field and the mechanical electrons.

The dualistic aspect of Lorentz's theory was distasteful to those monistically inclined, and attempts were made to secure a fusion of the

mechanical and the electromagnetic categories.   In the original attempts
the field magnitudes were treated as fundamental and the mechanical
ones as derived therefrom.   According to this program the corpuscular
electrons and protons were to be built up from field magnitudes alone,
and quite generally the mechanical laws were held to be consequences of
the laws of electromagnetics.   Any theory which professes to interpret
all the categories in terms of non-mechanical field magnitudes is called a
"pure field theory."   Consequently, the attempts we are here mention-
ing represented an endeavour to transform Lorentz's dualistic field theory
into a pure field theory.

Abraham, to whom we owe the first attempt of this kind, concentrated
on the mechanical concept of mass.   He assumed that an electron was a
rigid structure and he showed that the entire mass of an electron in
motion could be built up from the field magnitudes.   Mass, according to
this view, was fundamentally an electromagnetic manifestation.   Since
it was the electron's motion through the field which generated its mass,
an electron at rest should have no mass.   Lorentz himself proceeded along
different lines.   For reasons connected with certain experiments in electrodynamics, he postulated the so-called FitzGerald contraction for bodies
moving through the ether.   Thus an electron, which was assumed to be
spherical when at rest, became flattened when in motion.   In this assumption consists the hypothesis of "the contractile electron."   When the
problem of mass was investigated for the contractile electron, calculation
showed that though a part of the mass was electromagnetic in nature,
there remained a residue which could not be interpreted in terms of the
field.   A remnant of mechanical mass was therefore inevitable.

Experiments were soon devised for the purpose of deciding between
the opposing theories.   The most precise ones performed at the time were
those of Bücherer; and they were favorable to Lorentz's anticipations.
Abraham's pure field theory was abandoned in consequence.   But even
had Abraham's views been supported by experiment, other difficulties
would still have barred the way to a pure field theory.   For instance,
the electric charges on the surface of an electron, or in its interior, being
of the same sign, should repel one another, and hence the electron should
explode.   To account for the electron's stability, we must postulate some
intense extraneous pressure forcing back the escaping charges.   But this
pressure (Poincaré pressure) cannot be built up from the field quantities
of Maxwell and Lorentz; and so it must be of non-electromagnetic origin.
The electromagnetic field by itself is thus incapable of accounting for
the stability and the existence of the electron.   For these reasons and

many others, any pure field theory founded on the Maxwell-Lorentz electromagnetic field seems doomed to failure from the start. A more successful attempt was made at a later date by Mie, and at one time Weyl favored Mie's theory; but he has since abandoned it. In view of these failures physicists were thrown back on the dualistic theory of Lorentz, and it is the development of this theory and its connection with the theory of relativity that we shall now examine.

The theory of relativity, whether in the Lorentzian or the Einsteinian form, arose from the consideration of a group of phenomena known under the name of the electrodynamics of moving bodies and of moving fields. Lorentz's electron-field theory, even when supplemented by the FitzGerald contraction, was unable to account for the experimental results. In order to bring theory and practice into agreement, Lorentz was compelled to postulate revolutionary relations between space and time. These relations, which automatically embody the FitzGerald contraction, are expressed by the celebrated Lorentz transformation. Had Lorentz realized that the space and time, which he had subjected to the new relations, were the real space and time of physical science, he would have been the creator of the theory of relativity. But the progress of science is gradual, and before revolutionary changes are accepted, attempts are usually made to interpret the facts at our disposal in terms of those classical notions which have proved their worth in other situations. Lorentz accordingly viewed his transformation as purely formal and as having no bearing on the space and time of physics.

Then a year later, Einstein adopted the more revolutionary course, contending that the Lorentz transformation expressed the relations between physical space and time. This departure marked the start of the special theory of relativity. The ether, which had already been deprived by Lorentz of all its mechanical attributes, was now shorn of its ability to define a frame of reference in which velocity could be measured. The new views received a tremendous impetus from Minkowski's discoveries. Minkowski showed that if Einstein's contention were correct, this world of ours could not be the familiar world in which space and time are separate, but must be a strange non-Euclidean 4-dimensional world, since called *space-time*.

With the introduction of space-time, all physical magnitudes, formerly represented in the 3-dimensional space and separate time of classical science, now had to be expressed as 4-dimensional magnitudes of the new world. The new representation thereby entailed the crediting of additional components to the classical 3-dimensional magnitudes. The extra components turned out to be well-known magnitudes, so that the novelty of the space-time form of representation resided more particularly in its fusion of physical magnitudes formerly viewed as distinct. The net result was that many such physical magnitudes revealed themselves as the various components of the same 4-dimensional existents. The simplest illustration of this fusion is that of space and time, but there are others. Thus momentum and energy are unified, and the separate laws of the conservation of momentum and of energy are seen to express but one law of conservation in a 4-dimensional world. Force and work are also merged; so are electric current and charge.

It is particularly instructive to examine the fusion in the case of the electromagnetic-field magnitudes. As we have seen, in classical 3-dimensional space the electromagnetic field may be viewed as due to the superposition of two separate fields, the electric field, which is a vector field, and the magnetic field, which is a tensor field. Both fields are of course 3-dimensional since they are situated in 3-dimensional space. We also mentioned that these two fields are not mutually independent, for variations in the one entail modifications in the other. Maxwell's four equations express the mutual relations between the two fields. We must now transcribe the field magnitudes into 4-dimensional space-time. Calculation shows that the 3-dimensional electric vector and the magnetic tensor merge into a single 4-dimensional tensor, called the electromagnetic tensor. In other words, the 3-dimensional electric and magnetic fields become fused and appear as mere partial aspects of the same underlying 4-dimensional tensor field—the electromagnetic field. The interdependence of the electric and magnetic fields (which in 3-dimensional space had to be accepted as a fact of experience) is thus an immediate consequence of the new outlook. At the same time the exact quantitative expression of the connection between the two partial fields can now be obtained, even in the more involved situations which deal with the electrodynamics of moving bodies. Maxwell and Lorentz had to guess at this connection in the more complicated cases, and in point of fact guessed wrongly. Finally, we may mention that Maxwell's four 3-dimensional equations are reduced to two 4-dimensional ones.

In the special theory of relativity, Minkowski's space-time is viewed as flat (uncurved), rigid, and indeformable; it is thus the 4-dimensional counterpart of Newton's absolute space. But it has over Newton's space the advantage of showing clearly why uniform motion cannot be detected. This point may be understood when we note that in space-time the history of a particle, whether it be in a state of rest or of uniform motion, is represented by a straight world-line,* and that there is no means of differentiating one straight world-line from another. We also understand why acceleration may be distinguished from rest or uniform motion; it is because, in the event of acceleration, the space-time point describes a curve and no longer a straight line. Though the general concept of an ether, filling space-time, may be retained, no apparent justification can be advanced in its favor, and so the electromagnetic field is usually regarded as symbolizing a peculiar condition (non-mechanical) existing throughout space-time itself. The special theory of relativity is as dualistic as Lorentz's theory, because a successful interpretation of matter and electrons in terms of the field magnitudes seems as remote as ever.

The special theory is not, however, confined solely to electromagnetic and to mechanical actions; it also sheds some light on the phenomenon of gravitation. For example, a feature of Minkowski's space-time is that a velocity exceeding the velocity of light *in vacuo* cannot be attained, a fact which suggests that no physical action can be transmitted with a speed greater than that of light. Action at a distance even in the case of gravitation, seems therefore to be ruled out by the basic tenets of the special theory. The condition of Lorentzian invariance, which is characteristic of the special theory, also restricts the form of all natural laws and thereby affords a clue to the correct law of gravitation. Poincaré set himself the problem of amending Newton's law of gravitation so as to bring it into agreement with the Lorentzian restrictive condition. He obtained a slightly modified form of the law of gravitation, a form in which the action is propagated with the velocity of light. Incidentally, Laplace's error (see page 75) was disclosed in this investigation, and the way opened for a field interpretation of gravitation. If Poincaré's treatment is accepted, we must assume that Minkowski's rigid space-time may be permeated by a gravitational field (presumably of a non-mechanical nature), which may be superposed on the electromagnetic field.

Einstein, in his general theory of relativity, followed a different line of attack. He discarded his original conception of a perfectly rigid, flat space-time and assumed that the space-time structure, though normally flat, would become warped in the neighborhood of matter. It was this warping which was responsible for the presence of a gravitational field in the ordinary sense. The gravitational field, in contradistinction to the electromagnetic field, thus received a geometric interpretation.

---

* The successive points in space occupied by a particle over a period of time are represented in 4-dimensional space-time by a succession of space-time points. The aggregate of these space-time points defines the world-line of the particle.

So as to obtain a clearer understanding of the new views, let us revert to our hypothetical illustration of the sun suddenly coming into existence. For convenience we shall express the situation in terms of the more familiar categories of ordinary space and time. The sudden creation of the sun generates a warping of the immediately surrounding continuum. The warping then spreads in all directions with the velocity of light, decreasing in intensity as it proceeds. When the warping reaches a planet, the planet acts as though it were attracted towards the sun. It is, however, the warping of space-time in the immediate neighborhood of the planet which is responsible directly for the force. As in all field theories of gravitation, the sun is thus only indirectly responsible for the action.

In point of fact the 4-dimensional gravitational field, which Einstein identifies with the space-time structure, is of a far more general nature than is a gravitational field in the ordinary sense. In common parlance we should not speak of a gravitational field in a region of space infinitely distant from matter. Yet, inasmuch as in a distant region, space-time still has a definite structure (flat), Einstein's gravitational field is still present. To avoid confusion on this score, it is usual to refer to Einstein's 4-dimensional space-time field as the ''metrical field'' or the ''guiding field.'' The name ''metrical field'' is justified by the fact that Einstein's field determines the geometry which we obtain when we perform measurements with rods or with rays of light. The name guiding field is given because a free body, *e.g.*, a planet, follows a geodesic world-line through the space-time structure, as though it were guided by a groove.

Einstein's celebrated field equations express the general law of varying (or permanent) structure which space-time may assume; a special case of this structure is the flat structure, which holds far from matter. In the language of ordinary space and time, the equations embody the laws which control the permissible distributions and changes of the gravitational field. As such, Einstein's gravitational equations are the gravitational analogues of Maxwell's equations of the electromagnetic field.

We now examine in what respects Einstein's law of gravitation differs from Newton's. One way of illustrating the difference is to express Einstein's law in terms of the artificial, though more familiar, categories of ordinary space and time. But before doing this, we must mention one of the formulations of Newton's law of gravitation. Suppose we are considering the space surrounding the sun. In Newton's treatment, no appeal is made to a field as connoting a condition of space, but we may nevertheless speak of a field of force in a mathematical sense. Let us then consider the field of force which surrounds the sun. Since the force at any point is represented in magnitude and in direction by an arrow, we are dealing with a vector field. Instead of considering the gravitational force, we may introduce a magnitude which is related to the force and which is called the "Newtonian potential." Newton's gravitational field is then represented equivalently by the distribution of the Newtonian potential. The potential, in contradistinction to the force, is expressed by a simple number, *i.e.*, a scalar. The field defined by the potential distribution is therefore a scalar field. The mathematical relation which determines the possible distributions of the potential around matter is given by a celebrated equation studied by Laplace (Laplace's equation).

With this preliminary information disposed of, we may return to Einstein's 4-dimensional gravitational field. Let us picture it in ordinary 3-dimensional space. If we assume that the field is permanent, as it would be around a permanent body such as the sun, the representation of the space-time field may be regarded as due to the superposition of two separate 3-dimensional fields: a scalar field and a tensor field. If the mass of the body (the sun) generating the field is not too great, this scalar field is found to be none other than the field of the Newtonian potential. We conclude that the first field of the theory of relativity, if it existed alone, would entail Newton's law of the inverse square. We now pass to the second relativistic field: the tensor field. If, as before, the mass generating the gravitational field is not too great and if, in addition, the velocity of the planet is not too high, the second field has only a small influence. To the second field are due the small deviations between the anticipations of Einstein and of Newton. In particular, the secular advance of the perihelion of Mercury must be attributed to it. To avoid misconceptions, we must mention that this splitting of 4-dimensional space-time into 3-dimensional space and a separate time, which we have here considered for motives of simplicity, is not only artificial but also inaccurate. Furthermore, it obscures the fact that the two gravitational fields mentioned form a single 4-dimensional unit, just as do the two fields in electromagnetics.

The entire general theory revolves around Einstein's 4-dimensional gravitational field-equations, and hence around the structure, or field, of space-time. For this reason the general theory of relativity is a field theory. But it is not a pure field theory, because, in addition to the field, we have to consider matter under the corpuscular form of electrons, protons, and the like. Insofar as the theory of relativity is concerned, the atomicity of matter is accepted as an empirical fact, just as is the rotundity of the earth; no theoretical reason for this atomicity is even suggested. Although a pure field theory would require that we account for the atomicity of matter in terms of the field magnitudes, this aim has yet to be realized. The situation does, however, appear more promising than it did in electromagnetics, for now we have two fields at our disposal—the electromagnetic field and the gravitational, or metrical, field.

Various attempts were made by Mie and others to construct a pure field theory, but they were only partly successful. Einstein then attempted the reverse course, seeking to account for the field in terms of matter distributed more or less uniformly throughout space. He was thus led to the hypothesis of the "cylindrical universe" and to the introduction of the cosmological constant λ, which is proportional to the mean density of matter. In view of the existence of competitive cosmological theories and owing also to recent astronomical discoveries, this latter part of the relativity theory is still extremely speculative, and we shall not dwell on it. Turning to the more solidly established aspects of the theory, we may say that it is as dualistic as Lorentz's. Its primary concern is the field, but, for empirical reasons, electrons and matter atomistically constituted must be introduced subsequently.

Unified Field Theories—The general theory of relativity deals with two separate fields, the electromagnetic field and the metrical (or gravitational) field. The electromagnetic field acts on electrified bodies and is itself influenced by their presence. Similarly, the metrical field acts on masses and is influenced by them. Up to this point, the analogy between the two fields is very great; but there are also important differences which we shall now consider.

In the first place, the metrical field receives a geometric interpretation since this field is a mere expression of the geometrical structure of space-time. The electromagnetic field, on the other hand, cannot be attributed to the space-time structure. It is true that the presence of an electromagnetic field modifies the structure, but we cannot regard this modification as accounting entirely for the presence of the field. The electromagnetic field must therefore be regarded as resulting from the presence in space-time of another type of physical condition, the electromagnetic tensor. Thus, whereas the metrical field is a characteristic of space-time itself, the electromagnetic field appears as a foreign intrusion.

Let us also observe that the entire relativity theory would collapse were space-time to have no structure: in the absence of a structure the distinction between space and time would vanish; there would be no geometry, no geodesics, and hence there would be nothing to regulate the paths and motions of bodies. The absence of a space-time structure being incompatible with the theory of relativity, we conclude that a metrical field is a necessity in this theory and cannot possibly be non-existent. But the electromagnetic field, being divorced from the structure, may quite well be absent without endangering the theory. A further dualism is thus revealed : the metrical field is essential and the electromagnetic field contingent. The unity of the theory of relativity is marred by this dualism.

The subsidiary rôle played by the electromagnetic field, as contrasted with the fundamental nature of the metrical field, entails interesting consequences. Thus, when we wish to determine the path and motion, say, of a planet, we first determine the space-time curvature around the sun and we deduce therefrom the lay of the geodesics (straightest lines). The world-line of the planet will then lie along the geodesic corresponding to the initial position and velocity of the planet. When the geodesic is known, the path and motion of the planet are obtained. This is what is meant when we say that the space-time structure guides the planet. But the situation is entirely different when, in place of a massive body moving in a gravitational field, we consider an electrified particle moving in an electromagnetic field : the world-line of the electrified particle no longer coincides with a geodesic and, indeed, exhibits no connection with the space-time structure. A considerable gain in unity would be obtained if we could establish some connection between the world-line of the electrified particle and the space-time structure, for then the world-lines of the massive body and of the electrified particle would both be dependent on the same underlying space-time structure. Now this unification would be secured if we could merge the two fields, identifying them both with the structural peculiarities of space-time. We should then have what is known as a "unified field theory."

The earlier development of the theory of relativity gives a clue to the method whereby the unification may be obtained. We saw that, in 3-dimensional space, the electromagnetic field may be regarded as resulting from the superposition of two fields, which are different though causally related. We then found that when 3-dimensional space is replaced by 4-dimensional space-time, the fusion of the two fields is realized. This circumstance suggests that any attempt to unify the electromagnetic and the metrical fields will presumably involve an extension of the fundamental 4-dimensional continuum. In Kaluza's theory, the extension is secured by the addition of a fifth dimension ; in Weyl's, no extra dimensionality is invoked, but, as against this, the continuum is held to be susceptible of a richer variety of structures. Weyl's theory was subsequently extended by Eddington. Kaluza, as a result of his introduction of a fifth dimension, was able to identify a geodesic of the fundamental continuum, not solely with the path and motion of a mass in a gravitational field, but also with the path and motion of an electron in an electromagnetic field. Einstein himself, after investigating a different generalization from those suggested by Weyl and Kaluza, is now said to be working on a modified form of Kaluza's suggestion. Though retaining 4-dimensional space-time, he nevertheless introduces field magnitudes that were thought to be possible only in a 5-dimensional world.

Any one of these attempts, if successful, would furnish a unified field, both gravitational and electromagnetic. The next step would then presumably be to secure a pure field theory, by interpreting matter and electrons in terms of the field magnitudes. The foremost theoretical physicists, however, have expressed the belief not only that a pure field theory is impossible, but also that field theories in general can give only a macroscopic picture of reality. The fact is that all field theories are rigorously deterministic, or at least have always been assumed as such. But quantum mechanics leads to the suspicion that rigorous determinism is a myth, and that the determinism accepted by classical science (and the theory of relativity) refers merely to average results, not to individual processes. If this be so, the field theories of the past can be compatible only with a macroscopic survey of phenomena.

In addition to these objections of a general nature, to which all field theories are open, there are others which pertain to Maxwell's theory in particular. According to Maxwell's theory of the electromagnetic field *in vacuo,* any disturbance emitted from a point-source should be propagated equally in all directions. Consequently, radiation from a point-source of light should consist of spherical waves. Maxwell's field theory is thus in conflict with the corpuscular nature of radiation, which recent experiment appears to demand. On the other hand, Maxwell's equations must certainly contain a large part of the truth, for their mathematical consequences express physical results which are nearly always verified by experiment. The difficulty may be explained away if we suppose that Maxwell's field theory represents an approximation to reality—an approximation which is sufficient for our needs so long as we view phenomena from a macroscopic standpoint.

One of the first successful attempts to interpret Maxwell's continuous electromagnetic field as the macroscopic manifestation of fundamentally atomistic processes is the theory developed by Dr. Leigh Page in 1914.

He postulated particles moving with the velocity of light and showed that a macroscopic condition which could be represented by Maxwell's field equations would arise. Dr. Page writes:

"Instead of being propagated through an elastic medium in the manner of material waves, electromagnetic waves are to be thought of as carried by discontinuities emitted from the source with the velocity of light." *

With the discovery of Heisenberg's uncertainty relations, the conflict between the continuous field and the discrete corpuscles of radiation assumes a novel aspect. The continuity of the field is connected with the theoretical uncertainty which prevents the precise localization of particles. In place of discrete particles we have blurred regions of space which give the impression of continuity. But it should be added that the entire subject is still extremely obscure.

---

* Leigh Page. The Emission Theory of Electromagnetism. Transactions of the Connecticut Academy of Arts and Sciences, 26, Jan. 1924; pp. 213-243.

CHAPTER XI

PHENOMENOLOGICAL THEORIES

THE source of our knowledge of the physical world resides in the sensations of light, heat, sound, touch, and the like, which we experience directly. The world of the infant consists in bare sensations. Then gradually, through the inferences to which cumulative experience gives rise, sensations are sorted out, connected, and formed into a mental picture of the outside world. The more or less conscious efforts of our infancy are soon forgotten or ignored, and unless we be inclined towards introspection and analysis, we find it natural to say that concrete objects, such as the table in our room, are observed directly.

It is useless to quibble over the meaning of words, and so we shall agree that, when no conscious inferences are made, a phenomenon will be said to have been observed directly. But suppose now that on peering through a microscope we see a strange pattern of rods and circles. Any person acquainted with a microscope will contend that he has observed the texture of the preparation placed in the slide. The inferential nature of this conclusion is, however, undeniable, for it would not be made by a savage. If, then, we insist on speaking of "direct observation" in connection with the microscope, we are certainly extending the previously accepted meanings of words.

Let us proceed a step further. The quantum theorists regard ultra-violet light as observable. Yet ultra-violet light is invisible. When we speak of ultra-violet light as observable, what we really mean is that certain observable occurrences (*e.g.*, those affecting a photographic plate) are attributed to the presence of ultra-violet light. But this conclusion is manifestly inferential, and to class ultra-violet light among the observable existents is to accept a still more extended meaning of words than was done in the example of the microscope.

The preceding analysis shows that a large number of the observables in physics are revealed through inference. If the inference is so plausible that it would be accepted without question by any man regardless of his training, the distinction between the observables of physical science and of ordinary life is not so striking. But when the inference forms a characteristic feature of some sophisticated theory, the magnitudes which are observable in principle according to this theory may cease to be so when, at some subsequent date, the theory is rejected.

As an example, let us consider the kinetic theory of gases, in which elastic molecules are assumed to be rushing hither and thither in accordance with the laws of mechanics. The pressure which the gas exerts against the walls of the enclosure may be observed directly, for instance, by the effort we have to exert on a piston to compress the gas. The inference characteristic of the kinetic theory is to ascribe this pressure to the impacts of the invisible molecules. In virtue of this inference the molecular impacts become observable, at least in much the same indirect sense that ultra-violet light is claimed to be observable. On the other hand, if we reject the kinetic theory, the impacts cease to be observable.

A further illustration is afforded by the concept of absolute acceleration. Both the classical theory and the general theory of relativity recognize that the so-called forces of inertia can be detected and measured. But whereas the classical theory ascribes these forces to an acceleration through absolute space, the general theory of relativity connects them with the contingent structure of space-time. The net result is that, from the classical viewpoint, absolute acceleration is revealed through the forces of inertia; whereas, from the standpoint of the theory of relativity, these forces betray something entirely different and absolute acceleration is meaningless.

Planck expresses the situation in the following passage:

"For the question whether a physical magnitude can in principle be observed, or whether a certain question has a meaning as applied to Physics, can never be answered *a priori*, but only from the standpoint of a given theory. The distinction between the different theories consists precisely in the fact that according to one theory a certain magnitude can in principle be observed, and a certain question have a meaning as applied to Physics; while according to the other theory this is not the case. For example, according to the theories of Fresnel and Lorentz, with their assumption of a stationary ether, the absolute velocity of the earth can in principle be observed; but according to the theory of Relativity it cannot; again, the absolute acceleration of a body can be in principle observed according to Newtonian mechanics, but according to Relativity mechanics it cannot. Similarly the problem of the construction of a *perpetuum mobile* had a meaning before the principle of the conservation of energy was introduced, but ceased to have a meaning after its introduction. Hence it is not sufficient to describe the superiority of Quantum-mechanics as opposed to classical mechanics, by saying that it confines itself to quantities and magnitudes which can in principle be observed, for in its own way this is true also of classical mechanics. We must indicate the particular magnitudes or quantities which, according to Quantum-mechanics, are or are not in principle observed; after this has been done it remains to demonstrate that experience agrees with the assertion." *

Planck's lucid presentation stresses the difficulty of deciding *a priori*, in the absence of a theory, whether a postulated magnitude is observable or not. We must remember, however, that in this discussion the word *observable* is used in a highly sophisticated sense. If, on the other hand, we take the word "observable" with its commonplace connotations, neither absolute space nor ultra-violet light is observable. The usual practice is to adopt neither the highly sophisticated definition of the word "observable" nor its commonplace meaning. A compromise is accepted, and the distinction between magnitudes that are observed and those that are not observed is taken to rest on the more or less intuitive appeal that the inferences from the accepted theories would have for the average man. Ultra-violet light, for instance, will be said to be observable, whereas absolute space and the motions of the molecules in the kinetic theory will be classed as unobservable. Some degree of vagueness is inevitable, as in all situations where a compromise is made, but in the main no serious disputes need arise. At all events, it is in this last, half-way sense that the word observable will be applied in the remainder of this chapter.

The purpose of a physical theory is to establish causal relations between the facts observed, so that, from given facts, others may be predicted. But a causal connection between two observed magnitudes, A and B, need not express a direct relation; more often we have a causal chain extending from A to B and involving connections between intermediary magnitudes. Here a difference in method may arise.

1. We may stipulate that all the magnitudes in the causal chain must be observable. Theories in which this restriction is imposed are called "phenomenological"; thermodynamics is the outstanding example. In this science the quantities that appear in the equations, or causal relations, are the observable magnitudes: pressure, volume, quantity of heat, temperature; and also auxiliary magnitudes, such as entropy, built up from the former.†

2. We may disregard the foregoing restriction and accept intermediary magnitudes which are not observed, but which are postulated so

---

* Max Planck. The Universe in the Light of Modern Physics. W. W. Norton and Co., New York, 1931, p. 49.

† Another magnitude which enters into the equations of thermodynamics is "internal energy," and it is questionable whether this magnitude can be said to be observable.

as to secure a causal connection between the facts that are observed. Theories of this type involve hidden occurrences and are obviously more speculative, for the evidence in favor of the precise hidden occurrences postulated is necessarily indirect and inferential.

Illustrations of the latter type of theory are found in the mechanistic theories and in the field theories discussed in the preceding chapters. For instance, the kinetic theory of gases postulates invisible occurrences; they consist in the disorderly movements and collisions of the molecules. Einstein's theory of the gravitational field affords another example. Thus, if we confine ourselves to what is actually observed, all we can say is that a stone, when released, falls to the ground. Our attitude is then equivalent to supposing that there is a direct relationship between the earth and the stone. But in the general theory of relativity the causal connection is indirect: the earth causes a warping of 4-dimensional space-time; and this warping, which is not observable, is then assumed to be the cause which directs the stone in its fall to the earth.

In many cases the same group of phenomena may be investigated either by a phenomenological theory or by one which introduces hidden occurrences (*e.g.*, both thermodynamics and the kinetic theory can be applied to gases). In the phenomenological theories, our attention is restricted to the macroscopic properties which appear on the commonplace level of experience. These macroscopic properties are viewed as fundamental, and no attempt is made to analyze them further. In the opposing theories, the macroscopic properties are dissected, as it were, and usually (but not always) microscopic entities are postulated to represent the fundamental reality behind the macroscopic appearances. For this reason we may often refer to the latter theories as "microscopic theories."

Lorentz, in connection with electromagnetic phenomena, developed both kinds of theory. Let us first examine the macroscopic aspect of electromagnetic phenomena in the presence of matter. These phenomena are represented by electric currents, charged bodies, and fields of force. Matter appears under the form of conductors and non-conductors (dielectrics). A macroscopic theory would involve systems of relations connecting these magnitudes directly. Lorentz, however, in his attempt to construct a macroscopic theory, proceeded indirectly by first concentrating on the microscopic theory. He assumed that, from the microscopic standpoint, an electromagnetic phenomenon would only include electrons, protons, and fields. No fundamental unit of magnetism was invoked, so that magnetism was viewed as a macroscopic effect, much as temperature is shown to be in the kinetic theory of gases. Similarly, the electric currents of the macroscopic level were ascribed to streamings of electrons. Lorentz's microscopic theory, represented by his microscopic equations, expressed the relations connecting the electrons and fields. Lorentz then obtained the macroscopic magnitudes by submitting the microscopic ones to an averaging process. Magnetism and electric currents reappeared, and the differences in the behavior of conducting and of nonconducting media were accounted for. Lorentz's macroscopic laws or equations were derived in a similar fashion from the microscopic equations by an averaging process, the microscopic equations being regarded as fundamental.

Mach and Ostwald in the course of the last century insisted that theories which postulate hidden occurrences should be rejected and that phenomenological theories alone should be retained. A somewhat similar philosophy has been defended in recent years by the quantum theorists, but the incentives of the quantum theorists differ so greatly from those of Mach that it will be advisable to examine them separately. We first consider the views of the Machian school; those of the quantum theorists will be discussed towards the end of the chapter.

According to the school of Mach and Ostwald, theories which introduce hidden occurrences smack of metaphysics and should be discarded in principle lest we fall into the errors of the metaphysicians. It seems difficult to accept this view, for the repugnance which the average man of science experiences towards the writings of the metaphysicians is due, not to the speculative spirit manifest in their schemes, but to their disregard or ignorance of the most elementary facts of mathematics and of physics. Inasmuch as Mach and Ostwald did not clarify their idea further, their criticisms appear to have been dictated by preference rather than by reason. At all events their views found favor with many physicists who, though of first rate ability in the experimental field, never evidenced any particular liking for mathematics. Many of them have expressed their mathematical dislike openly, and one cannot help suspecting that the more difficult mathematical technique of the microscopic theories may have influenced their philosophy.* The fact, however, that the critics most averse to mathematics usually sponsor thermodynamics as a theory suiting their taste would seem to mitigate the force of their criticisms, for if we accept thermodynamics we are automatically accepting mathematics as an adjunct to physical research. Under these conditions the degree of difficulty of the mathematics incorporated in a theory should play no part in a philosophical discussion, especially so since the difficulty of mathematics depends largely on the amount of study we have

---

* Ostwald was a chemist and Mach made no contribution to theoretical physics.

devoted to the subject. The only fair way to determine the merits of the two opposing kinds of theories is to examine the influences they have exerted on the development of science. We shall find that theories of both types have been of extreme importance and that they have supplemented one another. We consider the phenomenological theories first.

Thermodynamics, which is the prototype of these theories, requires for its development the most elementary kind of mathematics; a few superficial notions of the calculus suffice. Furthermore, owing to its neglect of microscopic differences, thermodynamics deals with relatively simple situations. All this has an important bearing on the nature of the problems that can be solved by thermodynamical methods. Many discoveries which have been arrived at with comparative ease in thermodynamics would have been arrested by hopeless complications had a microscopic theory been adopted. Gibbs's celebrated phase rule is a case in point.

Because of its rejection of speculative hidden occurrences, a phenomenological theory is on safer ground than a microscopic one. This may be an advantage from a certain point of view, but it would not seem to constitute any decided superiority. Scientific progress is not realized by following dogmatically the motto, "Safety First." If it were, the best course would be to do nothing and to speculate on nothing; and this would lead to utter stagnation. Besides, the safety and reliability of a phenomenological theory is usually more apparent than real.

For example, so long as we are guided solely by the methods of thermodynamics, we have no reason to cast doubt on the law of entropy. Yet when this law is investigated by means of the kinetic theory of gases, a microscopic theory, its validity is seen to be only statistical. We may readily understand why thermodynamics failed to reveal the true situation, and why similar limitations are to be expected in all phenomenological theories. Thus when we restrict ourselves to a phenomenological, or macroscopic, theory, the only relations and properties that we are concerned with are those disclosed by direct observation. The mathematics of the theory then enables us to derive the necessary consequences of these observed relations, and by this means predictions can be made. But the drawback to this procedure is that the mere refusal on our part to take unobservable occurrences into consideration does not suppress the possibility of their existence; and if such occurrences do exist, they too will entail phenomena, some of which may be perceptible. If these latter perceptible phenomena are generated only on very rare occasions, our original direct observations may fail to detect them; and the net result will be that our macroscopic theory will leave us in complete ignorance of their existence. For these reasons a phenomenological doctrine is not necessarily as reliable as we might suppose.

Other illustrations may be given. The theoretical values of the chemical constants of thermodynamics, the reasons why some chemical reactions progress more slowly than others, catalytic effects and the like, are beyond the province of thermodynamics as such. Here speculative assumptions which are certainly not observable (in the sense we have adopted) must be introduced.

A similar situation occurs in the quantum theory. Physicists were seeking the law of equilibrium radiation. Thermodynamical arguments proved of considerable help by imposing a restriction on the form of the unknown law.* But thus far and no further could they go. Planck's microscopic assumptions were required to derive the radiation law which bears his name.

We have mentioned the most conspicuous advantages and disadvantages of the phenomenological theories. Let us now pass to the microscopic theories. Much light is cast on the various problems at issue when we examine the reasons that gave rise to these theories. We recall that physical science originated from the accumulation of observational and experimental facts. Since it was the magnitudes and conditions observed on the commonplace, or macroscopic, level which forced themselves most immediately on the attention of physicists, the earlier facts necessarily pertained to macroscopic properties; and as a result the empirical laws secured by connecting these facts were likewise of the macroscopic variety. But investigators did not rest content with this mode of procedure, and even in Newton's day we find microscopic theories of the corpuscular and also of the continuous type developed in connection with optical phenomena. In the eighteenth century Daniel Bernoulli conceived the kinetic theory of gases. Bernoulli was undoubtedly guided to his theory by the empirical facts which suggested a corpuscular constitution for matter. Even so, in view of the hypothetical nature of his assumptions, we may well inquire what advantage he saw in taking this leap in the dark. Why should Bernoulli not have been satisfied with the empirical laws and causal relations established by direct experiment? Presumably, his desire was to obtain a better understanding of things—to peer behind the scene with the hope of tracing the various macroscopic properties to a common cause. The appreciation of what constitutes the most satisfac-

---

* The restriction leads to the conclusion that the law of radiation must have the general form expressed by Wien's relation (page 453).

tory form of understanding is, however, largely a personal matter which differs from one individual to another. We shall not dwell therefore on the motives that may have dictated Bernoulli's course. Instead, let us judge his kinetic theory by its results.

Thanks to the mathematical development of the kinetic theory at the hands of Maxwell and Boltzmann, unsuspected peculiarities of gases were revealed; for instance, Maxwell showed that the viscosity of a gas is independent of its pressure. More important still, as we mentioned previously, the law of entropy was shown to have but a statistical validity. The usefulness of microscopic theories is further revealed in electromagnetics, for Lorentz's microscopic theory enabled him to obtain his macroscopic equations and to supplement the deficiencies of Maxwell's equations in the presence of matter. To this microscopic theory also is due the prediction of the Zeeman effect and other phenomena.

From this analysis we conclude that both kinds of theories have their advantages and their disadvantages, and that the type of theory to be preferred depends largely on the viewpoint we propose to adopt. When we deal with matter in bulk, thermodynamics is sufficient for our needs, the exceptions to the principle of entropy occurring on such rare occasions that as yet no one has ever observed them. But if our concern is with minute volumes, the microscopic theories cannot be dispensed with. The fame acquired by an investigator in developing a theory of one type or the other is not necessarily a proof that his preferences lean in one direction. More often it is the circumstances surrounding his discoveries which govern his course. Indeed many of the greatest theoretical physicists have devoted their efforts to the development of both kinds of theories. Maxwell, Gibbs, Planck, and Einstein are among the most distinguished.

The New Quantum Theories—The new quantum theories, developed by Heisenberg, Born, Pauli, Dirac, and Bohr, are usually classed as phenomenological. When Heisenberg first formulated his Matrix Theory, considerable stress was placed by him on the necessity of introducing none but observable magnitudes in the equations. Heisenberg's attitude might be thought to be the same as Mach's. But a closer examination of the new theories and of their recent developments shows that this belief would be unwarranted.

In the first place, we must remember that the quantum theorists are among the foremost theoretical physicists of the present day, so that the mathematical difficulties which may have colored the philosophy of the Machian school need not be reckoned with. Furthermore, we shall see that the quantum theorists did not start out by rejecting hidden occurrences on principle. In fact their initial attempts were directed towards the microscopic theories, and only after repeated failures was a more phenomenological outlook adopted as a last resort. Finally, we are scarcely justified in calling the new quantum theories phenomenological; for though their equations connect only observable magnitudes,* these magnitudes are not connected directly, but only indirectly by means of an underlying network of intricate mathematical relations. The relations are of a degree of abstraction never before encountered in physical theories, and they cannot be visualized in terms of familiar notions. What we are witnessing is an extreme mathematicizing of physical Nature. It would therefore seem that the new quantum theories are more adequately expressed by the appellation "mathematical theories." From the standpoint of the disciple of Mach and Ostwald, the present situation is worse than ever before.

The motives and characteristic features of the new theories will best be understood if we follow the historical order of their appearance. This leads us back to Bohr's mechanistic theory of the atom, because it was the failure of Bohr's theory to give a satisfactory account of subatomic phenomena that precipitated Heisenberg's matrix method. When Bohr initiated his theory, the empirical evidence indicated that an atom was similar to a miniature solar system. Thus an atom was supposed to be represented by a central nucleus (built up of electrons and protons) which had a net positive charge and around which electrons were circling in accordance with the laws of mechanics. In this model the electrostatic attraction of the nucleus on the electrons played the part of the sun's gravitational attraction on the planets. The simplest of the atoms was the hydrogen atom, in which a single electron was circling around a proton as nucleus. The model just described represents the hidden microscopic occurrences, the radiations emitted from the atom constituting the observed events which must receive their interpretation from the invisible motions. The foregoing conception of the atom was adopted by Bohr, so that his theory was a mechanistic theory of the corpuscular type.

The application of the classical laws of mechanics and of electrodynamics to this model did not, however, account for the precise radiations observed. Even in the atom of hydrogen, which because of its simplicity offered the best hope of success, the distinctive spectral series could not be interpreted. The originality of Bohr's theory consists in imposing so-called quantum restrictions on the classical mechanical laws,

---

* These include ultra-violet light.

and also in rejecting classical electromagnetics in its entirety. Thanks to Bohr's restrictions and postulates, the spectral series of hydrogen were accurately predicted, and since Bohr's theory was the only theory then known which was able to secure this result, it was accepted. The restrictions imposed by Bohr were not exactly hypotheses *ad hoc* postulated for the sole purpose of interpreting the hydrogen spectrum. They were natural extensions of the quantum assumptions previously introduced by Planck in a totally different field of investigation; namely in the study of black-body radiation. Bohr's results served therefore as an indirect confirmation of Planck's original assumptions and strengthened the belief that, with quantum phenomena, a new body of physical occurrences had been uncovered.

At first sight it would seem that two different sets of mechanical laws should be assumed: firstly, the classical (or relativistic) laws which hold for the planets of the solar system and more generally for bodies moving outside the atoms; secondly, the quantum-restricted mechanical laws which regulate motions inside the atom. In itself this dualism is unsatisfying; besides, since an atom has no definite boundary, we cannot decide under what conditions an electron will be moving inside or outside a given atom. Bohr proved, however, that the dualism was more apparent than real. He showed that the effect of the quantum restrictions would tend to vanish when the electron was describing a larger orbit round the atomic nucleus, or when the dimensions of the mechanical system were increased.* According to this view, classical mechanics supplemented by the quantum restrictions constituted the correct mechanics; classical mechanics in the absence of the restrictions was a mere approximate doctrine tending to become valid only on the macroscopic level of experience.

Thus far we have been concerned with the hydrogen atom, but the situation became hopelessly complicated when the heavier atoms were considered and when a more careful examination was made of the atomic spectra. It was found necessary to introduce, as and when needed, additional quantum restrictions on the mechanical laws. The expression of some of the mechanical magnitudes had to be modified and the mechanical principles revised accordingly. Even when all these changes were made, Bohr's theory was unable to predict with precision the phenomena actually observed. The belief generally entertained at this stage may be expressed in the following statement:

There is but one mechanics holding within and without the atom, and it differs considerably from the classical variety formerly thought to be

---

* A similar situation was shown to hold for the electromagnetic modifications which Bohr had postulated.

correct. On the macrascopic level we may accept classical mechanics without danger, for the refinements of the new mechanics then cease to exert perceptible effects; but if we are dealing with a microscopic mechanism, the refinements become perceptible, and we obtain totally incorrect results unless the correct mechanics is applied. Unfortunately the precise form of the correct mechanics is unknown, so that we cannot construct a mechanistic theory of the atom. Thus, though in principle a mechanistic theory of the hidden-occurrence type is conceivable for the atom, it must be rejected in practice.

This brief summary gives an idea of the situation that was responsible for the rejection of Bohr's mechanistic theory and that prompted Heisenberg to devise a new method of attack. Heisenberg, in his matrix theory, did not deny that an atom was constituted as had been assumed by Bohr, namely, by electrons circling round a nucleus. But he pointed out that since the laws regulating the electronic motions were utterly mysterious, the best course was to disregard these motions and to concentrate on the radiations emitted from the atomic system. Accordingly in Heisenberg's method, the positions, motions, and orbits of the electrons in the atom were not mentioned, and the aim was to connect the radiations by appropriate mathematical schemes. Thus, it was under the stress of circumstances, and not by choice, that Heisenberg and the other quantum theorists came to formulate a phenomenological theory of the atom. No *a priori* rejection of the hidden-occurrence theories is found in the writings of any of these men. This last point is sufficiently obvious when we note that in addition to Bohr himself many of the other quantum theorists, chiefly Born, Pauli, and Heisenberg, had devoted considerable time to Bohr's original mechanistic atom.

The following passage, written by Born in 1926, exhibits the idea we are expressing.

"We therefore stand before a new fact which forces us to decide whether the electronic motion or the wave shall be looked upon as the primary act. After all theories which postulate the motion have proved unsatisfactory we investigate if this is also the case for the waves." *

A study of Heisenberg's matrix method shows that it is not so phenomenological as it is sometimes claimed to be. Thus, when we treat the problem of the hydrogen atom by the matrix method, we assume that the atom is formed of a proton and an electron which attract each other

---

* M. Born. Problems of Atomic Dynamics. Massachusetts Institute of Technology; 1926; p. 70.

In the above passage Born uses the word "waves" in place of "radiations."

according to Coulomb's law of the inverse square. To this extent, invisible occurrences are postulated exactly as they were in Bohr's theory. In the second place, we find that although Heisenberg never introduces the position and momentum of the electron in his equations, he nevertheless appeals to magnitudes which play exactly the same rôle as do the mechanical magnitudes in Bohr's theory.* If, then, by a phenomenological theory we mean one which postulates no hidden occurrences, Heisenberg's matrix method cannot be said to be of this type. Besides, it is questionable whether all the magnitudes which Heisenberg introduces in his equations are truly observable. The phases of the vibrations, for instance, cannot be observed, and yet we cannot avoid considering them in the application of the theory. Finally, there is a last point which differentiates the matrix theory from the commonly-accepted standard of a phenomenological theory, such as thermodynamics. In thermodynamics the observed magnitudes are linked directly; but, as we mentioned on a previous page, in the new quantum theory the observed magnitudes are connected by an elaborate mathematical substructure of the most abstract type. It can be visualized only when we represent it in a space which has an infinite number of imaginary dimensions.

A question that might be asked is whether the matrix theory has been successful in disclosing the unknown mechanical laws of the subatomic world. The answer is in the affirmative, but at the same time it assumes an unexpected form, for the matrix theory and the other quantum theories lead to the conclusion that there are no exact microscopic mechanical laws; the true mechanical laws merely restrict uncertainties. One of the mistakes in Bohr's theory was to assume rigorous microscopic laws and then attempt to guess at them. The new theories also show that the mechanical restrictions which Bohr introduced in his theory of the atom are mere manifestations of the fundamental natural uncertainties discovered by Heisenberg.

At about the time Heisenberg was developing his new theory, de Broglie suggested a theory in which mysterious waves (not to be confused with waves of radiation) play a controlling part. De Broglie accepted the simultaneous presence of waves and particles; the office of the former was to guide the latter. In the hydrogen atom, de Broglie retained Bohr's mechanical model of an electron circling around a proton, but he did not rely on the classical mechanical laws to determine the motion; instead, he appealed to the controlling waves. The waves were regarded as physic-

* We refer to the matrices q and p and to the adoption of the classical Hamiltonian function of the problem.

ally real, so that de Broglie's original theory was a compromise between a mechanistic theory based on corpuscles and a theory involving continuity. Furthermore, since neither the electrons and their motions, nor the waves, are observable, we must conclude that de Broglie's theory is not a phenomenological one. Schrödinger developed some of the implications of de Broglie's waves and came to the conclusion that these waves must in many cases be supposed to be propagated in multidimensional spaces. This circumstance in itself prevents the waves from being physically real, as de Broglie had assumed them to be. Gradually, and in spite of de Broglie's resistance, the wave theory lost more and more of its physical attributes and became an abstract theory in which the substructure was entirely mathematical. Finally, Schrödinger's wave theory was proved to be mathematically equivalent to Heisenberg's matrix theory. In our opinion the fusion of these two theories, which were initially conceived in a totally different spirit, is of considerable importance. It shows that the individual preferences of the various investigators (whatever they may have been) have had no effect on the final results. The highly mathematical scheme to which both theories lead would thus appear to spring from the nature of things rather than from some passing vogue.

CHAPTER XII

A RETROSPECT ON THE DIFFERENT KINDS OF THEORIES

In the preceding chapters we have examined the more important features of mechanistic theories, of field theories, and of phenomenological theories. As for the modern quantum theories, they are best described by the qualification "mathematical." Jeans, commenting on the status of modern physical science, illustrates this idea in the words: "The Great Architect of the Universe now begins to appear as a pure mathematician."

A characteristic feature of the mechanistic theories, whether of the corpuscular or of the continuous type, is that they postulate underlying occurrences which are relatively simple to visualize. This feature is only natural since the occurrences postulated were suggested by phenomena which are familiar to us on the macroscopic level, for instance by swarms of particles or by elastic jellies. We cannot so easily form a mental picture of the field theories, especially when we are dealing with a field in 4-dimensional space-time. In particular, attempts to visualize the curvatures of space-time in the general theory of relativity would probably lead to incorrect conclusions. Nevertheless the difficulty we experience is due solely to the four dimensions of the warped continuum: if space-time were two-dimensional, its warpings could easily be imagined. The practical impossibility of representing the situation is therefore more one of degree than of essence.

In the phenomenological theories of the classical period, e.g., in thermodynamics, the urge to visualize did not come into consideration because the equations of the theory expressed direct connections among the magnitudes observed. The situation is different in the modern quantum theories. Here we have an underlying mathematical scheme which we might wish to render more concrete by picturing it as the expression of physical occurrences. However, efforts in this direction have failed, and we must be content with mathematical symbols. It is true that quantum phenomena have been interpreted by means of waves and that the wave form of representation might seem to afford a more physical model. But the similarity of the underlying occurrences to waves is deceptive, for these waves differ in too many respects from those we see moving over the surface of a pond.

Whatever may be said of the urge to visualize, we must agree that it is a natural psychological trait, issuing from the desire to interpret the unfamiliar in terms of the familiar. Any man, when faced by an unfamiliar situation, first tries to cope with it according to the familiar methods which have been of service to him on other occasions. He may subsequently discover the inefficacy of his measures, but only by testing them can he convince himself of this fact. Now in the physical world, mechanical categories are the ones with which we are best acquainted. We are surrounded by them on the commonplace level, and they cannot well escape our notice. Mechanics, moreover, was the first of the sciences to attain a considerable mathematical development, and so the earlier theoretical physicists, on looking around for hidden occurrences on which to base their theories, naturally selected mechanical ones. Indeed, no others were known before the days of Maxwell.

Today, for reasons which will appear presently, mechanistic theories are discredited. But Jeans and Eddington, in their popular writings (Jeans especially), convey the impression not only that mechanistic theories must be rejected in view of the discoveries in modern science, but also that these theories are totally wrong departures and never should have been considered. The claim is made that these mechanistic theories have led physicists astray, forcing them into a blind alley whence science has had to retrace its steps.

Jeans's main contention is undoubtedly correct, for there is little doubt that mechanistic theories are gone, never to return. But this does not necessarily imply that *a priori* cosmological beliefs were in any sense responsible for the mechanistic theories of the past, and that, had the earlier physicists been blessed with greater wisdom, theories of this type would never have arisen. Mechanistic theories found protracted favor simply because they seemed to account, with a high degree of accuracy, for the phenomena observed. In the kinetic theory of gases, for example, theory and experiment were in remarkable agreement in the days of Maxwell and Boltzmann. Only recently, thanks to the improvement in experimental technique, has the behavior of gases at exceedingly low temperatures been studied and have important discrepancies been observed. When these discrepancies, and also others relating to kindred mechanistic theories, were established beyond any reasonable doubt, mechanistic theories were immediately abandoned by an overwhelming majority of theoretical physicists without any particular sign of reluctance—manifest proof that no deep-seated beliefs were being discarded.

But there is also another aspect to the problem. The recent developments of the quantum theory show that as we near the commonplace level of ordinary experience, the mathematical scheme underlying the new quantum theories tends to merge into the mathematics derived from the classical mechanistic theories. This is but another way of saying that as we near the macroscopic level, the underlying occurrences may be represented by schemes that are mechanistic in the ordinary sense.* Indeed it is surprising to find how far the classical mechanical analogies may be accepted without involving us in perceptibly incorrect anticipations.

Compton for example, in his original interpretation of the effect which bears his name, assimilated the collision of a photon and electron to that of two perfectly elastic billiard balls. The Compton effect is no longer treated in this crude mechanistic way today; nevertheless, Compton's mechanistic interpretation gives very approximate results and it prompted him to further discoveries. The important rôle played by the mechanistic theories in the progress of science must also be borne in mind. The mechanistic methods developed by Boltzmann and by Gibbs in the classical kinetic theory of gases were transplanted into the new gas theories, so that to this extent the classical theory served as a model for the modern theories. It would be difficult to imagine how, in a single leap, science could possibly have passed from the macroscopic gas laws of Boyle and Charles to the extremely abstract mathematical schemes of the new gas theories. Similarly, the modern quantum theories of the atom could scarcely have arisen had it not been for Bohr's prior mechanistic theory.

From this analysis we conclude that the mechanistic theories of the past should be viewed as first approximations, and that subsequent discovery has served to refine but not altogether to destroy them. Furthermore, dogmatic assertions on the finality of the modern theories are out of place. For all we know, the modern theories in turn may be only approximations which will appear strangely crude twenty years hence. The field theories, in particular, cannot be the last word—they may be accepted only provisionally till a higher stage of approximation is reached. The progress of science must be viewed as the forging of a chain; each link has its place and cannot be suppressed without breaking the continuity of the chain; and the mechanistic theories represent one of these indispensable links.

Although we believe that Jeans's condemnation of the mechanistic

---

* The essence of this discovery is embodied in Bohr's Correspondence Principle.

theories is too severe, his stand becomes comprehensible when we take into consideration the general attitude of English physicists in the last century. The possibility of devising a familiar model of underlying physical occurrences was long regarded in England as an argument in favor of a theory. Kelvin once said that he could not understand a phenomenon unless he could devise a mechanical model of it. With these promptings, he proceeded to construct models of atoms by postulating vortices in the ether; the idea was suggested to him by Helmholtz's discovery of the great stability of vortex motions. Even more recently we find Larmor interpreting the electromagnetic field as due to motions of the continuous ether. Finally, at the Solvay Congress of 1911, the only two scientists to dissent strongly from Planck's quantum-interpretation of radiation, and to suggest in its place theories based on familiar models, were the two English representatives, Jeans and Lord Rayleigh.* But mechanistic interpretations did not find the same favor among the scientists of other countries. These thinkers leaned more towards analytical (mathematical) theories, and although viewing models as having a certain utility, claimed that on no account should the models be taken too seriously. Boltzmann expresses this view when he writes in the introduction to his Kinetic Theory of Gases:

"In England mechanistic ideas on the nature of heat, and on atomism in general, have been adhered to. But on the Continent the same is not true...."

It is also instructive to examine Boltzmann's own attitude in connection with his kinetic theory. In the introduction previously mentioned he writes:

"Who can prophesy the future? Let all ways remain open therefore, and away with all dogmatic statements, either for or against atomistic ideas! Besides, when we present ·the theory of gases as a complex of *mechanical analogies,* we are in fact indicating, by the choice of this expression, how far we are from admitting, in any certain fashion and as a reality, that bodies are from all standpoints constituted by very small particles."

Elsewhere, discussing the significance to be attached to models, Boltzmann continues:

"The question of the utility of atomistic conceptions is of course quite independent of the fact on which Kirchhoff has insisted, that our theories bear the same relationship to Nature as symbols do to the entities they

---

* Lord Rayleigh could not attend the congress but wrote a letter supporting the stand taken by Jeans.

stand for, or as the letters of the alphabet bear to the voice, or as notes do to music. It is also independent of the question of ascertaining whether it would not be better to call these theories simple descriptions, so as to remind us always what their status is with regard to Nature.''

In a similar vein, Poincaré writes: ''Maxwell's theory is Maxwell's system of equations.'' Poincaré's statement implies that the mathematical relations embody what is essential in a theory and that the conceiving of a model does not contribute anything of importance. The reader is referred to Poincaré's chapter on ''Hypothesis''.

We have mentioned these points because, in our opinion, the revulsion against mechanistic theories heralded by Jeans and other English physicists arises from a belated recognition on the part of English science that mechanism is not the final solution.

CHAPTER XIII

PSYCHOLOGICAL DIFFERENCES AMONG PHYSICISTS

In the vernacular, the term physicist is applied both to the experimenter and to the theoretical investigator; and the awarders of the Nobel prize make no distinction between the two. Among the physicists to receive the prize in recent years are Chadwick, M. and Mme. Joliot, Anderson, Hess, Heisenberg, Dirac, Schrödinger, and de Broglie. The first five are experimenters, whereas the last four are purely theoretical men who have never contributed to the experimental field.

There are significant differences between the qualifications of these two types of investigators. The successful experimenter must be gifted with infinite patience, dexterity, and mechanical ingenuity in the devising and constructing of apparatus. On the other hand, he need have but little knowledge of mathematics. The theoretical investigator may be manually clumsy; a pencil and paper are his sole instruments. He must, however, be fully conversant with the experimental results obtained in all spheres of physics and must have a wide knowledge of those parts of mathematics which are of use to him in his work. His interest is primarily in coordinating results, if necessary by means of the introduction of auxiliary hypotheses. To the philosophic syntheses he creates, the name "natural philosophy" rightfully applies. As may be gathered, the qualifications necessary for the experimenter and the theoretical physicist differ so widely that very few men have been gifted and learned enough to distinguish themselves in both fields. Fresnel and Hertz were among the privileged few.

During the course of the present century the gap between the two groups of investigators has rapidly widened: on the one hand, successful experimentation is making ever-increasing demands on laboratory technique; on the other, the more recent physical theories make use of mathematical doctrines which are far from elementary. Thus, whereas we may study thermodynamics or even Newton's theory of gravitation with a minimum of mathematical knowledge, we could never expect to gain any profound insight into the theory of relativity or into the quantum theory without an extensive mathematical preparation demanding years of preliminary study. When it comes to determining which of the two types of investigators has contributed most to the advance of natural philosophy, no answer can be given; theorists and experimenters supplement each other. Without the experimenters, or gatherers of facts, there would be no facts on which to speculate; and without the theoretical investigators, or amalgamators of facts, there would be no possibility of interpreting the various facts and of establishing their relations.

As may well be imagined, the experimenters as a whole evince little sympathy with the modern theories of mathematical physics. The advanced mathematics involved, and the impossibility of obtaining concrete pictures of the underlying occurrences, act as a deterrent to many of the laboratory workers. The layman who wishes to gain a general idea of the reactions of the foremost physicists to the modern theories should therefore be careful to ascertain whether the authority, whose opinions he is weighing, has acquired his reputation in the experimental or in the theoretical field.

But even if we restrict our attention to the theorists, we find sharp divergences. When confronted with two possibilities, some theorists will pursue the first alternative while the others will follow the second. We have encountered several illustrations in the preceding chapters. Kelvin favored mechanical models, whereas other physicists preferred the more mathematical theories. Planck and Einstein refuse to countenance the rejection of the causal doctrine, whereas Bohr, Heisenberg, and Dirac do not regard this doctrine as unassailable and experience no particular qualms in abandoning it. Similarly, absolute space was not repugnant in principle to Newton, Euler, Lagrange, Laplace, and many others including Planck among the moderns; whereas it was abhorrent to Mach. Action at a distance seems to have had no defenders among the leaders in natural philosophy, though the classical theory of gravitation accepted it implicitly. Finally, there have always been investigators who have preferred the phenomenological outlook of thermodynamics to the more speculative one which prevails in the kinetic theory of gases.

In this matter of psychological differences, it is of interest to note the marked monistic tendencies of Einstein. By his monistic tendencies, we mean his desire to interpret what appear to be different physical categories in terms of some common category. Of course, this tendency is general among natural philosophers, but it is more apparent in Einstein than in others: dualism seems to be definitely repugnant to him. For example, when discussing the coexistence of material corpuscles and of Newton's corpuscles of light, he writes:

"Moreover, the introduction into the theory of the two entirely different kinds of material particles that are necessary for the representation of ponderable matter and of light is in itself most unsatisfactory; and later there appear the electric corpuscles as a third kind of particle, with again entirely different properties." *

In another passage where he discusses the coexistence of the mechanical and field categories in electromagnetics, he says:

"But this change led to a fundamental dualism which in the long run was insupportable." †

Einstein's formulation of the general theory of relativity in which he identifies inertial mass with gravitational mass is another case in point. So also is his attempt to rid the general theory of the dualistic conception it confers on the structure of space-time—a structure which appears to be part due to matter and part due to the nature of space-time itself. We may recall that it was in furthering this attempt that he imagined the cylindrical universe.

In this connection, Einstein writes:

"The physical properties of space would not then be wholly independent, that is, uninfluenced by matter, but in the main they would be, and only in small measure, conditioned by matter.   Such a dualistic conception is even in itself not satisfactory." ‡

On the other hand, neither de Sitter nor Eddington have objected to the dualism to which Einstein refers.   In his latest attempt to fuse the metrical and the electromagnetic fields, Einstein is furnishing another illustration of his monistic tendencies.

In some cases the conflict in the points of view of the various thinkers is striking.   The following passage, taken from the records of the Solvay Congress of 1927, illustrates Lorentz's reactions to the new concepts of the quantum theory which were there under discussion:

"We wish to obtain a representation of phenomena and form an image of them in our minds.   Till now, we have always attempted to form these images by means of the ordinary notions of time and space.   These notions are perhaps innate; in any case they have been developed by our daily observations.   For me, these notions are clear, and I confess that I am unable to gain any idea of physics without them. . . .
"For me, an electron is a corpuscle which at any given instant is situated at a determinate point of space, and if I believe that at the following instant this corpuscle is situated elsewhere, I attempt to imagine its path, which is a line in space.   And if this electron meets an atom and

---

* James Clerk Maxwell; Cambridge, 1931; p. 68.
† Einstein. Sidelights on Relativity. E. P. Dutton and Co., 1920; p. 8.
‡ Einstein. The Meaning of Relativity. Princeton, 1923; p. 109.

penetrates in its interior and, after several adventures, leaves the atom, I attempt to construct a theory in which this electron has retained its individuality . . .

I would like to retain this ideal of other days and describe everything that occurs in the world in terms of clear pictures.'' *

Contrast this passage with Dirac's statement:

''The only object of theoretical physics is to calculate results that can be compared with experiment, and it is quite unnecessary that any satisfying description of the whole course of the phenomena should be given.†

The conflict between Lorentz and the quantum theorists brings to light an interesting point. It is the extraordinary rapidity with which science has developed in the last forty years. The theory of relativity was the work of two men in their twenties, Einstein and Minkowski. With the sole exception of Poincaré, no outstanding theoretical physicist past fifty years of age would countenance it. Those who did not oppose it outspokenly, complained that they were too told to be interested in such revolutionary ideas, and that the task should be entrusted to the younger generation. In those days Einstein was viewed as an extreme radical in science. As time passed, the theory of relativity was accepted, and today no natural philosopher of repute has expressed his dissent or suggested any modification of it (except in the most speculative part pertaining to the universe as a whole).

Then came quantum mechanics, also the work of men in their twenties, and history has repeated itself. Planck and Einstein, the extreme radicals of yesterday, now appear as conservatives, taking little part in the recent developments though opposing the new doctrines. It is not that Planck and Einstein have changed; it is merely that the new theories are so revolutionary that the supposedly revolutionary theories of thirty years ago (theory of relativity and earlier quantum theory) now appear to be almost classical by contrast. If the past is any guide to the future, we may safely predict that, thirty years hence, Heisenberg and Dirac, the extreme radicals of today, will in turn be regarded as conservatives and will resist the still newer doctrines of tomorrow.

On the whole, the psychological differences that we have noted among natural philosophers have not occasioned incompatible scientific developments. For unless the divergences relate to problems which are insoluble

---

* Electrons et Photons, p. 248. Solvay Congress of 1927.
† Dirac. The Principles of Quantum Mechanics. Oxford, 1930; p. 7.

(such as we suspect the doctrine of determinism to be), there eventually comes a time when new discoveries or some crucial experiment settle the dispute. No one conversant with modern science would suggest a return to the mechanistic theories; we doubt whether even Kelvin, if he were alive, would champion them.

Closely related to the personal preferences of the different investigators is the emphasis these men place on one or another of the various features that may characterize the same conceptual notion. For instance, a notion, such as absolute space, may under ordinary conditions be associated with different attributes A, B, and C. Suppose now the attribute A is lacking. The writer who feels that A is less important than B or C may not deem it necessary to change his terminology and may still speak of absolute space. Another writer who views the attribute A as essential will hold that, in its absence, space can no longer be called absolute. For those who are in a position to follow the mathematical explanations which accompany the presentation of a physical theory, differences of terminology are of minor importance, because the meaning the writer wishes to ascribe to his words is sufficiently clear. But the layman and the metaphysician, who may be interested in science and who must rely on popular books, are likely to be misled by seemingly contradictory statements made by the various theoretical physicists.

The theory of relativity, even more so than the quantum theory, appears to have provoked much confusion in this connection. For lack of space, we shall not discuss exhaustively the various meanings that are attributed to such words as relative, absolute, force, real, fundamental. However, as an illustration of the confusions that may occur, we shall examine some of the seemingly contradictory statements that have been made by leading physicists on the nature of 4-dimensional space-time. All theoretical investigators agree that the 4-dimensional space-time of the special theory of relativity is a four-dimensional amplification of Newton's absolute space. But whereas Weyl views the space-time of the general theory as deprived of all absolute features, Planck still stresses its absolute nature. Einstein's attitude is somewhat different and we shall examine it later. A few quotations will clarify the situation. We start with the space-time of the special theory. Einstein writes:

"The principle of inertia, in particular, seems to compel us to ascribe physically objective properties to the space-time continuum. Just as it was necessary from the Newtonian standpoint to make both the statements, *tempus est absolutum, spatium est absolutum,* so from the standpoint of the special theory of relativity we must say, *continuum spatii et* *temporis est absolutum.* In this latter statement *absolutum* means not only "physically real," but also, "independent in its physical properties, having a physical effect, but not itself influenced by physical conditions." *

Weyl expresses himself in equivalent terms. Thus he writes:

"Where do the centrifrugal and other inertial forces take their origin? Newton's answer was: in absolute space. The answer given by the special theory of relativity does not differ essentially from that of Newton. It recognizes as the source of these forces the metrical structure of the world and considers this structure as a formal property of the world." †

We now pass to the space-time of the general theory. To quote Weyl:

"The falling over of glasses in a dining-car that is passing round a sharp curve and the bursting of a fly-wheel in rapid rotation are not, according to the view just expressed, effects of 'an absolute rotation' as Newton would state but whose existence we deny; they are effects of the 'metrical field' or rather of the affine relation associated with it." ‡

The significance of this statement is to rule out absolute rotation and thereby the absolute continuum with respect to which rotation would acquire a meaning. But let us see what Planck has to say. Planck, commenting on the space-time of the general theory of relativity, writes:

"If from space and time we should take away the concept of the absolute, this does not mean that the absolute is thereby banished out of existence, but rather that it is referred back to something more fundamental. As a matter of fact, this more fundamental thing is the four-dimensional manifold which is constituted by the welding together of time and space into a single continuum. Here the standard of reference and measurement is independent of the observer." ¶

Weyl and Planck appear to be at loggerheads; and yet, when the mathematics of the theory is understood, we find that Weyl and Planck are merely stressing different aspects of the same situation. The fact is that the attributes of an absolute continuum are varied. We leave out of account the mystical significance attributed to absolute space in the writings of Henry More and other metaphysicians, for it has nothing in common with the situation we are here considering. The features of

---

* The Meaning of Relativity, p. 61.

† H. Weyl. Space-Time-Matter. Methuen, London; p. 218.

‡ Space-Time-Matter, p. 221.

¶ Planck. Where Is Science Going? W. W. Norton and Co., New York, 1932; p. 197.

an absolute continuum (such as Newton's absolute space or Lorentz's stagnant ether) that are relevant in the present discussion are:

A. Its ability to exert physical effects.
B. The rigidity of its structure; and by this we mean the impossibility of our influencing its structure in any way through the action of matter or of other physical agents.
C. The existence of geometric properties which transcend the frame of reference or the point of view of the observer who surveys them.

The attribute A does not appear to reflect so much on the absolute nature of the continuum as on its physical existence. A continuum which would exert no physical effects would be a mere void having the sole attribute of extension; it would fail to manifest itself physically, and there would be no particular reason to postulate its existence in any physical sense. In Newton's absolute space, in Lorentz's stagnant ether, and in the 4-dimensional space-time of the special theory of relativity, the attribute A is illustrated by the forces of inertia that the continuum exerts on any accelerated body.

No particular comment is required on the attribute B; its significance is clear.

The significance of the attribute C is best understood by an illustration. The length of a shadow cast on a plane varies with the inclination given to the plane. On the other hand, the distance between two points in space is determined solely by the geometry or structure of space; it thus betrays an intrinsic property which a change in our frame of reference cannot affect. It seems trivial to mention this feature in the case of Newton's space, for it appears obvious. But Minkowski's discovery of the property of invariance for the distance between two points in 4-dimensional space-time was very important. By welding together space and time, it gave reality to the 4-dimensional continuum. The 4-dimensional continuum thus became a physical existent endowed with a structure transcending the contingencies of space and of time, that result from the varying point of view of the observer. The geodesics of space-time, in particular, appear as intrinsic features of the continuum.

Since the space-time of the special theory of relativity has the three attributes mentioned, no question has arisen as to its absoluteness. Einstein, however, objects to the simultaneous presence of the attributes A and B, and he expresses the belief that the attribute B should be discarded. To quote Einstein:

"In the first place, it is contrary to the mode of thinking in science to conceive of a thing (the space-time continuum) which acts itself, but which cannot be acted upon." *

---

* The Meaning of Relativity, p. 62.

Einstein postulates therefore in his general theory of relativity not only that the space-time structure acts on matter, as had previously been assumed, but that, conversely, matter must exert an action on the structure of space-time. The attribute B is thus withdrawn; the attributes A and C alone remain. Weyl, who views the attribute B as the one truly characteristic of an absolute continuum, is accordingly consistent in his contention that space-time has lost its status as an absolute; whereas Planck, to whom the attribute C appears the most important, is justified in holding that space-time is still absolute.

Lastly, we come to Einstein's contention. He expresses the opinion that the space-time of the general theory, having lost the attribute B while retaining the attributes A and C, is a hybrid concept which is neither absolute nor relative. Einstein's monistic tendencies urge him to remove this dualism by withdrawing also the attributes A and C. We must understand exactly what this implies. In the general theory the structure of space-time is affected by the proximity of matter, but, even at an infinite distance from matter or in a universe totally devoid of matter, a definite structure still remains—the flat structure of the special theory. Suppose, then, we assume the world to be empty of matter, a small test body alone being present. The flat structure will manifest itself by acting on the test body, and the forces to which the body is submitted can no longer be attributed to the action of other masses. Thus, in an empty world, space-time would still retain the attributes A and C.

Commenting on this dualism of the space-time of the general theory, Einstein, in a passage previously quoted, writes:

"The physical properties of space would not then be wholly independent, that is, uninfluenced by matter, but in the main they would be, and only in small measure, conditioned by matter. Such a dualistic conception is even in itself not satisfactory."

If we are to remove these attributes, as Einstein here aspires to do, we must assume that, in a world empty of matter, space-time would lose all trace of a structure. Matter would then be the sole cause of the structure of space-time, and space-time in the absence of matter would lose all its attributes and would cease to have any physical existence. It was to further this idea that Einstein devised his theory of the cylindrical universe.

The analysis we have given of the different aspects of an absolute continuum is typical of many concepts that might be mentioned. In mathematics also we find examples of a concept which has various attributes and for which the popular terminology is retained even when the more familiar attributes are lacking. The concept of perpendicularity is a case in point. We all have a clear understanding of what is meant by two perpendicular lines; and we should deem it absurd to speak of a line that is perpendicular to itself. The mathematician, however, is not concerned with the visual aspect of perpendicularity. His interest lies in the mathematical relation that is satisfied by perpendicular lines; and so, from his standpoint, the realization of this relation constitutes the true characteristic of perpendicularity. Now there is a certain kind of line, called a minimal line or a null-line, for which the above relation is satisfied, not in connection with a second line, but in connection with the null-line itself. The mathematicians therefore feel justified in saying that a null-line is perpendicular to itself. Null-lines play an important part in Minkowski's space-time; they define the world-lines of particles (photons) moving with the velocity of light.

PART II

PHYSICAL THEORIES OF THE CLASSICAL PERIOD

CHAPTER XIV

MATHEMATICAL PRELIMINARIES

(Functions and Groups)

In the previous pages we mentioned that the progress of physical theories is dependent on the mathematical development of the day. Thus the general theory of relativity could never have arisen had it not been for the prior discoveries of Gauss and Riemann in non-Euclidean geometry; and almost all theories draw upon the calculus. For these reasons a few general considerations on various aspects of mathematics may not be out of place.

The Greeks were highly proficient in geometry, furnishing mathematical geniuses of the first rank, notably Euclid, Appolonius, and Archimedes. But ignorant as they were of the more powerful methods of analytical geometry and the calculus, their demonstrations were laborious, and the majority of the theorems they discovered can be established today by much simpler means. In point of fact analytical geometry and the infinitesimal methods were both foreshadowed by Archimedes, but the lack of a simple notation and other causes precluded the great Syracusan from deriving full benefit from his ideas. Not until the seventeenth century did these two great instruments of research become available; and, thanks to them, the scope of mathematical investigations was extended to a degree which would have seemed unbelievable to the ancients.

The outstanding mathematician of the scientific Renaissance in the seventeenth century was Newton. The eighteenth century produced three giants, Euler, Lagrange, and Laplace. The nineteenth century witnessed a much larger number of masters. In the chronological order of their birth, the great mathematicians of the nineteenth century were Gauss, Cauchy, Abel, Jacobi, Hamilton, Weierstrass, Hermite, Riemann, Sophus Lie, and Poincaré. To this list we might add Poisson and Cayley. A host of other mathematicians, such as Fourier, Dirichlet, Liouville, also made discoveries and evolved methods of the highest importance; and indeed these mathematicians, taken as a whole, have probably contributed more than any single one of the great masters.

The chronological order in which the names of the masters has been written is, very approximately, the order in which the student of mathematics will come across their discoveries. Thus Euclid is the first of the great mathematicians whose theorems we study; whereas it would require many years of hard work before we could become conversant with the contributions of Poincaré, or even understand in what they consist. The fact is that each successive creative mathematician stands, so to speak, on the shoulders of his predecessors, utilizing their discoveries.

In a broad way we may classify mathematical advances under two headings. Firstly, new methods are discovered, methods which render possible the solving of problems formerly insoluble and which open up new avenues of research. Secondly, considerable advances may ensue from the application of known methods. Analytical geometry, the calculus, the theory of groups, vector analysis, and tensor analysis illustrate new methods. Newton's deduction of the law of the inverse square from Kepler's laws is an example of the successful application of a known method to a particular problem.

We might suppose that the mathematician who originates a new method deserves greater credit than the one who applies it. But this is not necessarily so, for the application of a known method may make greater demands on the mathematical ingenuity of the investigator than may the bare discovery of the method. Fourier, for instance, constructed the celebrated series which bear his name, and the importance of these series is today so considerable that we could scarcely do without them. Yet no mathematician would think of classing Fourier with the greatest masters. Luck may also play a part in the fame a new method or theory receives. Of two methods equally deep, one may prove relatively sterile whereas the other has a number of applications. Hadamard, commenting on the theory of integral equations developed by Volterra, Fredholm, and Hilbert, writes:

"The theory of integral equations, born yesterday, is already classical. It has been introduced in several university courses. There is no doubt—perhaps after further improvements—that it will soon impose itself as of current use in mathematics. This is a rare piece of good fortune for a mathematical doctrine, for mathematical doctrines so often become museum exhibits."

The great masters have been prolific not only in the application of known methods, but also in the discovery of new ones. The caliber of a mathematician is usually determined by the volume of his productions and by the diversity of the subjects he has treated. This statement may

seem strange when we recall that many of the masters have compared
their science to an art; and an artist is not gauged by the quantity of
his productions, but by their quality. The paradox is attenuated, however, when we realize that there can be no such thing as mediocre mathematics. A mathematical demonstration is either right or wrong, and if
it is wrong, it is not mathematics and finds no place in a mathematical
publication.* To be sure, there may be a slip in a demonstration, even
when furnished by a master like Riemann; but the slip is eventually
detected and remedied, or else, if it vitiates the whole, the demonstration
is condemned and forgotten. On the other hand, it would be misleading
to say that, in mathematics, quality plays no part at all. In many cases
a theorem may originally be demonstrated in a cumbersome way and then
later receive a more simple demonstration characterized by that subtle
quality which is referred to as "mathematical elegance." Nevertheless
in such cases, the theorem is usually named after the original discoverer;
to him is given credit for the discovery rather than to the investigator
who has simplified the demonstration.

Pure and Applied Mathematics—A distinction is usually made between pure and applied mathematics. According to the prevalent classification, the subject matter underlying a mathematical doctrine is assumed
to establish the distinction. When this subject matter pertains to the
physical world, the mathematics is said to be of the applied variety; and
when it is selected without regard for the world of our physical experience,
the mathematics is called pure. On this basis theoretical physics and
mechanics must be regarded as branches of applied mathematics, whereas
the theory of functions and the theory of groups are representative of
pure mathematics. Euclidean geometry was at one time classed with
pure mathematics because its basic postulates were thought to express
requirements of the human mind. But the discovery of the non-
Euclidean geometries has since dispelled this belief; it has shown that
the Euclidean postulates are mere idealizations of the results of our
physical experience.

The accepted distinction between pure and applied mathematics is
far from satisfying. In the first place, it permits no permanent classification. As an example let us consider the doctrine of classical mechanics.
When the basic postulates of classical mechanics were estalished by

---

* We are not referring here to the discussions on the nature of mathematics and
of mathematical reasoning. In this latter sphere of investigation, appropriately
christened "metamathematics," it seems to be a case of every man for himself.

Galileo and Newton, they were thought to express physical characteristics of the world. Classical mechanics was thus regarded as a branch of applied mathematics. But today, as a result of the theory of relativity, we know that the classical postulates do not correspond to physical reality. Strictly speaking, we should therefore reverse our former stand and view classical mechanics as an abstract doctrine pertaining to pure mathematics. From this illustration we see that the prevalent classification of the mathematical doctrines can never aspire to any degree of permanency since it may be modified by every new discovery.

More important than these objections is the fact that the prevalent classification pays no heed to one of the essential characteristics of mathematics: namely, in a mathematical doctrine the subject matter is of no importance. What the mathematician is concerned with are systems of relations, and his aim is to extract from given relations their necessary consequences. In celestial mechanics, for example, we may be concerned with determining how two or more bodies will move under their mutual gravitational attractions. The physical part of the problem is exhausted when we have utilized the mechanical postulates and have transcribed our physical problem into the mathematical form of a system of differential equations. The answer to the physical problem is then obtained if we can solve these equations. In this attempt the physical significance of the equations may be lost sight of entirely; the difficulties in our task become purely mathematical, and they are not dependent upon the fact that the equations may or may not be the mathematical expressions of physical occurrences. The purely mathematical difficulties in obtaining the solutions of seemingly simple mechanical problems are often so great that they have challenged the efforts of the foremost mathematicians. Many of the most abstract methods in mathematical research have been devised for the purpose of obtaining these solutions; and we find that the men who have contributed most to the development of mechanics and celestial mechanics are the same masters who have made some of the greatest discoveries in pure mathematics.

In our opinion the distinction between pure and applied mathematics should rest not on the subject matter of a mathematical doctrine, but on the method of procedure of the investigator. The true difference, according to this view, must be sought between pure and applied mathematicians rather than between pure and applied mathematics; and the name "applied mathematician" should be given to the investigator who utilizes established mathematical results without perhaps knowing, or even caring, how they are arrived at. For instance an applied mathematician may, in the course of his work, be confronted with some differential equation which he does not know how to solve. He turns to a textbook or calls upon some pure mathematician and asks him to solve the equation. His sole interest lies in the cut and dried solution and not in the complicated theory that may be involved in obtaining the solution.

Usually, theoretical physicists are of this type or, at least, they used to be until recent years. But in the new physical theories, some of the mathematical problems raised are of a novel kind, and they cannot be solved by the classical methods. For this reason the modern theoretical physicist has been thrown back on his own resources and, under the force of circumstances, has been compelled to become a pure mathematician. Nevertheless, in the majority of cases, the requirements of the theoretical physicist even today are not those of the pure mathematician. As an example, let us examine the most celebrated equation of wave mechanics, i.e., Schrödinger's equation of the hydrogen atom.

Schrödinger sought to solve this differential equation. He proceeded along standard lines by splitting the equation into two simpler ones. The first of these is a classic to mathematicians and is solved by means of functions introduced by Laplace. The second equation is of a standard form investigated by Fuchs in the last century. Schrödinger solved it by applying a method devised by Laplace. Subsequently, he noticed that the solution of this equation could be obtained more simply by solving another equation previously studied by the mathematician Laguerre. The solution of Laguerre's equation being well known, the problem of the hydrogen atom was solved automatically. From this account we see that the solution of Schrödinger's wave equation did not involve any new mathematical discoveries, but was arrived at in a routine way by a direct application of the methods previously established by the pure mathematicians.

The foregoing analysis is not meant to disparage Schrödinger's mathematical ability; it merely proposes to show that the contribution of the mathematical physicist, as such, resides not in the solution of purely mathematical problems, but in the construction of physical theories which enable him to transcribe physical processes and phenomena into the form of mathematical equations.

Functions of a Real Variable

Functions—The Greeks, in addition to their study of conics (on which they stumbled when considering the plane intersections of a circular cone), invented a number of other curves: the cissoid, the strophoid, the cycloid, épicycloids, spirals. Many of these curves were introduced for the purpose of solving some particular geometrical problem, *e.g.*, the trisection of an angle. Then, when in the seventeenth century analytical geometry was invented by Descartes, an infinite variety of new curves was brought to the attention of mathematicians. At a still later date, these curves, which had been accepted as mathematical by the investigators of the seventeenth and eighteenth centuries, were found to form only an insignificant minority among the mathematical curves. It is with the evolution of the concept of a mathematical curve that we shall be concerned in the first part of this chapter.

We must first examine what is meant by a function. In a broad sense, if one thing depends on other things, the former is said to be a function of the latter. Thus the growth of a plant is a function of the nature of the soil, of its moisture content, and of the amount of sunshine the plant receives. The dependences here involved are, however, vague and ill-defined; and only when the relationships are precise and susceptible of numerical expression are we properly dealing with a mathematical function. Nevertheless it is certain that man was acquainted with the concept of function long before this concept was introduced explicitly into mathematics during the seventeenth century. Thus the notion of function, like so many others in mathematics, is plainly taken over from the world of common experience.

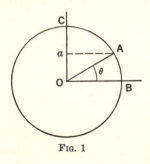

Fɪɢ. 1

Geometry affords many illustrations of functions. As an example, consider a circle. Draw any radius OA making any arbitrary angle θ with the horizontal OB.

Project OA on the vertical OC. We thus obtain the length Oa. The magnitude of this length will depend on the value of the angle θ; and to each angle θ will correspond a definite length Oa. We say that the length Oa is a function of the angle θ. This relationship is perfectly definite; there is nothing vague about it; and so a definite functional relationship is determined. Symbolically, the function is written sin θ, and the functional relationship is expressed by

$$\text{length } Oa = \sin \theta \times \text{length } OA.$$

Quite generally, if we call y the magnitude which depends on the value of x, and if we wish merely to express the existence of a functional relationship without having to know its exact nature, we write:

$$y = f(x).$$

Under this form the exact functional relationship is left unspecified. The magnitude y, whose value is determined by the value ascribed to x, is called the *dependent variable*; whereas x is called the *independent variable*. It may happen that y depends on more than one independent variable, *e.g.*, on three, which we may represent by x, u, v; the functional relationship would then be written

$$y = f(x, u, v).$$

A physical illustration of this more complicated type of function is given by the dependence of the apparent height of the sun on the time of the day, the season of the year, and the latitude of the observing station. In what follows, we shall confine ourselves to functions of a single independent variable.

A graphical representation of a function of one variable is obtained when we follow the Cartesian method of plotting the value ascribed to x along a horizontal and the corresponding value of y along a vertical. When a given value is ascribed to x, we thus obtain a point P, as is seen in the figure. If now we repeat this operation for all values of x in the range considered, *e.g.*, from $-\infty$ to $+\infty$, we obtain an infinite aggregate of points in the plane. This aggregate of points gives a graph, or geometrical representation, of the function. If the points determine a continuous curve, the function is said to be continuous.

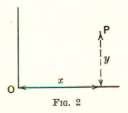

Fig. 2

Next we must examine how a function will be defined in mathematics. According to the modern conception, y is viewed as a known function of x when we have stated clearly the quantitative law of correspondence between the two variables. The law may be expressed purely verbally, according to our fancy; it need not be defined by known mathematical operations. But the mathematicians of the seventeenth and eighteenth centuries were more exacting. The only functions which they accepted as susceptible of mathematical study were those which could be defined by means of well-known mathematical operations. We shall refer to such functions as *mathematical functions*. All functions which did not satisfy this condition were regarded as beyond the scope of mathematics. For example a function such as

$$y = \frac{\sqrt{x^2 + 1}}{x^3 - 4} - x + 2$$

was accepted as mathematical, because the expression on the right involves mere algebraic operations. We need scarcely mention that the curve which represents this function is obtained when we make x run over a continuous range of values; then calculate the corresponding values of y; and, finally, proceed according to the Cartesian method. The curve is a continuous one. Similarly, the relationship expressed by

$$y = \sin x,$$

since it can be determined by a geometric construction, was also accepted as defining a mathematical function. The curve obtained is called a sinusoid.

Since a continuous curve was associated with each of the continuous functions that could be imagined, and since these functions were legion, the mathematicians of the seventeenth century were conversant with a vast collection of continuous curves. Among these, all the curves known to the Greeks readily found a place. For instance,

$$y = x^2 \text{ and } y = \pm \sqrt{R^2 - x^2}$$

represent, respectively, a parabola and a circle of radius R. It was impossible to coin a name for each one of the new curves which any mathematician could invent, and so the majority of curves went unnamed. Some few, however, received names, either on account of their simple geometric properties or by reason of their connection with the solution of some conspicuous problem. Thus we have the oval and the folium (Descartes), the lemniscate (Bernoulli), Pascal's curve, and at a later date Gauss's curve of errors.

Towards the end of the seventeenth century, infinite series were studied by Newton and Leibnitz, and, by utilizing these series to define new functions, a considerable extension was given to the class of continuous curves accepted as mathematical. This leads to a digression on series.

Series—Suppose we take the algebraic sum of a billion finite numbers. We assume that the numbers are positive or negative, or that some are positive and others negative. The sum is obviously finite. But suppose that in place of a billion numbers, we consider an infinite succession of numbers. We now have an *infinite series*. We cannot speak of the *sum* of an infinite series, since infinity is never exhausted; but we may consider a magnitude which plays the same rôle with respect to an infinite series as a sum does to a finite succession of terms. This magnitude is the one towards which the sum of the first n terms of the infinite series tends when n is made to increase indefinitely. It is called the *limit* of the infinite series.

Important differences distinguish the sum of a finite number of numbers from the limit of an infinite number of numbers. Let us mention a few. We know that the sum of a billion numbers, some of which are positive and others negative, has exactly the same value whether we effect the summation in the order given or in any other order. Yet, as was proved by Dirichlet in the nineteenth century, this conclusion is not necessarily correct when we are dealing with infinite series: the limit may have one value or another according to the way in which the terms are ordered. Here is a first important distinction between the sum of a finite number of terms and the limit of an infinite number of terms. But there are others. Thus the sum of a finite number of numbers is always finite and, as just explained, is perfectly well determined. On the other hand, the limit of an infinite series may be finite or infinite, and sometimes the series has no limit. In the two latter cases the series is said to *diverge*. We may illustrate these various situations by elementary examples. If we add the number 1 a billion times in succession, we obtain a perfectly definite finite sum, *i.e.*, one billion. But if we have the unending series

$$1 + 1 + 1 + 1 + 1 + \ldots\ldots,$$

the limit is infinite—the series diverges. Similarly, if we have a finite sequence of terms such as

$$1 - 1 + 1 - 1 + 1, \text{ or } 1 - 1 + 1 - 1 + 1 - 1,$$

the sum will have the value 1 or 0 according to whether the number of terms is odd or even. But if we are dealing with an infinite series of this type, there is no definite limit, for we have a kind of oscillation between 1 and 0—the series diverges.

As an illustration of a series the limit of which is well defined and finite, we may mention the series

$$1 + \frac{1}{2} + \frac{1}{4} + \frac{1}{8} + \ldots\ldots$$

Its limit is 2. Such a series is said to *converge*. The series

$$1 - \frac{1}{2} + \frac{1}{3} - \frac{1}{4} + \frac{1}{5} - \ldots\ldots$$

also converges, but here we must be careful not to change the order in which the terms are written. If we change the order of the terms, we may find that the series diverges. These brief illustrations on numerical series will suffice for our purpose. What we shall now be concerned with are series of functions.

Series of functions differ from numerical series in that the successive terms are functions of a magnitude x whose precise numerical value is not specified. Thus the infinite series

$$f_1(x) + f_2(x) + f_3(x) + \ldots \ldots,$$

where $f_1(x)$, $f_2(x)$, $f_3(x)$, . . . are known functions of x, represents a series of functions. Let us assign the same arbitrary numerical value to x in each one of these functions. The functions now define numbers, and our series of functions becomes a numerical series. By changing the value ascribed to x, we obtain one numerical series after another. In most cases the numerical series converges if certain values of x are substituted, whereas it diverges (*i.e.*, has no limit or becomes infinite) when other values of x are taken. Suppose for argument's sake that the series converges for all values of x between 0 and 1, and diverges for all other values of x. The limit of the series then has a well-determined value only when x is situated between 0 and 1, and usually this value will depend on the value assigned to x in the interval. The limit of the series thus defines a function of x. But note that it is only when x is situated between 0 and 1 (where the series converges) that the series defines a function; elsewhere the series is meaningless and defines nothing.

We thus realize the importance of determining for what range of x (if any) a series of functions converges. At the same time we see that if a series converges for a certain range of x, the series *via* its limit enables us to define a function, of which we may obtain a graphical representation by Descartes' method. This introduction of series as a means of defining functions enables us to obtain new functions and therefore new curves. The problem of determining for what range of x any given series of functions converges is one which even today is often of extreme difficulty. The mathematicians of the seventeenth and eighteenth centuries, handicapped as they were by their ignorance of the properties of infinite series, were restricted to the simplest kinds of series. Such series were the elementary geometric progressions and the more general *power series* or *Taylor series*.

Now the functions defined by Taylor series are functions of a very special kind. These functions, today named *analytic functions*, form but a very small sub-class among the functions that can be defined by other types of series and by other kinds of mathematical expressions. But this fact was not realized by the mathematicians of the seventeenth and eighteenth centuries; and so they made no distinction between mathematical functions and analytic ones. They erroneously believed that any mathematical function (any function that could be defined by a mathematical expression) was necessarily analytic, *i.e.*, susceptible of being expressed by means of a Taylor series.* The mistake made by the earlier mathematicians is comprehensible when we are told that all those mathematical functions whose properties they were able to investigate happened to be analytic.

Taylor Series—In view of the importance of Taylor series in the historical development of mathematical analysis, it will be advantageous to examine their principal characteristics. Consider the two following infinite series:

$$(1) \qquad a_0 + a_1 x + a_2 x^2 + a_3 x^3 + \ . \ . \ . \ .$$

and

$$(2) \quad a_0 + a_1(x - k) + a_2(x - k)^2 + a_3(x - k)^3 + \ . \ . \ . \ . \ ,$$

in which a_0, a_1, a_2, a_3 . . . . . and k are arbitrary, fixed finite numbers, and x is a variable. It will be noted that both series contain the successive integral powers of $(x - k)$, the only difference between the two series is that, in the first, k is assumed to have the value zero. Series of this kind are called "power series."

Now a power series, such as (2), always converges for the value $x = k$, since it then reduces to its first term, a_0. If the power series diverges for all other values of x, it merely defines a point, *i.e.*, the point $x = k$. But the power series may also converge for some other value of x. Mathematical analysis shows that in this case the series will necessarily converge for a continuous range of values of x and that the point $x = k$ will occupy the centre of the range. Within this range of convergence (which may be limited or may comprise all values of x), the power series defines a function of x which can be shown to be a continuous function, all of whose derivatives are also continuous.

---

* It was in this connection that the appellation "analytic" originally arose; it was regarded as synonymous for mathematical. Subsequently, when it became apparent that not all functions that could be expressed mathematically were susceptible of being expressed by means of Taylor series, the word "analytic" was retained only for the more restricted type of mathematical functions.

A power series of this last kind constitutes a Taylor series,* so that a continuous function defined by a power series is what we have called an analytic function of x.

As mentioned in a previous paragraph, all the functions known to the mathematicians of the eighteenth century happened to be analytic; and we have said that this fact prompted the idea that all mathematical functions were analytic. An example may now be given. The function

$$y = \sin x$$

expresses a relation between x and y, which is defined by means of a geometric construction. Such a function is therefore mathematical. But it was soon discovered that this geometric function could be expressed equivalently by means of the Taylor series

$$y = x - \frac{x^3}{3!} + \frac{x^5}{5!} - \frac{x^7}{7!} + \ldots \ldots,$$

and that as a result the function was analytic. In all other cases examined by the mathematicians of the seventeenth and eighteenth centuries, the same situation was found to occur.

Analytic Curves—If we plot, according to the Cartesian method, the x and y of a function defined by a Taylor series (*i.e.*, an analytic function), we obtain a curve which we also call "analytic." We have said that the function defined by a Taylor series in the interval of convergence is always a continuous function with all its derivatives continuous. This implies that when we plot the function by Descartes' method, the analytic curve obtained will be continuous and will have no sharp bends in it, and that the radius of curvature will also vary continuously as we move along the curve. In a crude way, we may say that analytic curves are smooth curves. But there are other important features to analytic curves; these features are elucidated when we revert to the Taylor series that define the curves.

As a first illustration, let us examine the trigonometric function

(3) $$y = \sin x.$$

---

* From this account it would appear that a Taylor series is always an infinite series. As a special case, however, we may suppose that the series contains only a limited number of terms. In this case it reduces to a polynominal and it converges for all finite values of x.

Let us construct the curve by plotting along a horizontal the values we ascribe to x, and along a vertical the corresponding values for y. We obtain a snake-like curve (Figure 3) extending from $-\infty$ on the left

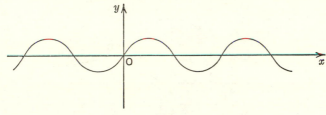

Fɪɢ. 3

to $+\infty$ on the right; the curve is called a sinusoid. We have seen that this function may be expressed by the Taylor series

$$(4) \qquad y = x - \frac{x^3}{3!} + \frac{x^5}{5!} - \frac{x^7}{7!} + \ \ldots \ldots \, ,$$

and that this fact proves the function, or curve, to be analytic. Now the series just written converges for all values of x; hence we conclude that the series also represents the curve over its entire domain of existence, from $-\infty$ to $+\infty$. The two expressions (3) and (4) of the function are equivalent in all respects. By this we mean that if in (3) and in (4) we give the same current value to x, exactly the same value is obtained for y in the two cases. The Taylor series just written is not the only one that can represent the function; an infinity of such series can be obtained. For instance,

$$(5) \qquad y = \frac{1}{\sqrt{2}} + \frac{1}{\sqrt{2}} \left(x - \frac{\pi}{4} \right) - \frac{1}{2\sqrt{2}} \left(x - \frac{\pi}{4} \right)^2$$
$$- \frac{1}{6\sqrt{2}} \left(x - \frac{\pi}{4} \right)^3 + \ \ldots \ldots$$

is a Taylor series in power of $\left(x - \dfrac{\pi}{4} \right)$, which represents the same

function. Like the series (4) it converges for all values of x.

Let us pass to a second illustration. The function

$$(6) \qquad y = \frac{1}{x^2 + 1}$$

is analytic, as is obvious since it is defined by mere algebraic operations. If we construct the curve by giving all values to x and calculating from (6) the corresponding values for y, we obtain a curve (Figure 4) extending from $-\infty$ to $+\infty$.

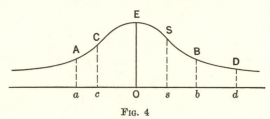

Fig. 4

Since the function, or curve, is analytic, we may also express it by means of a Taylor series. Two such series are

$$(7) \qquad y = 1 - x^2 + x^4 - x^6 + x^8 - \ \ldots \ldots$$

and

$$(8) \quad y = \frac{4}{5} - \frac{16}{25}\left(x - \frac{1}{2}\right) - \frac{16}{125}\left(x - \frac{1}{2}\right)^2 + \frac{384}{625}\left(x - \frac{1}{2}\right)^3 - \ \ldots$$

Both of them represent the curve; but there is an important difference between the properties of these two Taylor series and those of the series (4) and (5), that we considered in the previous example. The two series (4) and (5) converged for all values of x and therefore represented the entire curve. With the two series (7) and (8) a new situation arises, for these series converge only when x is situated in certain ranges. Thus the series (7) converges only when x is taken between -1 and $+1$ (*i.e.*, from a to b in the figure); whereas the series (8) converges only when x is

within the interval defined by the two values $\dfrac{1 - \sqrt{5}}{2}$ and $\dfrac{1 + \sqrt{5}}{2}$. The

second interval of convergence is marked by the points c and d in the figure (Figure 4). Owing to the peculiarities just mentioned, the first series represents only the arc AB of the curve, and the second series represents only the arc CD. Neither series defines the whole curve; hence neither series is completely equivalent to the function defined by (6). Incidentally, we note that our two series are in powers of $(x - k)$, where k

has the values $k = 0$ and $k = \frac{1}{2}$ respectively, and that in either case the point $x = k$ is at the midpoint of the interval of convergence. (In the figure these two centres of convergence are represented by the points O and s respectively). The positions of the two points $x = k$, at the centres of the intervals of convergence in the two foregoing Taylor series, illustrate the general rule mentioned on a previous page.

We may wonder why these Taylor series do not always converge. In particular, why does not the first series converge to the right of b? The point b does not appear to constitute any fundamental obstacle, since the second series converges beyond it. These questions could not be answered by the mathematicians of the eighteenth century. Not until the complex variable was systematically introduced by Cauchy, was the reason for the strange breaks in the convergence fully understood. We shall revert to this problem on a later page. For the present we wish to examine a different point.

Let us suppose that we are not given the equation (6) of the curve or any of the series which define stretches of the curve. We shall assume, however, that we are informed with mathematical precision of the lay of a small arc of the analytic curve—an arc which may be as tiny as we choose. Mathematical analysis then shows that it will be possible to construct the Taylor series which defines the curve over equal stretches to the right and left of the given arc. To take a definite example, let us assume that the tiny arc is situated at the central point marked E (see Figure 4). In this event, the Taylor series (7), which defines the curve from A to B, can be constructed. As a result, the curve is known in the interval AB, and hence it is known in particular over any tiny arc within this interval. Let us then fix our attention on the arc in the neighborhood of the point S. Since this arc is now known with mathematical precision, we may proceed as before and construct the Taylor series that converges over a stretch of which S, or more properly s, is the centre. The Taylor series in question is the series (8). The curve is thus known from C to D, and hence altogether from A to D (since the part from A to C is already known). Let us now consider a tiny arc between B and D. Proceeding as before, we may construct a third Taylor series converging symmetrically about the new arc and extending to the right of the point D. Thanks to this third series, the construction of the curve to the right is extended. We may continue this method indefinitely, deducing one Taylor series from another, and by this means we may ultimately construct the entire curve in piecemeal fashion from A to a point infinitely distant to the right.   In a similar way we might construct the curve to the left.*

The method of progressive construction, just explained, is called *analytic continuation*.   If we recapitulate what we have done, we see that, starting as we did from an assumed precise mathematical knowledge of an arc (however small) of the analytic curve, we have been able to construct the entire curve.   No other curve than the one obtained could have been constructed by this method.   This discussion brings to light an important characteristic of analytic curves.   It shows that an analytic curve is completely determined (subject to the reservations mentioned in the note) when any arbitrarily small part of the curve is specified.†

The foregoing properties of analytic curves are reminiscent of those of physical systems evolving in a rigorously deterministic world.   In Chapter VII we mentioned that the specification of an initial state of a physical system sufficed to determine the entire history of the system. The specification of an initial state of a physical system is therefore similar in its significance to the giving of a small arc of an analytic curve, and the determinism associated with physical processes is seen to be shared by analytic curves.

An immediate consequence of the properties of analytic curves is that if two analytic curves coincide over an arc, however small, they will necessarily coincide over their entire lengths.   A straight line is the simplest illustration of an analytic curve, and it may be referred to as such in spite of the contradiction in the popular meanings of the words curved and straight.   Since two analytic curves cannot coincide along an arc without coinciding everywhere, we conclude that if even a

---

* The foregoing explanations, which we have illustrated in connection with an analytic curve, assume a more rigorous form when we reason in terms of analytic functions.   Thus the giving of a small arc of an analytic curve is equivalent to the specifying of the values of an analytic function $f(x)$ for all values of the variable x within a small interval.   This information enables us (in theory) to determine the values of all the derivatives of the function for any value x_0 of the variable x within the given interval.   We may then construct the Taylor series, in powers of $(x - x_0)$, which represents the function, and from this series determine the values of all the derivatives of $f(x)$ for any other value x_1 of x within the interval of convergence of the series.   This latter information in turn permits us to construct a second Taylor series, in powers of $(x - x_1)$, which may converge beyond the original one, and so on.

† If a singular point is present on the real axis (*e.g.*, if the function becomes infinite for a certain value of x), the giving of an arc of the curve does not enable us to construct the entire curve, but only a part thereof.   This is because the Taylor series will not converge beyond the singular point.   However, this restriction can be removed by the introduction of the complex variable (see page 146), so that the general statement made in the text is correct.

tiny arc of an analytic curve is straight, the entire curve must necessarily be straight.

Consider then two straight segments forming a V. In view of the previous statement, the V-curve cannot possibly constitute a segment of an analytic curve. This fact entails an interesting consequence, illustrated in the problem of a vibrating string. As d'Alembert remarked:

"The ordinary way, indeed practically the only way, to set a string into vibration, is to hold it at a point, and then pull. And this yields two straight lines which form a broken line."

Whence d'Alembert concluded that the initial position of the string did not define an analytic curve, so that more generally a vibrating harp string could not always define an analytic curve at any current instant during its vibration. The consequences of this conclusion will be examined presently.

We may also understand why a curve traced at random with a pencil cannot in general be analytic. For if the curve were analytic, its entire length would be determined by the initial arc we might trace; and this is absurd, because, having started a curve with our pencil, we are free to continue it to our fancy. Nor need many other curves encountered in physics be analytic. For example, there is no reason why the graph traced by a registering thermometer should be analytic, for there is no particular reason why the sequence of temperatures over a given small interval of time should determine the temperature at all future instants.

We have dwelt at some length on analytic curves, because they were the only curves or functions which the mathematicians of the seventeenth and eighteenth centuries thought it possible to express by means of mathematical relations. All other curves, in the opinion of these early mathematicians, were incapable of mathematical representation and were ruled out as non-mathematical and as beyond the scope of mathematical investigation.

Non-analytic Curves—The inability of the earlier mathematicians to deal with non-analytic functions and curves precluded these investigators from attacking problems in which such functions appeared. Fortunately, the functions which they encountered in mechanics and in celestial mechanics were analytic, and because of this circumstance many mechanical problems were solved in the eighteenth century. But in general physics, the situation was different: non-analytic functions were often met with; and so mathematicians believed that the mathematical instrument was inapplicable to the majority of physical processes. As an illustration, we recall that the shape of a vibrating string at any given instant is not usually represented by an analytic function. Accordingly, the mathematicians of the eighteenth century came to the conclusion that the problem of a vibrating string could not be investigated in its full generality by mathematical means.

It was therefore a momentous event in the history of mathematical physics when Daniel Bernoulli, towards the middle of the eighteenth century, claimed to have achieved the impossible. He constructed mathematical series which, so he maintained, were able to define the curve traced by a harpstring at any instant of its vibration. Needless to say, these series were not Taylor series, since the latter define none but analytic curves. The astonishment of Euler and d'Alembert may well be imagined; it was tempered, however, by the fact that Bernoulli's demonstration was lacking in rigor, so that the results he thought he had established remained in dispute. We may wonder why the dispute was not settled then and there; for is not a mathematical demonstration correct or incorrect? The trouble was that although Bernoulli had obtained series which formally defined the shape of the string, he was unable to establish the convergence of his series—and we have seen that unless a series converges, it defines nothing.* Not until the following century were Bernoulli's results vindicated by Fourier and by Dirichlet. His series (whose terms are the trigonometric functions *sine* and *cosine*), are today known as Fourier series.

The advance made by Fourier consisted in the formulation of a general method for constructing the Fourier series which, in his opinion, would represent any arbitrarily assigned function, analytic or not. Unfortunately Fourier, like Bernoulli, failed to give a rigorous proof of the convergence of his series, so that it was questionable whether they truly represented the functions they were supposed to represent. Owing to this lack of rigor, Fourier's memoir, though it received the prize offered by the French Academy, was nevertheless severely criticized by the judges, Lagrange, Legendre, and Laplace. Towards the middle of the nineteenth century, however, Dirichlet stated conditions under which Fourier series would certainly converge, and he thereby established the kinds of functions that could be represented by these series. Dirichlet's conditions have since been refined by Camille Jordan and others. Many of the masters, including Riemann, spent considerable time in investigating the various properties of Fourier series.

---

* In modern mathematics, however, divergent series are not entirely useless.

Although Fourier gave no rigorous proof of the convergence of his series, he proceeded on the assumption that, in all the cases contemplated, convergence would be realized, and that the series would thus define curves, or functions, whether analytic or non-analytic. These considerations were then extended by him to series of several variables. At the same time, Fourier established the empirical laws of heat conduction and showed that problems of heat conduction and of temperature equilibrium could be solved by means of his series. He also treated in more amplified form the problem of vibration, extending it from the case of strings to that of membranes and elastic solids. Thanks to Fourier, theoretical physics in its modern sense was born.

Before we proceed to a discussion of Fourier series, two important points may be noted. In the first place, the discovery of these series arose from an attempt to solve a physical problem; many other mathematical discoveries have been brought about in this indirect way. Indeed, the statement has often been made that the progress of mathematics would soon have come to a standstill had it not been for the suggestions derived from the physical world. Another feature illustrated in the discovery of Fourier series is the procedure whereby great advances in mathematics have often been obtained; namely, by stumbling and guessing rather than by pursuing a rigorous chain of deductive reasoning. Though rigor is essential in mathematics, yet in view of the limitations of the human mind it is often preferable not to demand too much of it in the early stages. A premature appeal to rigor might stifle discovery. Its effect might be to cloud the vision of the whole by concentrating our attention on some minor detail. History shows that detail may wait; the important thing is to explore the territory first and to establish its general outline. In Poincaré's words: "Of what use is it to admire the work of the mason if we cannot appreciate the design of the architect?" At a later date, subsequent mathematicians may advance along the grand avenues opened by the pioneers and concentrate on the details left in abeyance. Sometimes the working out of the details reveals errors invalidating the whole structure, and in this event the structure must be abandoned or considerably modified. But more often than not, the faith of the pioneers has been found justified. Besides, the subsequent attempts to vindicate or to invalidate their results have often led to other discoveries.

Let us pause to consider the kinds of functions that Fourier series can represent. Fourier himself believed that all continuous curves could be expressed by his series. In this, however, as was subsequently proved by Dirichlet and by Jordan, he was mistaken; for although Fourier series can represent all analytic curves and also a large class of non-analytic continuous curves (*e. g.*, curves scribbled at random with a pencil), many continuous curves cannot be represented. But the interest of Fourier series is not restricted to continuous functions. These series can also express a large class of discontinuous curves, or functions. Discontinuous curves are curves which have breaks in them, *e.g.*, the curve traced in the figure. The number of discontinuities (or breaks)

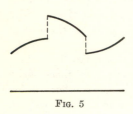

FIG. 5

over a finite stretch may be so large that a graphical representation by means of a succession of tiny arcs is impossible in practice. In such cases the word "curve" is no longer appropriate, and we must adhere to the more technical appellation, "function." A Fourier series cannot, however, represent any arbitrary discontinuous function; restrictions must be imposed on the nature of the discontinuities.

Following Fourier's discoveries, mathematicians sought to construct functions by utilizing other developments in series. Weierstrass constructed a series more general than that of Fourier and showed it to define a continuous function, or curve, which exhibits the remarkable property of having no tangent at any of its points. The Weierstrass curve is described most simply when we say that it contains waves within waves *ad infinitum*; obviously it cannot be drawn with a pencil. We have here an illustration of a continuous function which cannot be represented by a Fourier series. In modern mathematical physics, *e.g.*, in the theory of heat conduction, theorists utilize constantly the more general series devised by the mathematicians who followed in Fourier's steps. All this development of theoretical physics would have been impossible had our mathematical knowledge been restricted to that of the mathematicians of the early eighteenth century.

The gradual incorporation of all continuous curves into mathematics led to a study of continuous curves the properties of which baffle our geometric intuition. Here rigor is essential. We have already mentioned Weierstrass's curve which has no tangents; and it

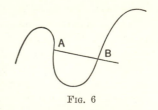

FIG. 6

may be of interest to understand how such a curve can exist. As is well known, the tangent to a continuous curve at a point A is defined by the limiting position of the straight segment drawn from A to a current point B on the curve, when this point B is made to approach A indefinitely. In order, then, that the curve should have a tangent at A, this segment must tend to a limiting position. (The concept of limit previously referred to in connection with series thus crops up again.) But suppose that in the neighborhood of the point A, the curve contains waves within waves *ad infinitum*. In this case, when B moves towards A, the segment AB oscillates incessantly without tending to a limiting position; and as a result there will be no tangent to the curve at the point A. Now this is precisely the situation that holds at each point of the Weierstrass curve, and so we understand why this curve has no tangents. The foregoing discussion shows that a curve must exhibit a certain smoothness if it is to have a tangent, or slope, at every point.

Weierstrass's curve is but one among the strange curves. Thus we may conceive of continuous curves which are sufficiently smooth to have a tangent at every point and yet which have no definite radius of curvature at every point. We may also mention Peano's curve, which passes through every point of a given area and thereby fills the area completely. The collapse of our geometric intuition that is betrayed when we seek to visualize these strange curves, led mathematicians to question the reliability of this intuition in other cases. And so they raised such questions as: Is it always possible to attribute a definite length to an arc of a curve limited by two points A and B? May we always view as determinate the area contained within a closed continuous, plane curve? Does a closed continuous, plane curve (which has no multiple point) necessarily divide the plane into an outside and an inside part?

Geometric intuition would prompt us to give an affirmative answer to all these questions. But past experience shows that we must refrain from jumping at conclusions. Thus Camille Jordan has shown that the length of a curve, between two points A and B taken arbitrarily on the curve, may be indefinite. The Weierstrass curve is of this type. It can also be shown that a closed curve need enclose no definite area (in the ordinary sense of the word). On the other hand, even though a closed curve may have no definite length, it may enclose a definite area. As for the question of deciding whether a closed continuous curve which does not intersect itself will always divide the plane into two parts, it has been settled by Jordan affirmatively. In this case therefore our geometric intuition has not led us astray. But to obtain this assurance Jordan was compelled to resort to a lengthy demonstration, and the problem has many pitfalls. The fact that alternative demonstrations have been given by de la Vallée Poussin and by Brouwer shows the importance mathematicians attribute to this seemingly self-evident theorem.

Thus far we have been concerned with continuous curves. But we have said that Fourier series can also represent a certain class of discontinuous functions, or curves. Thanks to this possibility, discontinuous functions have acquired a mathematical status which was formerly denied them. Following Fourier's discoveries, mathematicians sought to express a wider variety of discontinuous functions by means of series of one kind or another. In this connection we must mention that a discontinuous function $y = f(x)$ can have an infinite number of discontinuities corresponding to a finite range of values credited to the variable x. But here important differences arise according to whether the type of infinity contemplated is the one exemplified in the infinite sequence of integers or is the one associated with the aggregate of all real numbers (the mathematical continuum). A point of historic interest is that the study of the discontinuities of discontinuous functions led Cantor to his classification of the various kinds of infinity and thereby to his celebrated theory of infinite aggregates.

In another direction also, the developments originally suggested by Fourier series precipitated an important generalization in our understanding of mathematical functions. We saw that the earlier mathematicians regarded as mathematical only those functions which were susceptible of being expressed by means of mathematical operations. Prior to the advent of Fourier series, this restriction narrowed down the mathematical functions to the analytic ones. Then, as a result of Fourier's discoveries and of the developments to which they gave rise, many nonanalytic functions were found to be mathematical in the sense just defined. But in the middle of the last century Dirichlet suggested that we generalize our conception of a mathematical function. According to Dirichlet, y will be a mathematical function of x whenever a precise law of correspondence between x and y can be stated clearly. Dirichlet no longer regards as indispensable that this law of correspondence be defined by mathematical operations; he also accepts any purely verbal definition provided it be precise.

An illustration of a function defined verbally is the function called Dirichlet's function. It is defined as follows: In the relationship, $y = f(x)$, the dependent variable y is assigned the value $+1$ or 0 according to whether the value of x is a rational or an irrational number. If, then, we assume that x is made to vary continuously between, say, 2 and 3, the value of y jumps back and forth between 1 and 0. It so happens that Dirichlet's weird function can receive a mathematical expression by means of a series, but this circumstance is accidental and in no wise restricts the general significance of Dirichlet's extended conception of a function. Discontinuous functions manufactured so to speak artificially were thus introduced into mathematics. The modern advance in our knowledge of discontinuous functions is due to René Baire. He classes them into two groups according to whether they are *pointwise* discontinuous or *totally* discontinuous. Since the considerations he develops involve an understanding of the theory of aggregates, we shall dispense with further details.

As may well be imagined, the introduction of the stranger curves and functions has necessitated a generalization of the concept of integration. The method of integration was invented for the purpose of evaluating the area bounded by any given plane closed curve. In the seventeenth and eighteenth centuries, when only analytic curves were known, the areas to be calculated were always bounded by curves of this type, and the method of integration was that of Newton and Leibnitz. No particular difficulty was experienced in its application. But with the advent of the more general curves of the nineteenth century, Newtonian integration often became impossible. Indeed the very conception of an area enclosed by a curve becomes blurred when the bounding curve is a discontinuous one. However, despite the breakdown of geometric intuition in such cases, it is possible to generalize our understanding of integration in such a way that areas limited by discontinuous curves acquire at least a formal significance. The first generalization was accomplished by Riemann. A second one of greater scope has more recently been elaborated by Lebesgue.

The reader may well wonder whether these excursions of the mathematicians into the weird world of discontinuous functions are of any practical interest in theoretical physics. In this connection Baire, in the year 1905, prior to the development of the quantum theory, expressed himself as follows:

"Some physical theories in chemistry and mineralogy bear a certain resemblance to mathematical discontinuity. In any case, in spite of the old dictum, happily discarded, there is nothing which allows us to assert that 'Nature makes no jumps'. Under these conditions, is it not the duty of the mathematician to begin by studying *in abstracto* the relationships between the two notions of continuity and discontinuity, since these, while opposed, are intimately connected? This is perhaps the best way to prepare the advent of a mathematical physics in which the rôle of hypothesis will be reduced to a minimum."

The Complex Plane

Analytic Functions of a Complex Variable—In view of the fact that all the functions investigated during the eighteenth century were analytic, we might suppose that, at the close of the century, the properties of analytic functions were well known. But this belief would be erroneous. As Volterra has remarked, the knowledge which the eighteenth century mathematicians had of analytic functions was comparable to the knowledge we might have of a book by merely viewing its cover.

Strictly speaking, the opening of the imaginary book in Volterra's analogy occurred in the eighteenth century, when the variable x in an analytic function $y = f(x)$ was allowed to assume complex as well as real values. In this way analytic functions of a complex variable were obtained. But no particular advantage was derived from this generalization till 1825, when Cauchy established a series of theorems on analytic functions of a complex variable and showed that the properties of these functions revolutionized our understanding of the analytic functions of a real variable. Cauchy's investigations opened up a new field in mathematical research, a field which is far from being exhausted even today. Some of the more important points established by Cauchy and his successors will now be examined.

We saw in connection with analytic functions of real variables, that if we allow x to assume all real values and then consider the corresponding values for y, we may represent the connection between x and y graphically

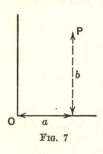

Fig. 7

by means of a curve. This method becomes impracticable when x is allowed to assume complex values, and so a different method of representation must be followed. We adopt the Gauss form of representation for complex numbers. Thus, if x is given a complex value $a + ib$, where a and b are both real numbers and where i stands for $\sqrt{-1}$, we agree to represent this complex value of x by a point P of abscissa a and ordinate b.* The figure is self-explanatory. If x assumes only real values, then $b = 0$, and a alone varies. The point P then describes the horizontal line in the figure. This line is called the *real axis*; and only when x has a real value is the representative point P situated on this axis. Now for each value ascribed to x, and hence for each point P of the complex plane, the functional relationship imposes a value (or sometimes several different values) on y. This value (or values) is usually also complex. We require therefore a second complex plane to represent the value of y. However, we need not represent the value of the function y geometrically, for we may confine ourselves to associating mentally a value (or values) of y with each point P of the first plane.

---

* The plane on which the point P is represented is called the ''complex plane.''

Cauchy restricted his attention to *analytic* functions of a complex variable. Such functions are easily obtained if we select any analytic function of a real variable, *e.g.*,

$$(9) \qquad y = \sin x, \quad y = \frac{1}{x}, \quad y = \frac{1}{1 + x^2},$$

and then agree that x may assume complex as well as real values. In some cases, when we make x (*i.e.*, P) move over the plane, we find that at certain points the corresponding value of y is infinite. Such points are called the *poles* of the function. For instance in the case of the second function (9), the point, $x = 0$, is a pole. This point is situated on the real axis, at the origin. The third function has two poles: they are given by the values of x which make the denominator $1 + x^2$ vanish, *i.e.*, the values $x = \pm \sqrt{-1}$, or $x = \pm i$. These two points are no longer situated on the real axis of the complex plane, but on the vertical at unit distances above and below the origin. As for the first function written, it has no poles.

In addition to poles, other important kinds of points must be considered. So called *branch points* may arise; these will be present only when to a given value of x correspond two or more values for y. Let us place ourselves in this case, and as an illustration let us consider the function

$$y = \sqrt[3]{x - (2 + 3i)}.$$

If in this function we give some value to x, we obtain, in general, three possible values for y. The point B, defined by $x = 2 + 3i$, at which the three values of y coalesce into one, *viz*, $y = 0$, is called a branch point of the function. To investigate the properties of branch points, let us suppose that we start from some definite value of x, say $x = 0$. We represent the value ascribed to x by a point P (here situated at the origin O). To this value of x correspond, as we have said, three values for y; we take one of them and disregard the others. Suppose now that starting from the point O we make x vary continuously over a range of complex values, and then make x return to O. In this event the point P, which represents the value of x at any instant, describes a closed path over the complex

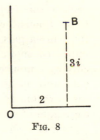

Fig. 8

plane; and while P moves continuously along the path, the value of y also changes continuously. If the closed path does not surround the branch point B, we shall find that when P has described the closed path and is back at its starting point O, the value of y will also resume its original value. On the other hand, if the closed path surrounds the branch point, the value of y will no longer be the initial value with which we started but will be one of the other two values that we disregarded. A second circulation round the branch point, in the same general sense of circulation and with a return to the same point O, makes y change values once again; and we obtain the third possible value. Finally, a third circulation restores the original value selected for y.

Branch points are thus present only when to any given value of x more than one value of y corresponds. Functions in which this situation arises are called *many-valued,* or *non-uniform.* When y has only one value for each value of x, the function is called *one-valued,* or *uniform.* There are no branch points in this latter case; and any closed path followed by x will always bring y back to the same initial value. Non-uniform functions introduce complications, for their non-uniformity interferes with our associating a definite value of y with each value of x, *i.e.*, with each point P on the complex plane. Even if we agree on the initial value for y when x is at some point P, and then make x follow an open path reaching some other point P', the value of y at P' will differ according to whether the path followed by x passes on one side or on the other of a branch point.

Cauchy sought to rid non-uniform functions of this ambiguity. So he made cuts in the plane, extending them from the branch points to infinity. The cuts act as barriers which cannot be crossed by continuous paths, and they thus prevent us from circling completely around a branch point. Thanks to this artifice, if we now make x describe any closed path which is not barred by the cut, we shall invariably return to our point of departure with the same value for y with which we started. But the drawback to this method is that having started with one of the possible determinations for y, we are unable to obtain the other determinations, so that only a part of the function can be represented at a time. Riemann improved on Cauchy's method. He replaced the Cauchy plane and its cuts by a number of superposed planes, or sheets, in which appropriate cuts were made. Strips passing through the cuts joined one sheet to the next and thereby acted in the capacity of staircases connecting the aggregate of superposed sheets. Suppose, then, we attempt to make the point x circle around a branch point. Owing to the communication strip, the cut radiating from the branch point no longer interferes with the circling; the point x passes through the cut, moves along the strip, and thus reaches the sheet situated immediately below the original one. During this process the value of y changes and we find ourselves with a new value for y at a point which is situated just underneath the original point of departure, and is on the next lower sheet. The net result is that to each point *on each sheet* corresponds only one value for y, so that when we take into consideration the individualities of the various sheets, the requirements of uniformity are satisfied. Riemann's device has thus for effect to restore uniformity, or one-valuedness, to the function; at the same time it enables us to obtain all the various determinations of y.

This famous invention of Riemann's, called the method of Riemann surfaces (not to be confused with Riemann spaces in non-Euclidean geometry), may appear trivial. In point of fact it has revolutionized the theory of functions. We may readily understand that once the branch points of a function are known, it is possible, in theory, to construct the corresponding Riemann surface with its communicating strips, and thereby render the function uniform. We also sense the close relationship between the peculiarities of the communicating Riemann sheets and the general theory of the connectivity of surfaces, *i.e.*, that branch of mathematics called Analysis Situs (initiated by Listing and developed by Riemann). We have here an unexpected correlation between the theory of functions and Analysis Situs, two branches of mathematics which at first sight might seem to have nothing in common. Incidentally, the method of Riemann surfaces, by giving so to speak a geometrical model for the elaboration of an abstruse theory, illustrates one of the characteristics of Riemann. He seeks to visualize. Nothing of this sort is found in the works of Riemann's great contemporary, Weierstrass, whose major achievements are also found in this same theory of functions.

Poles and branch points are called *singular* points. There is also another kind of singular point, studied by Weierstrass, but it need not detain us here.

Analytic functions of a complex variable elucidate many of the features which mystified the mathematicians of the eighteenth century. We recall that analytic functions of a real variable may always be expressed by Taylor series, and that these series usually have only limited ranges of convergence. The underlying reason for these limited ranges becomes clear when complex variables are introduced. To understand this point we must mention that analytic functions of a complex variable, similarly to those of a real variable, may be expressed by Taylor series, *i.e.*, by series the successive terms of which contain the increasing powers of $(x - k)$, where k is any number, real or complex. Cauchy showed that, if we are given an analytic function of a complex variable and express it by means of a Taylor series in powers of $(x - k)$, where k is any number, real or complex, then this Taylor series will converge only so long as x remains within the circle whose centre is situated at the point k and whose circumference passes through the nearest singular point of the function. This *circle of convergence,* as it is called, obviously contains no singular point in its interior. Of course, if our function has no singular points, all circles of convergences will be the same: all of them will then extend over the entire plane to infinity.

Let us apply these considerations to the analytic function

$$(6) \qquad y = \frac{1}{x^2 + 1},$$

which we have previously discussed from the standpoint of real variables.

Two singular points are apparent immediately: they are the two poles, $x = \pm i$, at which the function becomes infinite. On the complex plane these two poles, P and P', are situated at unit distance above and below the origin O, on the vertical passing through O. There are no other singular points. Let us consider the Taylor series in powers of $(x - k)$

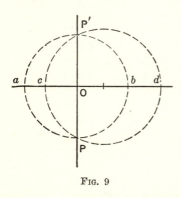

Fig. 9

that represents the function (6). According to Cauchy's rule this series will converge in a circle whose centre is the point $x = k$, and whose circumference passes through the nearer of the two points P or P'. In particular, let us consider the Taylor series in powers of x and the one in powers of $(x - \frac{1}{2})$. We have here two Taylor series in which k is taken to have the real values zero or $\frac{1}{2}$. The point, $x = k$, centre of the circle of convergence, is then situated at the origin O or at a point on the real horizontal axis separated from the origin by a distance $+\frac{1}{2}$. Circles of convergence having these points as centres may then be drawn, as in Figure 9.

If we confine our attention to real values for x, we see that the first of the two Taylor series converges only when x is between the points a and b, whereas the second Taylor series converges only when x is between c and d. The two Taylor series we are here considering are the two series (7) and (8), which we discussed on a previous page. In our earlier discussion we could give no reason for the strange limitations in the ranges of convergence of these series. The reason is now clear: the limited ranges of convergence are due to the presence of the singular points, P and P', situated on the complex plane. The mathematicians of the eighteenth century, by disregarding the complex values of x, were unable to understand the situation.

We also see why, in the case of the real function, $y = \sin x$, the Taylor series representing the function converges for all real values of x. It is because this function has no singular points (at finite distance), and so the one and only circle of convergence is the infinite circle, which covers the entire plane.

These elementary examples show how a knowledge of a function's singular points on the complex plane may yield important information on the behavior of the function, even when only real values of x are considered. In other words, the situation over the complex plane influences profoundly the occurrences along the real axis, and we cannot obtain any deep insight into the latter if we ignore the former. The limitations of the mathematicians of the eighteenth century now become obvious. All they knew was the behavior of their analytic functions along the real axis; they ignored the complex plane which extends above and below. As a result, they were in the situation of men wearing blinkers who walk along a road unable to see to the right or left. To such men the crossing of a bridge or the presence of a tunnel would be unpredictable events. Could they but look away from the road, they would perceive the river in the distance and the chain of hills, and they would expect the bridge and tunnel before reaching them. To the introduction of the complex variable, mathematicians owe the removal of the blinkers.

An important property of analytic functions of a complex variable is the following: If we are given the values of the function over an arc, as small as we choose, in the complex plane, we may deduce therefrom a knowledge of the function over its entire domain of existence.

Let us understand how the foregoing statement may be justified.

Consider a small arc of a curve in the complex plane, and suppose that we are informed of the value of an analytic function $f(x)$ at each point x of this arc. This information enables us to determine the values of all the derivatives of the function at any point A on the arc. The knowledge of the values of the function and of its derivatives at the point A permits us to construct a Taylor series which represents the analytic function, and which converges within a circle having the point A as centre. The circumference of this circle will pass through the nearest singular point of the function. If there is no singular point at finite distance, the circle will extend to infinity, covering the entire complex plane, and the function will thus be known in its entire domain of existence. But let us suppose that there is a singular point at finite distance, so that the circle of convergence has a limited radius. The function is then known only within this circle and is unknown at exterior points. Nevertheless, as we shall now see, the function can be constructed outside the circle.   Consider any point B inside the circle.   At this point B the values of the function and of all its derivatives are furnished by the Taylor series.   The knowledge of these values enables us to construct a second Taylor series which also represents the function and which converges within a circle having the point B as centre.   This second circle will usually overlap the first one and will thus determine the function beyond the limits of the first circle.   Continuing in this way, we may determine the function progressively over more and more extended regions, until eventually we have determined it over its entire domain of existence.   In this consists the method of *analytic continuation* as applied to analytic functions of a complex variable.

The domain of existence of the function may cover the entire complex plane or it may be restricted to a finite part.   If the latter situation arises, the boundary line of the domain of existence acts as a barrier preventing further analytic continuation.   This implies that the boundary line is formed of points all of which are singular—the boundary line itself is then singular.   Beyond this line the function is meaningless.

These considerations on analytic continuation in the case of analytic functions of a complex variable clarify certain points we mentioned in connection with analytic functions of a real variable.   Thus we mentioned that if a tiny arc of an analytic curve was given, we could deduce therefrom the entire curve by analytic continuation, provided, however, no singular point was present on the real axis (see note page 132).   But we added that the introduction of the complex variable enabled us to remove this restriction.   The justification for this statement is now clear.   The singular point on the real axis prevents the real Taylor series from converging beyond this point, so that if we are restricted to real variables, the method of analytic continuation will enable us to construct the curve up to the singular point but not beyond.   On the other hand, if we utilize the complex variable, we may proceed by analytic continuation around the singular point and then regain the real axis on the other side of the point.   Thus the singular point ceases to constitute an obstacle.   At the same time we see that the rigorous determinism which characterized analytic functions of a real variable (or analytic curves) finds its full expression in the analytic functions of a complex variable.

Applications of the Complex Plane—A particularly beautiful illustration of the utility of the complex plane, even when our interest is confined to functions of a real variable, is afforded by complex integration. For instance, consider the analytic function

$$(10) \qquad\qquad y = \frac{1}{(1 + x^2)^2} \cdot$$

We assume that x takes only real values, so that the function may be represented by a curve. We thus obtain the curve traced in Figure 10; it extends from $-\infty$ to $+\infty$. Suppose now we wish to evaluate the area comprised between this curve and the Ox axis. The straightforward method is to integrate the function along the whole extent of the real axis (the horizontal line). Integrations of this sort are often tedious, but, as we shall now see, the introduction of the complex variable simplifies the problem considerably.

Thus in the present example let us assume that x takes all complex values; our function must then be considered over the whole extent of the complex plane. We note that there are two poles present in the complex plane, *i.e.*, the two points $x = \pm i$. We now replace our

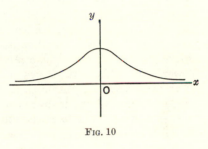

Fig. 10

original problem of integration along the real axis by a different problem. We integrate the complex function around a closed curve enclosing the pole $+i$. In particular, we select as closed curve the path defined by the entire real axis and by an infinitely extended semicircle having the point O as centre. The path of integration is illustrated in Figure 11, except that the dotted semicircle must be assumed infinitely large and hence the points A, B, and C infinitely distant.

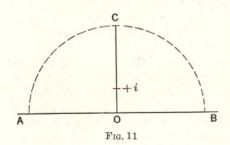

Fig. 11

Calculation shows that the integral along the circumference of the infinitely expanded semicircle has a zero value; and so the value of the integral around the entire closed curve has exactly the same value as the integral along the unlimited real axis AB. This latter integral, we recall, is the one we originally proposed to calculate. As a result, our problem will be solved if we can compute the integral around the entire closed curve. At first sight it might appear that we have merely replaced one problem by another equally difficult. But this is not so, for thanks to a theorem established by Cauchy, known as the "theorem of residues," the value of an integral along a closed curve surrounding a pole is obtained with little difficulty. In the present case the value is $\frac{\pi}{2}$. Here, then, is also the value of the integral along the real axis, and our problem is solved with a minimum of effort. Our example again shows how useful is the introduction of the complex plane, even in problems dealing with real variables.

The artifice we have just resorted to is called the *method of complex integration*; it is of current use for the calculation of integrals. Many of the integrations met with in theoretical physics, in optics, in celestial mechanics, and elsewhere are most easily calculated by this method; in particular, complex integration is utilized in the computation of the orbits in Bohr's atom. The complex plane has also been the source of powerful methods in the theory of differential equations. Applications of these methods are numerous in mathematical physics. For instance, in the problem of the hydrogen atom in wave mechanics, Schrödinger utilizes a theorem due to Fuchs and a method invented by Laplace, both of which involve the complex plane.

Elliptic Functions—We now propose to examine various special kinds of analytic functions of a complex variable. During the eighteenth century, mathematicians sought to calculate the lengths of the more important analytic curves. The length of an arc of a curve is furnished by an integral limited by the end-points of the arc. In particular if the curve is an ellipse, we obtain an integral of a special kind, which, in view of its connection with an ellipse, was called an *elliptic integral*. The calculation of the arcs of other curves may also lead to elliptic integrals, the curve known as the lemniscate of Bernoulli being a case in point.

Let us take on the lemniscate curve a fixed point A and a variable point B. The length of the arc AB depends solely on the position of B (since A is fixed). The elliptic integral which gives the length of the arc AB is therefore a function of the position of the variable point B. But we may reverse the relationship and view the position of B as a function of the length of the arc AB. This latter function, called an *elliptic function*, is of course not the same as the function of the end point defined by the aforementioned elliptic integral; the elliptic function is, so to speak, the inverse of the elliptic integral, and for this reason we say that it is obtained by inversion from the elliptic integral.

Euler, and principally Legendre, studied these elliptic integrals and functions, concentrating on the integrals rather than on the functions. The entire subject was revolutionized by a single remark of Abel, who pointed out that Legendre's procedure was unnecessarily laborious. Instead of deriving the properties of the function from the integrals, Abel maintained that the opposite procedure should be followed and that the study of the elliptic functions should be given priority. Abel and Jacobi soon put this idea into practice, and the results they obtained have entitled them to be regarded as the real founders of the theory of elliptic functions. Commenting on elliptic functions, Volterra writes:

"The story of elliptic functions is well known. It has often been told, for it is one of the most interesting chapters in mathematics. Surprise followed surprise as investigators passed from one stage of development to the next, which in turn brought new discoveries and new marvels."

Some of the more important properties of elliptic functions may easily be understood. Thus let

$$y = f(x)$$

represent an elliptic function. The variable x, as before, is allowed to move over the complex plane, and the corresponding value of y is then determined by the function. Only one value for y corresponds to each value of x, for there are no branch points and so the function is uniform, *i.e.*, one-valued. Elliptic functions betray important periodic properties, which we shall now examine. Elementary trigonometry makes us familiar with periodic functions. For instance, in the trigonometric function

$$y = \sin x,$$

where x is restricted to real values, we find that y (*i.e.*, the function) resumes exactly the same value when x is increased or decreased by 2π. Equivalently, we may say that y remains unchanged when $x \pm 2\pi$ is substituted for x. The magnitude 2π is called the *period* of the function; and since our function has only one period, it is called *simply periodic*. These results are expressed analytically by the equalities

$$y = \sin x = \sin(x \pm 2\pi) ;$$

and since the substitution of $x \pm 2\pi$ for x may be repeated any number of times in succession without the value of y being affected, we may write more generally

$$y = \sin x = \sin(x \pm m\, 2\pi),$$

where m is any integer, positive or negative.

Elliptic functions differ, however, from the trigonometric ones in that they are *doubly periodic*; by this is meant that they have two independent periods. These two periods, which we shall designate by ω and ω', may have any values, but not more than one of the two periods can be a real number. In general, both periods are complex. The double periodicity of elliptic functions is expressed by the statement: If $y = f(x)$ is an elliptic function, then y will resume the same value when $x + \omega$ or when $x + \omega'$ is substituted for x in the function. Inasmuch as these substitutions may be repeated consecutively any number of times, the double periodicity leads to the relations

$$y = f(x) = f(x + m\omega + m'\omega'),$$

where m and m' are any two integers, positive, negative or vanishing. Obviously, elliptic functions are, from the standpoint of periodicity, a generalization of the simply-periodic trigonometric functions.

The double periodicity of elliptic functions is susceptible of a geometrical representation on the complex plane. To understand it, consider a point P on the complex plane. This point defines some complex number. If we add another complex number $a + ib$ to the original number, we obtain a point Q situated at a distance a from P along a horizontal and at a distance b along a vertical. Figure 12 is self-explanatory. The length and slant of the segment PQ are thus determined by the value of the complex number we have added.

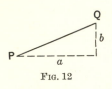

Fig. 12

We may now apply these considerations to the property of double periodicity of the elliptic functions. Let the point P (Figure 13) represent the value ascribed initially to x in the elliptic function $y = f(x)$. The two complex periods ω and ω' are represented by two different segments of definite slant and length. If, starting from P, we draw in all possible ways successions of segments equal and parallel to the period segments ω and ω', we obtain an aggregate of points in the complex plane. Each one of these points represents some numerical value for x, which differs from the value at P by integral numbers of time the magnitudes of the two periods. The aggregate of all such points defines the summits of a network of juxtaposed parallelograms, which extends over the entire complex plane; all these parallelograms are equal, their sides being equal and parallel to the period segments ω and ω'. Some of them are represented in Figure 13. The double periodicity of the elliptic function then shows that y resumes the same value when x is made to pass from the initial summit P to any one of the other summits. More generally, the value of the function y is repeated at all similarly-situated points in the various parallelograms. An immediate consequence is that if we know how the function behaves in any one of the parallelograms, we know thereby how it behaves over the entire complex plane.

In Figure 13 both periods ω and ω' are complex, but we have said that one of the two periods may be real. When one of the periods is real, the lower side of a parallelogram is parallel to the real axis (*i.e.*, the horizontal axis), as in Figure 14. In this case the elliptic function resumes the same value when we start, say, with a real value for x and then add to x an integral number of times the value of the real period. Only when we consider complex increments to the value of x is the second periodicity detected.

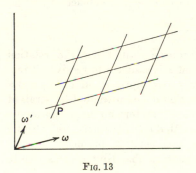

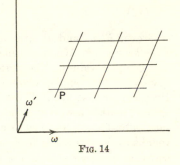

Fɪɢ. 13   Fɪɢ. 14

The elliptic functions known to Legendre were of the type last mentioned, and since Legendre confined his attention to real values, only one period (the real one) appeared to be associated with his elliptic functions. Later Abel and Jacobi, by introducing the complex variable, were led to the discovery of the second periodicity and gained thereby an understanding of elliptic functions, undreamed of by Legendre. Abel, Jacobi, Weierstrass, Riemann, and Hermite are the great names associated with elliptic functions. Various series and functions were introduced by these mathematicians in order to coordinate results, *e.g.*, the *Theta functions* of Jacobi and the *Zeta and Sigma functions* of Weierstrass. The range covered by the theory is so vast that voluminous treatises have been written on this single department of the theory of analytic functions.

Elliptic functions furnish the solutions of several mechanical problems. The angle of oscillation of a swinging pendulum, for example, is expressed at any instant by an elliptic function of the time. The problem of the spinning top also involves elliptic functions, and so does the calculation of the planetary orbits in the theory of relativity. But quite aside from their physical applications, elliptic functions have yielded the solution of many problems in geometry and in the various fields of pure mathematics. Furthermore, the methods devised in the study of elliptic functions served later as a guide for the elucidation of the properties of the more complicated functions that were soon to be discovered.

Modular Functions—Hermite, from a study of the elliptic functions, was led to the discovery of new uniform analytic functions with a more complicated kind of periodicity. These are the elliptic modular functions. If we designate an elliptic modular function by

$$y = f(x),$$

the function will resume the same value whenever x is replaced by

(12)
$$\frac{ax + b}{cx + d},$$

where a, b, c, and d are any real integers satisfying the relation $ad - bc = +1$.* We must remember that x is assumed to receive all real and complex values. The periodic properties expressed in (12) yield a division of the complex plane into repeating domains limited by straight lines and circular arcs. These domains are therefore no longer parallelograms; neither are they equal in area. Modular functions, like elliptic ones, afford additional means of solving mathematical problems. It was thanks to the introduction of modular functions that Hermite was able to solve the algebraic equation of the fifth degree. ·

Automorphic Functions—More general than the modular functions are the *automorphic* functions of Poincaré. The same substitution that causes a repetition in the value of a modular function still holds, but now the numbers a, b, c, d are no longer restricted to real integers. The discovery of these new functions, which Poincaré classes under the headings of Fuchsian and Kleinian in the order of their generality, has opened a new continent in mathematics, much of which is still virgin territory. The foremost contributor in this field, after Poincaré himself, was Klein.

Let us observe that both trigonometric and elliptic functions are particular cases of automorphic functions. Elliptic functions repeat when x is replaced by $x + \omega$ or $x + \omega'$, where ω and ω' are the two periods. That this property also holds as a special case for automorphic functions is seen by setting in

$$\frac{a\,x + b}{c\,x + d}$$

the values $a = 1$, $b = \omega$ or ω', $c = 0$, $d = 1$.

---

* We may also say that an elliptic modular function resumes the same value when x is replaced by $-\dfrac{1}{x}$ or by $x + 1$; for when we perform these substitutions in succession in all possible ways, the general substitution (12) is obtained.

In his discovery of the new functions Poincaré was guided by the theory of elliptic functions. The method of inversion, noted in connection with elliptic functions, and also the construction of theta series were carried over by Poincaré and generalized to suit the new conditions. The applications of automorphic functions in pure mathematics are numerous and varied. Poincaré showed that an important class of differential equations can be integrated when automorphic functions are utilized. Interesting relations between the new functions and problems of higher arithmetic and of non-Euclidean geometry were established. Finally, Poincaré observed that the properties of some of these functions suggested the possibility of solving by their means the algebraic equations of degree higher than five.

There is one further application that we should like to mention. It concerns the representation of algebraic curves. By an algebraic curve is meant a curve defined by any algebraic relation between x and y. If we assume real values for x and y, the curve may be represented graphically by plotting x as abscissa and y as ordinate. Thus

$$(13) \qquad x^2 + y^2 - R^2 = 0,$$

or equivalently

$$(14) \qquad y = \pm \sqrt{R^2 - x^2},$$

is an algebraic curve. It is of the second degree, since 2 is the highest power appearing in the equation (13). It represents a circle of radius R and with centre at the origin.

Now we see that for a given value of x between $-R$ and $+R$, there are two values for y; these correspond to two symmetric points, such as B and B', which appear above and below the Ox axis. In certain cases we may wish to avoid this many-valuedness or non-uniformity. Since it is the radical in (14) which, owing to its double sign, generates the non-uniformity, we must rid ourselves of the radical. This may be done in the case of the circle by introducing an auxiliary variable called a *parameter*. Thus with t as parameter, the equations

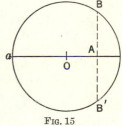

FIG. 15

$$(14) \qquad x = R\,\frac{1 - t^2}{1 + t^2}\,, \quad y = \frac{2Rt}{1 + t^2}$$

define exactly the same circle when t is made to vary from $-\infty$ to $+\infty$. When t is given the value $-\infty$, the values of x and y are $-R$ and zero respectively, so that we obtain the point a on the circle. Then, as t increases to $+\infty$, the point moves once around the circle in a counterclockwise direction. To each value of t, corresponds only one point on the circle; accordingly the ambiguity in the previous method of representation is removed. We now inquire: Can a similar procedure be followed for all algebraic curves? Can we, in every case, however complicated the curve, define a point of the curve without ambiguity when the value of some appropriate parameter is given?

Prior to the discovery of automorphic functions, such a representation was usually impossible. But by utilizing these functions, Poincaré showed that all algebraic curves could be given a parametric representation. As one commentator has said, this discovery gave Poincaré the keys to algebra. All these results are extremely important owing to their connection with the theory of Abelian integrals, which itself leads to one of the pinnacles of analysis.

Infinite Processes—From a psychological standpoint we first obtain an understanding of finite processes; only at a later stage are infinite processes considered. Thus the conception of the sum of a finite number of terms or of a finite number of areas preceded that of infinite sums, illustrated by series and integrals. The same is true of finite and infinite products. In this passage from the finite to the infinite, we cannot always extrapolate results uncritically from one domain to the other. The danger of a hasty extrapolation is apparent when we recall the important differences between the properties of the sum of a finite number of terms and those of the limit of an infinite series. We mentioned, in particular, that if a series which contains an infinite number of positive and negative terms converges, the limit of the series may change when the order of the terms is modified; the series may even be made to diverge. No such analogy holds for finite sums. A similar situation arises in the case of a convergent series of continuous functions. If there were only a finite number of continuous functions to be added, say a trillion, their sum would still define a continuous function. But this conclusion is not necessarily correct when we are dealing with an infinite series of continuous functions; the limit may now define a discontinuous function. It was by considering series of this latter type that Baire was led to his theory of discontinuous functions, referred to on a previous page. Many other properties that hold in the finite case may fail when infinity is considered. For these reasons, the passage from the finite to the infinite is always a delicate matter, and hasty conclusions are often found to be erroneous. Other examples of the passage to infinity are illustrated by Poincaré's study of infinite determinants and by Hilbert's investigations on infinite quadratic forms. Infinite determinants occur in the theory of integral equations (these equations are of frequent use in modern mathematical physics). As for the theory of infinite quadratic forms, it has proved to be at the basis of the new quantum theory of the atom. These points will be understood when we discuss Heisenberg's Matrix Method.

GALOIS' THEORY OF GROUPS

Groups of Substitutions—Galois appears to have been the first mathematician to realize the importance of the group concept in mathematics. A few hours before meeting his death in a duel, he jotted down a summary of his discoveries on groups of substitutions and showed that his ideas revolutionized the entire theory of algebraic equations. The following brief explanations will indicate the essentials of Galois' theory.

Let us consider a set of different mathematical operations, which we may designate by A, B, C, ... If any two of these operations performed in succession are equivalent in their effect to one of the operations of the set, then the set of operations is said to have the group property. But for the operations to form a group, the inverse of every operation must be present in the set, and one of the operations of the set must be the identical operation (*i.e.*, an operation which changes nothing).

Suppose, then, we have a number n of different objects a, b, c, . . . placed along a straight line. They may be ordered in $n!$ different ways. Each different method of ordering is called a *permutation*. For instance, if there are three objects a, b, c, there are six permutations, namely,

abc, acb, bac, bca, cab, cba.

To pass from the first permutation to the second, we must replace b by c, c by b, and a by itself. To pass from the first permutation to the last but one, we must replace a by c, b by a, and c by b. Each one of these sets of replacements is called a *substitution*. There are as many different substitutions as there are different permutations. The identical substitution, which changes nothing, is the one in which a is replaced by a, b by b, and c by c.

We may readily verify that the aggregate of the $n!$ different substitutions connected with n magnitudes a, b, c, . . . forms a group. Thus, if a substitution is performed on a given permutation, and if to the permutation obtained a second substitution is applied, we obtain a permutation which could have been derived directly from the original one by utilizing some appropriate substitution of the set. Furthermore, the set of substitutions comprises the identical substitution; and each substitution is accompanied by its inverse. The foregoing group, which contains $n!$ substitutions, is called the *symmetric group* of degree n.

The *degree* of the group is defined by the number of objects $a, b, c, \ldots$ on which the substitutions operate. The *order* of the group is defined by the number of distinct substitutions comprised in the group. Thus the symmetric group just considered is of degree n and of order $n!$

It may happen that some among the substitutions of a group can be withdrawn and that the remaining ones still form a group. The smaller group which is thus obtained is said to be a *sub-group* of the original group. For instance, the symmetric group of degree n always contains a sub-group, of degree n, represented by $\dfrac{n!}{2}$ of the original substitutions. This sub-group is called the *alternating group* of degree n. The alternating group in turn contains a sub-group. This sub-group may contain another, and so on.

Some sub-groups are characterized by a property described in the note. Such sub-groups are called *invariant sub-groups.** A given group may contain different invariant sub-groups. The invariant sub-group of maximum order is the one which comprises the greatest number of distinct substitutions.

Let us start with the symmetric group and consider its invariant sub-group of highest order (*i.e.*, the sub-group which involves the greatest possible number of distinct substitutions). The invariant sub-group in question turns out to be the alternating group. We then consider the invariant sub-group of highest order of the alternating group; and then the invariant sub-group of highest order of this sub-group; and so on. We thus obtain a succession of groups, each one of which is an invariant sub-group of maximum order of the preceding group. It is as though

---

* Let G_1 be a sub-group of G. Let S represent any substitution of G_1, and T any substitution of G. The inverse substitution of T is represented by T^{-1}. Suppose, then, that to any given permutation π we apply the substitution T^{-1}, obtaining thereby a new permutation π'. We then subject this new permutation to the substitution S and are thus led to a permutation π''. Finally, we perform the substitution T on π''. Our sequence of substitutions is represented by the product of substitutions $T^{-1} S T$. Since T^{-1}, S, and T are substitutions which are among those of the group G, it is certain that the product $T^{-1} S T$ is also a substitution belonging to this group.

Now, when S is any substitution of the sub-group G_1, and T any substitution of the group G, and when the substitution $T^{-1} S T$ always belongs to the sub-group G_1, then G_1 is said to be an invariant sub-group of G.

we had a collection of Chinese boxes, with each box (except the last one) containing another. The last member of this succesion of invariant sub-groups is represented by the identical substitution; for this substitution constitutes a group (though it is a very trivial illustration of one), and it also betrays the characteristics of an invariant sub-group.

To take a definite example, we shall consider the case of $n = 4$. The symmetric group contains $4! = 24$ distinct substitutions and ·hence is of order 24. The alternating group of order 12 is its invariant sub-group of maximum order. The alternating group in turn contains, as invariant sub-group of maximum order, a group of order 4. This sub-group contains an invariant sub-group of order 2; and the latter contains the invariant sub-group of order 1, represente 1 by the identical substitution. The orders of our succession of groups arc thus

$$24, \ 12, \ 4, \ 2, \ 1.$$

Galois considered the ratio of the order of each one of these groups to that of its successor. The ratios are given by the numbers.

$$2, \ 3, \ 2, \ 2.$$

He called this succession of numbers a *series of composition*. The present series arises when we start from the symmetric group for which $n = 4$. But the significance of a series of composition is the same whatever integral value may be ascribed to n.

For $n = 2, 3, 4, 5, \ldots n, \ldots$, the series of composition of the successive symmetric groups are the following:

$$(16) \quad \begin{array}{c|l} n = 2 & 2. \\ n = 3 & 2, 3. \\ n = 4 & 2, 3, 2, 2. \\ n > 4 & 2, \dfrac{n!}{2}. \end{array}$$

Sometimes a given group contains more than one invariant sub-group of maximum order. In this event, when we construct the series of composition, we are free to select any one of these sub-groups; and the series of composition may differ according to the choice we make. But it can be shown that the various series of composition will contain exactly the same numbers, and that the only difference that may arise consists in a change of the order in which these numbers appear in the series. Thus, all the series of composition of the same group contain exactly the same numbers, so that these numbers are characteristic of the group considered.

If we revert to the series of composition written in the table (16), we see that the first three series contain only prime numbers. On the other hand, when n exceeds 4, the number $\dfrac{n!}{2}$, which now always appears in the series of composition, is composite. Hence we conclude that, when $n > 4$, the series of composition for the symmetric group do not contain prime numbers exclusively. The importance of this observation will be understood when we consider Galois' theory of equations.

Algebraic Equations—The general algebraic equation of the nth degree (where n is any positive integer) is obtained by equating to zero a polynomial of the nth degree in x. It is thus of the general form

$$(17) \qquad x^n + c_1 x^{n-1} + c_2 x^{n-2} + \ldots \ldots c_{n-1} x + c_n = 0,$$

where the coefficients c are constants of arbitrary value. According to a theorem known as d'Alembert's theorem, the equation (17) is satisfied when the variable x is credited with any one among n appropriate values

$$x_1, x_2, \ldots x_n.$$

These values are called the *roots* of the equation. D'Alembert did not give a rigorous demonstration of his theorem, but it has since been furnished by applying the theory of functions of a complex variable. The values of the roots depend on the values credited to the coefficients $c_1, c_2, \ldots c_n$ in the equation, and hence the roots are functions of these coefficients. To solve the equation means to obtain the explicit expressions of the roots in terms of the coefficients.

At the close of the sixteenth century the general equations of the first four degrees had been solved. The solutions of these equations (*i.e.*, the formulae defining the roots) were obtained in the form of explicit algebraic functions of the coefficients,* so that no irrationals other than radicals appeared in these solutions. Owing to this circumstance, the equations of interest were said to be soluble by radicals, or, equivalently, to be soluble algebraically. At the time it was confidently expected that

---

* By an explicit algebraic function of the coefficients c is meant a function which can be constructed by performing on these coefficients a finite number of algebraic operations (additions, subtractions, multiplications, divisions, and extractions of roots). Hence the only irrationals that can appear in an explicit algebraic function are radicals. When therefore the roots of an equation can be expressed by explicit algebraic functions of the coefficients, we are justified in saying that the equation can be solved by radicals.

further study would furnish similar algebraic solutions for the general equations of higher degree; and during the next two centuries numerous attempts were made to secure such solutions. But unexpected difficulties were encountered, and no progress was made. Some mathematicians, notably Lagrange, expressed doubts on the existence of algebraic solutions for the equations of higher degree. The arguments in support of these doubts are readily understood when we recall that algebraic expressions, or functions, represent a very special class of functions, so that there is no *a priori* reason why those functions of the coefficients which define the roots should necessarily be explicit algebraic expressions in all cases. Doubts, however, even when plausible, do not constitute mathematical proofs, and so the search for the elusive solutions continued.

Finally, in 1825, Abel put an end to further discussion by proving in a celebrated theorem that the general equation of the nth degree ($n > 4$) cannot be solved algebraically. Shortly after, Galois, by applying his theory of groups, extended Abel's results. Galois established a general theorem which provided a test for deciding under what conditions any algebraic equation (not necessarily an equation of the general type)* can be solved by radicals, *i.e.*, algebraically. The application of Galois' theorem to the general equation of degree n then furnished an immediate proof of Abel's theorem.

The theorems of Abel and Galois merely established the impossibility of obtaining *algebraic* solutions for the general equations of higher degree; they did not imply that non-algebraic solutions were unobtainable. Indeed, some fifty years after Galois' memoir, Hermite solved the general equation of the fifth degree, and Poincaré indicated the possibility of expressing the roots of the equations of still higher degree. These two masters utilized modular, and automorphic functions respectively, and their solutions were, of course, non-algebraic. In this chapter, however, we shall confine our attention to algebraic solutions and to Galois' investigations.

Before we examine Galois' procedure, let us inquire why mathematicians should have evinced any particular interest for the precise form in which the roots of algebraic equations could be expressed. It would seem that the sole aim of the mathematician should be to solve a given problem as best he can. We should expect him therefore to abstain from imposing on the solution some artificially restricted form, which might render the existence of the solution impossible. Why, then, should algebraic equations have been classed as soluble or as insoluble

---

* If particular numerical values are assigned to the coefficients of the general equation, the equation is said to be non-general, or particular.

according to whether their roots could or could not be expressed by algebraic functions of the coefficients? The answer to this question and to similar ones must be sought in the historical development of mathematics. Usually, what appears today as a restriction on the form of a solution was not regarded as such by the mathematicians who first attacked the problem of interest. In their eyes the so-called restriction appeared as a necessary requirement for the solution to be rigorous, and so the restriction was accepted without question. With the progressive development of mathematics, other forms of solution were constructed, just as rigorous as the earlier ones but of a more general kind. Only then did the supposedly necessary requirement of earlier days reveal itself as artificial, and only then did it assume the status of a restriction imposed voluntarily. Even after this discovery, however, problems in which the restriction was retained were still given serious attention by mathematicians; the only difference was that these problems were now viewed in their proper light.

Let us see how these considerations apply to the solution of algebraic equations. From the standpoint of the mathematicians of the sixteenth century, the only explicit expressions which could be accepted as mathematical were algebraic expressions and trigonometric functions. Accordingly, unless the roots of an equation could be defined by such expressions, the equation was held to be insoluble—solutions in the form given by Hermite would have been incomprehensible. Of course the naïve outlook of the earlier mathematicians was not shared by Abel and Galois. These mathematicians undoubtedly recognized that explicit algebraic expressions formed a very restricted class. Nevertheless the problem of solving algebraic equations by means of radicals was deemed by them of sufficient interest in itself to warrant investigation.

Many illustrations of the same kind may be given. Thus we mentioned that, not until the close of the eighteenth century, did mathematicians realize the arbitrariness of their procedure in restricting mathematical functions to be analytic. Yet, even after analytic functions were recognized as forming only a small sub-class among the continuous functions, mathematicians still sought to determine under what conditions the solutions of given differential equations would be analytic. Similarly, the Greek geometricians insisted on the exclusive use of the straight edge and compass in the demonstrations of their theorems— though in so doing they presumably did not realize that they were imposing a restriction which was by no means necessary. Nevertheless, even when the restrictive nature of the condition imposed by the Greeks was understood, mathematicians still retained this restrictive condition in certain cases, and on this basis attempted to determine which of the geometrical problems were soluble or insoluble. For example, when we say that the squaring of the circle is impossible, what we really mean is that a square having the same area as a given circle cannot be constructed by means of the straight edge and compass. On the other hand, if we enlarge the field of permissible operations, the squaring of the circle may be performed, and the problem becomes soluble.

The Group of an Equation—Let us examine how Galois utilized his theory of groups in the study of algebraic equations. His first step was to connect a definite group with each equation. This group is called the *group of the equation*; its general significance will be understood from the following considerations.

Let x_1, x_2, . . . x_n represent the n roots of an equation of the nth degree,* and let $f(x_1, x_2, . . . x_n)$ be some function of these roots. If we knew the precise expressions of the roots in terms of the coefficients c of the equation, and if we replaced the roots in $f(x_1, x_2 . . . x_n)$ by these expressions, we should obtain the expression of the function f in terms of the coefficients.

Let us, then, consider all those rational functions † $f(x_1, x_2, . . . x_n)$ of the roots x_1, x_2, . . . x_n, which satisfy the following conditions: the coefficients they contain must be rational functions of the coefficients c_1, c_2, . . . c_n of the equation; furthermore, if in these functions the roots are replaced by their expressions in terms of the coefficients c_1, c_2, . . . c_n, the functions must become rational functions of the coefficients c_1, c_2 . . . c_n and must have rational coefficients. These functions form a class which we shall refer to as the class A for short. Suppose now that we consider one of the functions of class A at random, and that we submit its n variables to all the substitutions of one of the

---

* The equation is assumed to be irreducible. This means that it is impossible to split the polynomial on the left hand side of the equation into a product of two or more polynomials the coefficients of which are rational functions (with rational coefficients) of the coefficients c of the equation. For instance, if the coefficients c_1, c_2....c_n are integers and the equation is irreducible, it will be impossible to decompose the polynomial into a product of two or more polynomials the coefficients of which are rational numbers. Whenever an equation is irreducible, its roots are always unequal.

† Rational functions involve only the operations of addition, subtraction, multiplication, and division; and these operations may be repeated only a finite number of times.

groups of degree n. As an example let us suppose that we apply the substitutions of the symmetric group. These substitutions permute the positions of the variables x_1, x_2, . . . x_n in the function, and when account is taken of the numerical values of these variables (*i.e.*, the roots of the equation), some of the substitutions may be found to alter the numerical value of our function of class A. On the other hand, it may happen that the numerical value of the function remains unchanged under each one of the substitutions of this symmetric group. In such a case the function is said to belong to the symmetric group.

We first suppose that our function of class A belongs to the symmetric group. Obviously, then, the function will also remain unchanged in value under the substitutions of any one of the sub-groups of the symmetric group.

Next, let us assume that the numerical value of the function changes when the various substitutions of the symmetric group are applied. In this case we shall find that changes occur only for some of the substitutions, and that there will always be other substitutions of the symmetric group which do not change the value of the function. These particular substitutions that do not change the numerical value of the function always form a group, which is necessarily a sub-group of the symmetric group. Hence we conclude that there will always be a group of substitutions which leaves the numerical value of the function unchanged, and that this group is either the symmetric group or one of its sub-groups.

The considerations we have developed for one particular function of class A must now be extended to the whole aggregate of these functions. We are thus led to consider the group of highest possible order whose substitutions leave unchanged the numerical values of all the functions of class A corresponding to a given equation. This group may be the symmetric group or one of its sub-groups. It constitutes what Galois called the "group of the equation."

In short, all those rational functions of the roots x_1, x_2, . . . x_n, which can be expressed rationally in terms of the coefficients of the equation, are functions which retain fixed numerical values under the substitutions of the group of the equation.

Conversely, all those rational functions of the roots, which retain fixed numerical values under the substitution of the group of the equation, are functions which can be expressed by rational functions of the coefficients c_1, c_2, . . . c_n of the equation.

We do not propose to enter into a discussion of the methods whereby the group of the equation can be determined in any particular case. Suffice it to say, this group may always be obtained.

Galois' Theorem—We wish to ascertain whether a given irreducible algebraic equation (general or non-general) can be solved by radicals. Galois' method consists in obtaining the group of the equation and then proceeding to construct the series of composition which starts from this group. Galois' theorem then states:

The equation will be soluble by radicals if, and only if, all the numbers appearing in the series of composition are prime.

Let us apply Galois' theorem to the general equation of the nth degree. We first determine the group of this kind of equation. We find that it is the symmetric group of degree n. Next we consider the series of composition. Now the table (16) shows that the series of composition for the symmetric group of degree n contains none but prime numbers only when $n < 5$, and that composite numbers occur when n exceeds 4. We conclude that the general equations of degrees $n = 1, 2, 3, 4$ can be solved by radicals and that those of higher degree cannot.

Binomial Equations—A special class of non-general equations, of any degree n that can be solved by radicals is illustrated by the so-called binomial equations. For instance

$$(18) \qquad x^n - 1 = 0 \text{ (where } n \text{ is any positive integer),}$$

is a binomial equation. Galois' theory proves that equations of this type are soluble by radicals. However, we need not utilize Galois' theory to establish this point; it was well known to earlier mathematicians, to Gauss in particular.

Binomial equations have an important application in geometry. A simple demonstration shows that if the lengths defined by the roots of the equation (18) can be constructed, it will be possible to divide a circumference into n equal arcs. Obviously, if this can be done, the regular polygon of n sides inscribed in the circle is obtained by drawing chords to the successive equal arcs of the circumference. In short, we see that if the roots of the equation (18) can be constructed, the regular polygon of n sides can also be constructed.

Now we have said that the roots of the equation (18) can be expressed by radicals. By this we mean that the expressions of the roots contain only the coefficients (here ± 1), integers, and radicals. But in order to construct these roots geometrically, i.e., by ruler and compass, radicals other than square roots must be absent. Hence we conclude that the regular polygon of n sides can be constructed by the methods of geometry if, and only if, the roots of the equation (18) contain no radicals other than square roots in their expressions. The geometrical problem is thus transformed into the algebraical problem of determining the values of n for which the roots of (18) will have expressions of the type just stated.

Gauss solved the problem. He first restricted his attention to prime numbers for n. He then proved that the roots of (18) will have expressions of the required form when, and only when, the prime number n can be represented by the formula

$$(19) \qquad\qquad n = 2^k + 1,$$

where k is a positive integer. When in (19) we set $k = 1, 2, 3, 4, 5, 6, 7, 8,$ . . . , we obtain

$$n = 3, 5, 9, 17, 33, 65, 129, 257, \ldots .$$

Of the numbers written, only

$$(20) \qquad\qquad n = 3, 5, 17, 257, \ldots .$$

are prime.

Thus, Gauss's result is that the regular polygons of n sides (n prime) which can be constructed geometrically are those for which n is a prime number expressible by the formula (19), and hence is one of the numbers of the list (20).* The value $n = 3$ refers to an equilateral triangle and is of course trivial. The value $n = 5$ signifies that the regular pentagon may be constructed—a result known prior to Gauss. But the value $n = 17$ and the higher values were unknown. We have mentioned Gauss's theorem because it illustrates a relationship between seemingly different departments of mathematics; namely, geometrical constructions with ruler and compass, and the abstract theory of equations.

Other Applications of the Theory of Groups—To Galois goes the credit of having introduced the group concept into mathematics. But Galois considered only those groups of substitutions which permute

---

* More generally a regular polygon of n sides, where n is any positive integer, may be constructed provided n be of form

$$(21) \qquad\qquad n = 2^m \, p_1 \, p_2 \, p_3 \, p_4 \ldots ,$$

where m is any positive integer or zero, and where p_1, p_2, p_3, $p_4 \ldots$ may have the value 1, or else are prime numbers, all of which are different and can be expressed by the formula (19). From formula (21) we obtain

$$n = 3, 4, 5, 6, 8, 10, 12, 15, 16, 17, 20, \ldots .$$

for the numbers of sides of the regular polygons that can be constructed.

objects, whereas the concept of group may be extended to operations generally. For example, the rotations which transform a regular polyhedron into itself constitute a group. We shall illustrate this statement in connection with a tetrahedron.

A tetrahedron can be rotated about any one of seven different axes in such a way that the edges of the rotated solid occupy the positions vacated by other edges. The tetrahedron thus appears to resume its initial position in space at the end of each rotation, and for this reason we say that the rotations transform the solid into itself. Twelve different rotations about the seven axes have this property of transforming the tetrahedron into itself. These twelve rotations (one of which is a zero rotation leaving the solid unmoved) form a group, called the tetrahedral group. The group property issues from the fact that a composition of rotations about any two of the axes is a rotation about one of the axes. If instead of operating with a tetrahedron we operate with an icosahedron (regular polyhedron having twenty sides), we obtain a group containing sixty different rotations. Klein established the connection between this icosahedral group and Hermite's solution of the algebraic equation of the fifth degree.

Galois' groups, as also the groups which we have just considered, involve finite numbers of substitutions, or operations. But a group may contain an infinite number of substitutions. For example, the substitutions

$$x' = \frac{ax + b}{cx + d}, \ (ad - bc = 1)$$

where a, b, c, d are integers, are infinite in number, and they form a group. These substitutions, as we have seen, transform a modular function into itself, so that a modular function is a function which is reproduced under the substitutions of this group. Similarly, Poincaré's Fuchsian and Kleinian functions are transformed into themselves under the substitutions of groups which contain an infinite number of substitutions. Further examples of groups will be given in Chapter XX.

CHAPTER XV

MATHEMATICAL PRELIMINARIES *(Sequel)*

(DIFFERENTIAL EQUATIONS)

DIFFERENTIAL equations are probably the most important mathematical tools of theoretical physics. We shall examine only their most elementary properties. First, let us recall that when

$$y = f(x)$$

defines a continuous function y of a real variable x, the relationship between x and y can be represented graphically by means of a continuous curve (Cartesian method).

Derivatives—The rate of change of y when x varies is called the first derivative y' of the function. This first derivative usually varies in value when x is made to vary; hence it is itself a function of x. The value of the first derivative for any given value of x, corresponding to a point P on a curve $y = f(x)$, increases with the slope of the curve at P. It may be represented graphically by the trigonometric tangent of the angle θ contained between the Ox axis and the tangent AP to the curve at the point P (Figure 16), *i.e.*, by the ratio $\dfrac{BP}{AB}$. A continuous curve which, like Weierstrass's curve, has no definite tangent, or slope, at any point is thus equivalent to a continuous function having no first derivative.

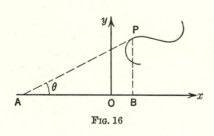

FIG. 16

The second derivative y'' of the function y is the first derivative of the first derivative; it represents the rate of change of the rate of change of y when x varies, or equivalently the rate of change of the slope of the curve as we proceed along the Ox axis. Obviously, if the curve is a straight line, so that the first derivative, or slope, remains unchanged, the second derivative will always vanish. Thus, the constant vanishing of the second derivative indicates that our curve is some straight line. We may consider higher derivatives, *e.g.*, the third derivative y'''; it is the first derivative of the second derivative. More generally the nth derivative is the first derivative of the $(n-1)$th derivative. These elementary notions will suffice for the present.

Ordinary Differential Equations—Consider the function

$$(1) \qquad y = \pm \sqrt{4 - x^2}.$$

If we plot x and y according to the Cartesian method, we obtain a circle of radius 2, with centre at the origin. Now at each point P of our circle there is a definite relationship between the slope of the circle and the position of this point P. Since the slope is determined by the first derivative y', there exists some definite relation holding, at all points of the circumference, between y' and the coordinates x and y of the point P. In the present example the relation is

$$(2) \qquad yy' + x = 0.$$

An equation of this kind is called a *differential equation* because, in contradistinction to the equation (1), the relationship (2) involves a derivative.

Let us examine the comparative significance of the two equations (1) and (2). We know that (1) represents the particular circle mentioned above, and that (2) expresses a relationship holding for this same circle. We therefore inquire: Are the two equations equivalent, *i.e.*, do they merely express the same relations in different forms? If they do, we must assume that (2), like (1), defines the former circle and nothing else. We find, however, that the two equations are not equivalent, for it can be shown that (2) defines *all* circles having the centre as origin. In mathematical language, all circles

$$(3) \qquad y = \pm \sqrt{R^2 - x^2},$$

regardless of the value of their radius R, satisfy the differential equation (2). We conclude that our differential equation defines not one particular circle, but a whole class of circles. This important property of defining a class, and not a particular member of the class, is characteristic of all differential equations.

Suppose, then, we are given some differential equation; it expresses a more or less intricate relationship connecting x, y, and the derivatives of y. To say that the differential equation has solutions means that there exist functions (or curves), $y = f(x)$, whose derivatives satisfy the relationship expressed by the differential equation. Our aim is to obtain these solution functions, as they are called. The problem of obtaining the solution functions of any given differential equation is called the problem of *integration*; it is one of the outstanding problems of analysis. To integrate a differential equation means to transform it in such a way that it is converted into a general mathematical expression which embodies all the solution functions. This general mathematical expression is called the *general solution,* or *general integral,* of the differential equation.*

Let us illustrate these considerations by reverting to the differential equation (2). All functions, or curves, that are solutions of this differential equation are expressed by the relation (3) in which any arbitrary fixed numerical value may be given to R. The function defined by (3) is thus the general solution of the differential equation. We obtain a particular solution by giving some specific value to R; for instance, the function (1) is a particular solution, which we obtain by setting $R = 2$. From a geometric standpoint the general solution represents all circles centred at the origin. A particular solution is any one of these circles.

Order of a Differential Equation—We may imagine as many differential equations as we wish: *e.g.,*

(4) $$yy' + x = 0$$

(5) $$y'' = 0$$

(6) $$y'''(1 + y'^2) - 3\,y'y''^2 = 0$$

are examples of differential equations. The order of a differential equation is by definition the same as the order of the highest derivative appearing in the equation. The equations written are thus of orders one, two, and three, respectively.

The equation (4) is the one just studied, and, as we have seen, it represents the class of all concentric circles that have the origin as centre; its general solution is given by (3), in which R is a constant of arbitrary value. The fact that one and only one arbitrary constant enters into the general solution shows that our aggregate of circles forms a simply infinite

---

* Cases may arise where the general solution does not comprise *all* the particular solutions, *e.g.,* when ''singular'' solutions are present.

family. The first order differential equation (4) thus defines a simply infinite family of curves, or functions.*

Let us now examine the equation (5). It expresses the vanishing of the second derivative; hence all functions that are solutions of (5) will be functions whose second derivatives always vanish. To interpret such functions geometrically, we recall that the vanishing of the second derivative implies *straightness*. Consequently, the differential equation (5) defines the class of all those continuous plane curves that have the attribute of straightness, *i.e.*, all the straight lines we may draw in the plane.†

Now all straight lines in the plane are expressed by the relation

$$(7) \qquad\qquad y = ax + b,$$

in which any fixed numerical values may be given to a and b. We conclude that (7) represents the general solution of the differential equation (5). All particular solutions, *i.e.*, all particular straight lines, are obtained by giving in turn definite values to the constants a and b. Owing to the presence of two arbitrary constants a and b in the expression (7), the class of all straight lines in the plane forms a doubly infinite family; hence the second-order differential equation (5) defines a doubly infinite family of curves, or functions.

The differential equation (6) is of the third order; it represents the class of all circles in the plane regardless of their centres and radii. The general solution which defines all these circles will contain three arbitrary constants. The presence of three constants (the radius and the two co-ordinates of the centre) merely expresses the fact that the class of all circles in the plane is a triply infinite family.

Quite generally, it may be shown that if the differential equation is of order n, the general solution will contain n arbitrary constants. Consequently, all those curves, or functions, which are solutions of a differential equation of order n will form, in their aggregate, a n-ply infinite family. Conversely, the differential equation which defines a n-ply infinite family of curves will be an equation of the nth order. As an example, let us consider the differential equation whose general solution is the class of all

---

* A physical illustration of a simply infinite family of curves, defined by a first order differential equation, is afforded by the so-called magnetic spectrum. Iron filings are spread thinly on a sheet of paper placed over a magnet. The filings set themselves end to end, forming curves connecting the extremities of the magnet. These curves, which define the lines of force, constitute in their aggregate the general solution of a certain first-order differential equation.

† Vertical lines must be excepted, since for these $y' = \infty$.

conics in the plane. Since five independent data are required to determine a conic, the class of all conics is a quintuply infinite family of curves; hence the differential equation which defines this family will be of order five.

These considerations show that the higher the order of a differential equation, the greater the variety of functions that are its solutions. In particular, then, if we increase the order of a differential equation by differentiating it (*i.e.*, by taking its derivative), we shall obtain a new differential equation which will define a broader class of functions. An elementary example is afforded by the differential equations

$$(7) \qquad\qquad y' = 0,$$

$$(8) \qquad\qquad y'' = 0,$$

$$(9) \qquad\qquad y''' = 0.$$

The second equation is obtained by differentiating the first; the third, by differentiating the second. It is therefore correct to say that the first entails the second and the third; and that the second entails the third. But the converse is not true, as is readily seen from the general solutions of these differential equations.

Thus the general solution of (7) is

$$(10) \qquad\qquad y = c,$$

where c is an arbitrary constant. It represents all the horizontal straight lines of the plane.

The general solution of (8) is

$$(11) \qquad\qquad y = bx + c,$$

in which b and c are arbitrary constants. It defines all straight lines in the plane and therefore defines a more extensive class of curves than does (10), which defines only horizontals.

The general solution of (9) is

$$(12) \qquad\qquad y = ax^2 + bx + c,$$

where a, b, and c are arbitrary constants. The class of curves here defined comprises all straight lines and also a restricted class of parabolas. We conclude that each one of the general solutions written above is more inclusive than the preceding one. For example, by assigning the value zero to a in (12), we eliminate the parabolas and are left with (11), *i.e.*, with the class of all straight lines. Similarly, by assigning the value zero to a and to b, we obtain (10), *i.e.*, the class of all horizontals. In short, all solutions of (10) are solutions of (12), but not all solutions of (12) are solutions of (10). We have thus verified that the differential equations (7), (8), and (9) define increasingly extensive classes of curves.

The theory of differential equations forms one of the most important chapters of analysis; and the problem of integration has ever been an outstanding one. Numerous methods of integration have been devised. Here also Cauchy's complex plane has proved useful. A fundamental problem has been to determine whether a given differential equation has solutions. Affirmative proof is called an *existence theorem*. Differential equations are intimately connected with the theory of functions (as is to be expected, since their solutions are functions). We mentioned that Poincaré, by constructing the automorphic functions, integrated an important category of differential equations. The significance of his discovery should now be clear: it is that the solutions of these differential equations can be expressed by automorphic functions.

Partial Differential Equations—The derivatives we have considered thus far are those of functions of a *single* variable x. We must now examine the derivatives of functions of several independent variables. Such functions are often encountered in mathematical physics. For instance, in a room with a curved ceiling, the height of the ceiling above a point P on the floor of the room varies with the position of P; the height is thus a function of P. If we define the point P by its Cartesian coordinates, x and y, and represent by z the elevation of the ceiling, then z is a function of both x and y. These two variables are the independent variables. We express this functional relationship by writing

$$(13) \qquad z = f(x, y).$$

The concept of derivation also arises for a function of several variables. But it is more complicated because the derivation may be performed in different ways. Thus we may conceive of the rate of change of z when x is varied while y is kept fixed; or again, we may keep x fixed and allow y to vary. Such derivatives, represented respectively by

$$(14) \qquad \frac{\partial z}{\partial x} \text{ and } \frac{\partial z}{\partial y}$$

are called the first partial derivatives of the function z with respect to x and with respect to y. These partial derivatives, like z itself, are functions of x and of y. Partial derivatives of higher order are obtained by taking the partial derivatives, with respect to x or to y, of the first partial derivatives. In particular, if the latter derivatives are differentiated only once, we obtain the second order partial derivatives, viz.

$$\text{(15)} \qquad \frac{\partial^2 z}{\partial x^2}, \; \frac{\partial^2 z}{\partial x \partial y}, \; \frac{\partial^2 z}{\partial y \partial x}, \; \frac{\partial^2 z}{\partial y^2}.$$

The two derivatives $\dfrac{\partial^2 z}{\partial x \partial y}$ and $\dfrac{\partial^2 z}{\partial y \partial x}$ are not necessarily the same, a fact which indicates that the order of derivation is not necessarily commutative. However, in theoretical physics the functions encountered are usually of such a kind that these two derivatives are identical. We have confined our discussion to derivatives of the first two orders, but derivatives of the third and higher orders may also be considered.

A partial differential equation is a relation connecting some of the partial derivatives of a function *inter se*; in addition, the relationship may comprise the function itself and the independent variables. The *order* of the partial differential equation is defined by the order of the highest partial derivative appearing in the equation. Explanation is made easier by considering a definite physical problem. Indeed it was for the purpose of understanding specific physical problems that partial differential equations were first studied; and we are following the method most acceptable to the human mind when we pursue the historical order. Let us suppose that we maintain the various points of the surface of some solid conductor at arbitrarily given temperatures. After a sufficient lapse of time the temperature distribution inside the solid becomes permanent. This permanent distribution will depend on the temperatures at which we keep the various portions of the surface (surface conditions), and also on the shape and size of the body—for instance, for a given body the temperature distribution will not be the same if we maintain the whole surface at the temperature of boiling water, or if we keep half the surface at this temperature and the other half at the temperature of melting ice. But though the precise temperature distribution must depend on the shape of the solid and on the surface conditions, there will always be a certain sameness in all temperature distributions: the temperature will always be finite and will always vary continuously from place to place; and so also will the gradient of the temperature taken in any direction; in addition, as Fourier has shown, the permanent distribution will always satisfy a second-order partial differential equation, called Laplace's equation. This equation, controlling as it does all permanent temperature distributions, represents the law of such distributions.

The mathematical significance of Laplace's equation is easily understood. Thus let us define the various points within the solid by their

Cartesian coordinates x, y, z; and let T denote the temperature at any point x, y, z. The value of T will usually vary from point to point, so that T will be a function of x, y, z. We may represent this function by $T(x, y, z)$. According to whether one temperature distribution or another is realized, the function $T(x, y, z)$ will be of one kind or another. The result obtained by Fourier from his speculations on heat conduction was that, regardless of what the precise temperature distribution might be, the function $T(x, y, z)$ would always have second-order derivatives satisfying the relation.

$$(16) \qquad \frac{\partial^2 T}{\partial x^2} + \frac{\partial^2 T}{\partial y^2} + \frac{\partial^2 T}{\partial z^2} = 0.$$

This relation constitutes Laplace's equation.*

Inasmuch as the equation (16) is to control the temperature distributions that will arise in all cases, regardless of the solid's shape and of the surface conditions, we conclude that the restriction embodied in Laplace's equation is insufficient in itself to determine one particular function $T(x, y, z)$, or temperature distribution: the equation merely determines a class of possible functions. Such functions (the solution functions of Laplace's equation) are called *harmonic functions*. In other words, whatever precise temperature distribution, or precise function $T(x, y, z)$, occurs in a particular problem, this function will always be one of the solution functions of Laplace's equation (16). The characteristic class-feature of the ordinary differential equations, examined previously, must therefore be extended to partial differential equations. Thus, partial differential equations define whole classes of functions or possibilities, and not any one particular function.

Guided by what we have said of the ordinary differential equations, we might suppose that to integrate a partial differential equation means to obtain a general formula which comprises all the solution functions. This procedure, however, is not usually feasible with partial differential equations, and piecemeal methods must be resorted to. We therefore proceed as follows:

We specify that the solution functions of the differential equation must satisfy additional restrictions. By an appropriate choice of these restrictions we may narrow down the original class of all the solutions to a sub-class or even to one particular solution. If, then, the correspond-

---

* The same equation is satisfied by the distribution of the Newtonian potential throughout empty space (see note, page 216), and it was in this connection that the equation was studied by Laplace.

ing solution or solutions can be constructed, the partial differential equation is said to be integrated in a special case.

The problem of determining the restrictions that eliminate all but one particular solution is a problem which the pure mathematician is called upon to solve. No general rule can be given because partial differential equations which appear to be nearly the same often have widely different properties. In the particular case of Laplace's equation, the restrictions which serve to eliminate all but one of the solutions may be guessed at from physical considerations. Thus we have said that Laplace's equation expresses the general requirements which all permanent distributions of temperature must satisfy in a solid body; and we know from experience that a definite continuous temperature distribution arises throughout a solid of arbitrary shape when the surface points of the solid are maintained at fixed temperatures. These empirical results, transcribed into mathematical language, imply that one particular solution function of Laplace's equation will always exist within a closed volume provided the following restrictions be imposed:

1—the solution function is to be one-valued, continuous, and with first and second derivatives continuous.

2—the solution function is to assume given values over the surface of the closed volume.

But, of course, the mathematician cannot be satisfied with such physical promptings; what he demands is a rigorous mathematical demonstration. In the case of Laplace's equation, the mathematical demonstration has been obtained; hence we are certain that the restrictions listed above will single out one particular solution. The mere assurance that a solution function exists does not, however, furnish this solution; and we are still confronted with the task of constructing it. If the solution function satisfying given restrictions can be constructed, Laplace's equation is said to be integrated in the particular case considered.

The method actually followed in obtaining the solution function of Laplace's equation varies considerably with the shape of the volume. According to whether the volume is a parallelepiped, a sphere, a cylinder, or an ellipsoid, it is simpler to seek the solution in the form of a Fourier series, or in the form of a series of Laplace functions, or Bessel functions, or Lamé functions. If the shape is irregular, Fredholm's method of integral equations is usually applied. From this analysis we realize how the shape of the volume, which at first sight appears to be a minor consideration, may lead to different methods of approach and to totally different mathematical problems,

Suppose, then, we are dealing with a solid of prescribed shape and size whose surface points are maintained at given temperatures. If we can construct the solution function of Laplace's equation that corresponds to the conditions of our problem, we shall automatically obtain the temperature distribution which will persist within the solid. Thus, our ability to predict the temperature distribution depends on our ability to integrate Laplace's equation in the particular case considered. Besides controlling the temperature distribution in solids, Laplace's equation controls many other physical phenomena, *e.g.*, phenomena involving gravitation, electrostatics, elasticity, and the permanent flow of liquids. If, then, we can integrate Laplace's equation in the various cases that present themselves in the different problems, we shall be in a position to predict not only the temperature distribution in a given solid, but also the behavior of systems in which gravitation, electrostatics, etc., are involved.

The Importance of Integration—Let us suppose that a physical system is controlled by Laplace's equation. We have just seen that our knowledge of this fact does not in itself permit us to predict the behavior of the system, and that, unless Laplace's equation can be integrated in the particular case which corresponds to our physical problem, the exact behavior of the system cannot be foreseen. These general conclusions, which we have derived from a discussion of Laplace's equation, apply to all differential equations and to the physical processes they may control. Since many of the laws which regulate physical processes are expressed by differential equations (*e.g.*, Maxwell's laws of electromagnetics and Einstein's gravitational equations), we may well appreciate the importance of integration in theoretical physics. Indeed, the theoretical physicist, in his attempts to draw definite conclusions from his premises, is utterly dependent on his ability to integrate differential equations. The following illustration which concerns the theory of relativity will serve to clarify the situation further.

According to the general theory of relativity, free bodies, such as the planets, follow the geodesics of 4-dimensional space-time. In the neighborhood of matter, space-time becomes curved, so that the geodesics become curved lines and hence free bodies pursue curved paths. In order to predict the paths and motions of free bodies in any given situation—*c.g.*, to predict the paths and motions of the planets round the sun—we must therefore determine the curvature of space-time from place to place, and then obtain the geodesics. Now we cannot expect the theory of relativity to define this curvature in any precise way, for the curvature will necessarily depend on the distribution of matter in the problem considered. What the theory of relativity furnishes is a system of relations which prescribes the connections which will always hold between the curvatures at neighboring space-time points. These relations are expressed by partial differential equations called Einstein's gravitational field equations. Einstein's equations, by the mere fact that they are of the differential type, do not define any specific space-time curvature from place to place; they merely furnish a whole class of possible curvatures. To obtain the exact curvature of space-time in any particular situation, we must integrate Einstein's gravitational equations and then restrict the solution so as to conform to the requirements of the situation contemplated. Let us suppose the solution has been found, so that we now know the curvature of space-time from place to place. We still have to determine the geodesics if we wish to predict the paths and motions of free bodies and of waves of light. The discovery of the geodesics requires further integrations, for these geodesics are themselves controlled by differential equations. In short, we shall be informed of the motions of free bodies in any particular situation only if we can perform several integrations, the first of which is the integration of the gravitational equations.

Einstein's gravitational equations, however, are so complicated that to the present day no one has been able to integrate them rigorously in the general case. Fortunately for the theory of relativity, the situation of major interest in astronomy happens to be one of the simplest imaginable. We refer to the situation which arises in the solar system, where a single spherical mass, the sun, may be considered alone, the masses of the planets being so small in comparison that we may neglect their effects on the space-time curvature. This circumstance simplifies considerably the problem of integration, for, in view of the symmetry of the conditions around the sun, we know that the solution we are seeking will be radially symmetric; all other solutions may therefore be ignored. By analogy with what we said of Laplace's equation, we must also stipulate the values of the solution over the bounding surface, here represented by spatial infinity. In the general theory of relativity the boundary values are furnished directly, thanks to Einstein's assumption that the space-time structure is flat at infinity.* The foregoing restrictions in the symmetry and in the boundary values being imposed on the solution functions, one definite solution is determined. Einstein himself was unsuc-

---

* The specification of conditions at spatial infinity was displeasing to Einstein; and one of his reasons for imagining the cylindrical universe was to render possible the determination of a solution without having to consider the conditions at infinity.

cessful in constructing the rigorous symmetric solution. But soon after, the gravitational equations were integrated by Schwarzschild in the special case considered, and the solution was constructed. The curvature of space-time round the sun being thus established, the geodesics were obtained by an easy integration and so the paths and motions of the planets and of waves of light were determined rigorously. It thus became possible to submit the theory to crucial observational tests.

The case of integrability just considered is one of the very few in which the integration of Einstein's equations has been performed; and, as we have said, this simplest of cases happens by sheer luck to be the one of major interest in astronomy. Other situations are also of interest, e.g., when there are two gravitating masses of the same comparative importance, a situation realized in a double-star system. But here integration has proved impossible, and we cannot tell exactly how the stars of a double-star system should move according to the theory of relativity. The conclusions we have arrived at for Einstein's differential equations are general. Until the equations have been integrated, they are practically useless. Many of the problems which have beset the pure mathematician have arisen from a desire to assist the theoretical physicist in his integrations.

Differential Equations and Physical Laws—With the exception of the statistical laws, all the physical laws formulated prior to the quantum theory can be expressed by means of differential equations of one type or another. It is with these laws that we shall be concerned for the present. First let us understand what is meant by a physical law.

The name law is given in physics to a restriction which controls all the phenomena belonging to a class. For instance, the laws of mechanics control all mechanical phenomena, and the laws of electromagnetism control all electromagnetic ones. As a further illustration of a physical law, we may mention Boyle's law of gases. It states that, for a given mass of gas at constant temperature, the product of the volume and pressure will always have the same value. Here we are expressing Boyle's law in the so-called finite form. But there is another form in which Boyle's law and all similar physical laws can be expressed. This alternative form, which is called the differential form, was rendered possible by Newton's mathematical discoveries. It offers many advantages because it is far more general than the finite form; indeed, many of the physical laws can be expressed only in the differential form.

When a law expressed in the differential form deals with magnitudes which can be represented in the space-time frame, the law defines relationships that must hold between one or more characteristic physical magnitudes at a current space-time point and these magnitudes at any space-time point in the immediate vicinity of the first one. In certain cases the phenomenon controlled by the law is static, so that the notion of time does not enter into account. Of this type is the permanent temperature-distribution in a solid. The law controlling this distribution (Laplace's equation) merely imposes a restriction on the manner in which the temperature may differ permanently from point to point. But in the majority of cases our concern is with processes of change. The laws regulating these processes then involve time.

Let us, then, consider some physical system which is subjected to change as time passes, and let us suppose that the law controlling the evolution of the system is known. The law expresses a permanent relation between the state of the system at any current instant and its state at the immediately following instant. The restriction which this relation imposes is not sufficient, however, to determine a definite course for the physical system; it leaves open the possibility of an infinite number of different courses. But if we supplement the law by specifying the initial state of the system, one definite course becomes inevitable (at least this is so if we accept the doctrine of rigorous causality). As for the meaning of a "state", it varies from one system to another. Quite generally, the state of a given physical system may be defined as that complex of data which, under the assumption of rigorous causality, suffices to determine the history of the system uniquely.

Further insight into the nature of physical laws is obtained when we examine the characteristics of the physical processes they control. Classical physicists assumed that all physical processes were continuous. Even in the collision of two billiard balls, where the courses of the balls appeared to change abruptly, the changes were claimed to be continuous though extremely rapid. The discovery of quantum phenomena has since led physicists to question the doctrine of the continuity of change. But for the present we shall not consider the quantum revolution and shall restrict our attention to the classical doctrine. In a continuous process, the sequence of states assumed by a physical system forms a continuous succession; by this we mean that the change in the state of the system, over an infinitesimal period of time, is itself infinitesimal. The laws controlling such processes will, therefore, exhibit the features of continuity.

From this analysis we see that a physical law (involving the passage of time) expresses a permanent relationship between the present state of a continuously evolving system and the state of the system at the following instant. More concisely, we may say that the law establishes a permanent relationship between the antecedent and the consequent; it defines the connection between the successive links in the causal chain.

An example of a physical law is afforded by Newton's second law of motion. According to this law the product of the mass and acceleration of a body is at all instants equal to the force acting on the body. Since the acceleration is the time-rate of change of the velocity, Newton's law is seen to define the necessary relation which must connect the velocity of the body at a current space-time point with the velocity at a point in the immediate vicinity of the first.

Next, we consider the mathematical expression of physical laws. The mathematical relations that will express a law must necessarily satisfy the following requirements: they must be susceptible of defining connections between magnitudes at neighboring space-time points; they must be sufficiently general to leave open a whole class of possibilities; and they must connote the continuity of change. All these requirements are satisfied by differential equations. Thus in a differential equation the derivatives establish a connection between the values of a magnitude at neighboring points. Furthermore, the presence of these derivatives implies that the feature of continuity is satisfied. Finally the feature of generality is assured by the fact that the general solution of a differential equation embodies a whole class of possibilities. We conclude that differential equations furnish the mathematical expression of the physical laws. Let us verify this conclusion in the particular case of Newton's second law of motion.

To take a simple illustration we shall assume that Newton's law is applied to a point-mass moving along a straight line under the action of a force. We denote by x the distance of the point-mass, at any instant, from a point O taken as origin on the straight line. At any instant t, the point-mass occupies some position on the line, and hence this position x is a function of the time t. Since the first derivative x' of x with respect to t is the time-rate of change of x when t varies, x' represents the velocity of the point-mass. Consequently, this derivative, or the velocity it measures, has one value or another according to the instant t considered. Similarly the second derivative x'', defining as it does the rate of change of the velocity at any instant, represents the acceleration of the point-mass at this instant. We shall assume that the force varies with the position x of the body, and we may therefore represent it by some function $F(x)$. Newton's law of motion is then seen to be represented by

(17) $$mx'' = F(x),$$

i.e., by a second-order differential equation of the simpler kind studied at the beginning of the chapter (for x is a function of only one independent variable, namely, the time t).

In this discussion we have supposed that the point-mass is moving along a straight line; but more generally it may be moving in space, and in this event we should have three differential equations similar to (17): one equation for each of the three coordinates x, y, z of the point-mass. If, instead of a single point-mass, we were dealing with a solid body of finite size or with several point-masses, Newton's law would be expressed by a larger number of differential equations, but these equations would still be of the same type as the equation (17). We therefore conclude that the equations of dynamics are second-order differential equations.

As we shall now see, the equation (17) does not prescribe any definite motion for the point-mass; instead, it defines a whole class of possible motions, the exact motion being determined only when the initial state of the point-mass is specified. To verify these assertions, we shall suppose for the sake of simplicity that the force acting on the point-mass is a constant force, so that it may be represented by a constant K. The differential equation (17) thus becomes

(18) $$mx'' = K.$$

We wish now to determine the motion of the body or, what comes to the same, its position x at any instant t. We acquire this information by integrating the differential equation, thereby obtaining its general solution. The general solution is

(19) $$x = \frac{K}{2m}\, t^2 + at + b,$$

in which a and b are arbitrary constants. By giving different sets of values to the constants a and b, and hence by considering different particular solutions of the differential equation (18), we obtain different motions for our point-mass. Each one of the possible motions is thus expressed by a particular solution of the differential equation, or mechanical law, (18). In the present case the differential equation admits a doubly-infinite family of particular solutions, so that it is compatible with a doubly infinite class of motions.

To obtain the particular motion of the point-mass in any given problem, we must specify the numerical values of the arbitrary constants a and b in the general solution (19). We may easily verify that these values are determined by the initial position of the point-mass and by its initial velocity.   Calling x_o and v_o the initial position and the initial velocity respectively at time $t = 0$, we obtain

(20) $a = v_o$ and $b = x_o$;

and our particular solution which defines the actual motion is

(21) $x = \dfrac{K}{2m}\, t^2 + v_o t + x_o.$

Since the initial position and the initial velocity determine the motion, these two items of information, considered jointly, constitute what we have called the "initial state of the point-mass."   Thus, the form of the differential equation that expresses the mechanical law has enabled us to establish the nature of the items of information which define a state of the mechanical system.   More generally, the state of any system whose history is controlled by a given differential equation, or law, is defined by the aggregate of those items of information which suffice to single out a particular solution from the general solution of the differential equation.

We have just seen that the equations of mechanics (for point-masses) are second-order differential equations of the ordinary type that were discussed at the beginning of the chapter.   But in the majority of cases physical laws are expressed by partial differential equations.   Even in mechanics these latter equations are often met with.   Let us, then, examine under what conditions the two different kinds of differential equations will occur in mechanics.

The position of a point-mass or of an extended solid is defined by magnitudes which vary solely with time.   Consequently the differential equations which control the motion will involve functions of a single independent variable (time).   We conclude that differential equations of the simpler type must be expected when we are dealing with the mechanics of point-masses or of solid bodies.   Partial differential equations in mechanics occur in connection with the motions of continuous deformable media.   A simple illustration is found in the vibrations of strings.

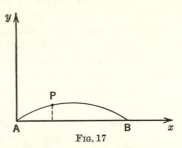

Fig. 17

Consider a vibrating string fixed at its two extremities A and B.   Let P be any point of the string.   We call x and y the coordinates of this point.   As time passes, the elevation of the point P varies, and hence y is a function of the time t.   But at any instant the elevation y also depends on the position of the point P along the string; hence y is also a function of x. Thus, y is a continuous function of both x and t. In order to determine how the string will vibrate, we must obtain the value of y for all values of x (over the length of the string) and for all values of the time t. The vibrations are caused by the restoring force developed by the tension of the string. When to this force we apply Newton's second law of motion, we obtain a differential equation involving the derivatives of the function y with respect to both variables x and t. We thus have a partial differential equation (called d'Alembert's equation), *viz.*,

$$(22) \qquad a^2 \frac{\partial^2 y}{\partial x^2} = \frac{\partial^2 y}{\partial t^2},$$

where a is a constant connected with the tension of the string. Incidentally, this is the equation which Bernoulli solved by expressing the solution function $y(x, t)$ as a Fourier series. We refer the reader to page 134, where the importance of Bernoulli's solution is discussed.

Similarly the equation of wave disturbances in a homogeneous elastic medium is given by d'Alembert's partial differential equation

$$(23) \qquad \frac{\partial^2 \theta}{\partial x^2} + \frac{\partial^2 \theta}{\partial y^2} + \frac{\partial^2 \theta}{\partial z^2} = \frac{1}{V^2} \frac{\partial^2 \theta}{\partial t^2},$$

where θ is the dilatation and V the velocity of propagation.

To summarize: All differential equations in which time is one of the independent variables express the continuity of change. The differential equations of the simpler type are associated with the motions of point-masses and rigid bodies; they furnish the proper mathematical instrument to be used if we adhere to an atomistic conception of Nature. Partial differential equations arise when we are dealing with continuous, deformable media. Interesting conclusions may be drawn from these statements. For instance, an elastic jelly is assumed to be formed, in the final analysis, of tiny molecules acting on one another and separated by empty spaces. From a macroscopic standpoint, however, the jelly appears to be a continuous substance. If we view the jelly as continuous, the equations that regulate the propagations of disturbances through its substance will be partial differential equations. If, on the other hand, we suppose that the jelly is atomistically constituted, we must consider the motions of each one of the individual corpuscles. With each corpuscle are then associated three differential equations of the ordinary type (we say three because of the three coordinates x, y, z defining the position of each corpuscle). The physical analogy between a continuous jelly and one atomistically constituted leads us to surmise that a partial differential equation may in certain cases be equivalent to an infinitely large number of differential equations of the simpler type. This assumption is indeed correct, but of course it must be proved mathematically and cannot be justified by mere physical intuition.

The conclusion to be derived from this example is important. It shows that even if a medium be discrete, provided it be composed of an enormous number of corpuscles in each tiny volume, we are perfectly justified from the standpoint of macroscopic treatment in viewing it as continuous, not solely on physical grounds but also from the standpoint of mathematical treatment. We are therefore justified in introducing partial differential equations in mechanics, even if we adhere to a corpuscular structure of matter. This fact simplifies considerably the mathematical investigations, for though a partial differential equation is more complicated than a differential equation of the simpler type, it is easier to manipulate a single partial differential equation than to treat the aggregate of an enormous number of the simpler equations.

The examples given thus far are drawn from mechanics, but quite generally we shall meet with partial differential equations whenever we are considering laws that regulate the continuous distribution of a magnitude through space. In the majority of cases the magnitude at a fixed point will also vary with time; and as a result the partial differential equation will also involve time-derivatives. Equations (22) and (23) are illustrations of partial differential equations involving time. Fourier's equation for the flow of heat is likewise of this type. It is expressed by

$$(24) \qquad \frac{\partial^2 T}{\partial x^2} + \frac{\partial^2 T}{\partial y^2} + \frac{\partial^2 T}{\partial z^2} = a\,\frac{\partial T}{\partial t},$$

where T is the temperature throughout the volume, and a is a magnitude characterizing the solid. Laplace's equation, which regulates the permanent distribution of temperature, is a restricted form of (24). Thus, if the temperature distribution is assumed permanent and hence does not vary with time at any fixed point of the volume, the derivative $\frac{\partial T}{\partial t}$ always vanishes. In this event, (24) becomes Laplace's equation (16).

Quite generally, the equations (or laws) of the field theories are partial differential equations. Maxwell's equations of electromagnetics and Einstein's gravitational equations are examples. The reason field theories lead to partial differential equations is that they deal with the continuous distribution of magnitudes throughout a continuum, such as space or 4-dimensional space-time. Indeed all the laws of the general theory of relativity are partial differential equations, and in this theory we are confronted with the simpler equations only when we wish to determine the motions of individual point-masses, such as the planets.

There are other physical laws, antedating the quantum theory, which cannot be expressed by means of differential equations of either of the two types. Laws of this kind have been revealed in the study of elasticity. Experiment shows that the evolution of an elastic system is not always determined solely by its initial state but also by all previous states: a kind of ancestral influence must be appealed to. Volterra, to whom the mathematical study of these elastic phenomena is due, has proposed calling the laws controlling them "hereditary laws." Such laws cannot be expressed by the differential equations mentioned in the present chapter; they require a new mathematical instrument, called the *integro-differential equation*, the study of which we owe chiefly to Volterra. As has, however, been remarked by Painlevé, the reason the knowledge of the initial state alone is insufficient to determine the evolution of the elastic system is that we do not define the word "state" with sufficient completeness. If, instead of viewing a state as defined by a small number of magnitudes, we assumed that it was defined by an appropriate choice of an infinite number of magnitudes, we should be justified in saying that the initial state does determine the evolution of the elastic system. Painlevé's attitude seems to be justified on mathematical grounds, for an integro-differential equation is equivalent to an infinite number of ordinary differential equations; and, if we replace the integro-differential equation by an infinite aggregate of ordinary ones, we may quite well claim that the evolution of the system is determined by the initial state alone. However, we must remember that the initial state now requires the enumeration of an infinite number of magnitudes. The physical significance of Painlevé's observation appears to be that, if we view the elastic body atomistically and therefore consider an extremely large number of ordinary differential equations regulating the motions of an extremely large number of corpuscles, we may dispense with the hereditary feature which experiment seems to suggest.

In the earlier days of the quantum theory, when the sudden jumps characteristic of quantum processes were first discovered, physicists were of the opinion that differential equations, because of their association with continuous processes, could no longer be applicable in theoretical physics. Other types of equations, known as difference-equations, were suggested as appropriate. These equations connect states that differ by finite, instead of by infinitesimal, amounts. In the more recent form of the quantum theory, where rigorous laws are no longer held to govern the courses of individual processes, differential equations are also useless. However, if we consider statistical results and not individual processes, continuity and rigorous determinism reappear. A continuous fluid of a fictitious nature, sometimes called "the probability fluid," is introduced in wave mechanics; and the motions of this fluid in space or in hyperspace (multidimensional space) are regulated by rigorous laws. In view of the fluid's continuity throughout space and time, partial differential equations express the laws. Schrödinger's wave equation of the hydrogen atom is of this type. As for the differential equations of the ordinary kind, they never arise in wave mechanics. The reason is clear: it is that the motions of individual particles are no longer assumed to be submitted to rigorous laws.

CHAPTER XVI

THE CONTROVERSIES ON THE NATURE OF MATHEMATICS

INASMUCH as mathematics is practically the only instrument that can be applied in the construction of physical theories, the nature of mathematical reasoning is a subject of interest to the physicist. Various definitions of mathematics have been proposed by different writers, but the consensus of opinion is that none of them is entirely satisfactory. The fact is that one cannot compress into a few words the definition of so vast and varied a subject. One of the definitions suggested is due to Benjamin Pierce. It runs: "Mathematics is the science which draws necessary conclusions." Although Pierce's aphorism is not patently incorrect, it may convey a misleading impression to the unwary.

In the first place, a conclusion can be termed necessary only in virtue of rules of reasoning which we may see fit to adopt. A proper definition of mathematics should therefore throw some light on the nature and origin of these rules. Pierce presumably supposes that they are the rules of logic, so that his definition is tantamount to identifying mathematics with logic. But we believe we are correct in saying that none of the leading mathematicians of the present day would subscribe to this identification. Finally, if mathematics is truly but the science which draws necessary conclusions, why is it that every man who has devoted his time to mathematics, and who presumably has a clear mind, has not revealed himself as a creative mathematician? We shall first seek an answer to this question.

There are many universities, and in each of these universities there are many men who deliver courses in elementary and in advanced mathematics. Many of them have written textbooks, which in view of their excellence have been translated into other languages. We may presume therefore that these men have a considerable knowledge of their special provinces of mathematics. And yet we may count on our fingers the number of creative mathematicians living today who have evolved new methods and established new theorems instead of merely commenting on the work of others. This is strange if, as Benjamin Pierce appears to believe, mathematics is nothing but the drawing of necessary conclusions. Also, how does it happen that an eminent logician like Bertrand Russell, who is constantly talking about mathematics, should never have furnished a single theorem, whereas Poincaré, whose hostility to the identification of mathematics and logic is well known, should have been one of the most prolific creators of all time? Are we to suppose that Mr. Russell is unable to draw necessary conclusions? We do not believe that this supposition would be warranted. In our opinion the answer to our question must be sought in a totally different direction.

Some hint is secured when we examine a mathematical demonstration for the first time. We read it over, understand it step by step, and are perfectly convinced of the validity of its proof. But suppose now we are required to repeat the demonstration, even a few hours later. We often find this impossible, and we excuse ourselves by saying that we do not recall the demonstration. Yet, if a mathematical demonstration involves nothing more than perceiving that p implies q, we may well wonder what part memory has to play in the matter. If A is greater than B and B is greater than C, we do not have to appeal to our memory to assert that A is greater than C. Anyone who wishes to test what we have just said can easily do so without any particular knowledge of mathematics. He has but to revert to Euclid's proof of the unending sequence of prime numbers or of the irrationality of $\sqrt{2}$. Both of these proofs are mentioned in Professor Dantzig's admirable book, "Number," which will long serve as a model of semipopular exposition in mathematical philosophy.*

This leads us to inquire: What do we have to recall in a mathematical demonstration in order to repeat it? On reverting to the demonstration which we have forgotten, we usually find that our memory has failed at a point where some expedient known as a mathematical artifice has been introduced. To take a simple example, Euclid's proof of the unending sequence of prime numbers starts with the introduction of factorials. Euclid shows that if n is any prime number, a greater prime number always exists. He establishes this theorem by considering the numbers defined by $n!$ and by $n! + 1$. Regardless of the sequel of Euclid's demonstration, his initial introduction of factorials is an artifice and is certainly not dictated by any logical necessity. Should the reader forget this first step, all the logic in the world would not enable him to reproduce the demonstration. Euclid does not tell us how he came to think of factorials. The idea may have come to him suddenly, like a flash, or it may have arisen as the result of prolonged reflection. These are the idiosyncrasies of the creator; they pertain to psychology, not to logic. Such

---

* T. Dantzig. Number. Macmillan, New York, 1930; pp. 48 and 102.

artifices are very frequent in the writings of the masters. In many cases they are devised because some seemingly insuperable difficulty blocks the way. An attempt is then made to transform the problem into an equivalent but easier one. Riemann, for instance, sought to solve a certain problem in the theory of functions when the area under consideration was that of an arbitrary continuous closed curve. But he did not solve the problem directly; he transformed it into a simpler one by transforming the arbitrary closed curve into a circle. With the circle, the solution is relatively easy and hence the original problem was solved. If we forget that Riemann made a transformation at this point, or even if remembering that a transformation was made we forget its precise nature, we shall probably be unable to reproduce the proof even though we once understood it. Mathematical transformations are often extremely ingenious, and on coming across one for the first time we are tempted to exclaim: "How did anyone ever think of that?" We cannot suppose that the great masters experimented with all conceivable transformations before hitting on the appropriate one; human life is too short to permit this procedure. Nor can we suppose that sheer luck invariably guided these men to the correct choice. What then is the secret of their success? Some call it genius; others intuition; but whatever it be, let us not be confused into thinking that it is merely an instance of *"p* implies *q."* If it were, we should have to admit that, except for a handful of creators, all the present-day mathematicians are logical imbeciles.

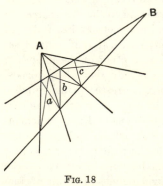

Fig. 18

A simple example of an ingenious transformation in elementary geometry illustrates the thoughts we have been expressing. The example is well known; it occurs in connection with the following problem.

Four straight lines issuing from a point *A* are intersected by two straight lines issuing from a point *B*. We draw the diagonals in the three quadrilateral figures thus formed, and we thereby obtain the three points of intersection marked *a*, *b*, and *c*. We are required to prove that these three points lie on the same straight line, *i.e.*, are collinear (see Figure 18).

This theorem may be proved by analytical geometry or by straightforward geometry, but in either case the proof would be laborious. However, there is a transformation which, if we are lucky enough to think of it, transforms our problem into an equivalent one of the utmost simplicity. The artifice consists in projecting the figure in an appropriate way onto another plane. Thus suppose the sheet of paper on which our figure is traced is set in a vertical position, with the line AB placed along a horizontal. Consider now a point-source of light S situated at the same elevation as the horizontal line AB. This point-source of light casts a shadow of the drawing on a horizontal sheet placed below the figure. The shadows of the straight lines will be straight lines, and as a result the four straight lines issuing from A will have as shadows four straight lines issuing from the shadow of A. But the shadow of A is at infinity; hence the four straight lines issuing from A will have as shadows four straight lines meeting at infinity; and this implies that the four shadow lines will be parallel. The same argument applies to the two lines issuing from B. Altogether, the shadow of our drawing will consist of four parallel straight lines intersected by two other parallel straight lines (see Figure 19).

Since the shadow of a straight line is a straight line, and vice versa, we see that the three points, a, b, c, will be collinear if their shadows, a', b', c', lie on a straight line. We shall therefore have solved our original problem if we can establish the collinearity of the points, a', b', c'. Thus our original problem is transformed into a new one. Now the new

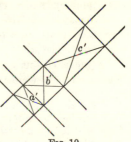

Fig. 19

problem, in contradistinction to the original one, is absurdly simple, for the points, a', b', c', obviously lie on a straight line. We conclude that a, b, c are collinear, and so our original problem is solved. Thanks to an ingenious transformation the solution has been obtained without difficulty.

Although we have expressed the artifice in physical language so that the non-mathematical reader may follow it more easily, the demonstration can be made perfectly rigorous. In short, by means of an artifice we have transformed a difficult problem into a simple one. In our opinion the student who would think of this simple artifice would prove to have greater mathematical ability than the one who might solve the same problem by more laborious but less ingenious methods.

These examples and many others would seem to indicate that the conception of the creative mathematician, as implied in Pierce's definition, is radically wrong. Between the creative mathematician and the logician there is a vast difference. The logician must be able to concentrate and to think clearly. So also must the mathematician; but, in addition, the creative mathematician needs imagination, or genius, or whatever we wish to call it. He must be an artist, not solely a worker with a one-track mind. Imagination to the logician, if not harmful, would at any rate be excess baggage. No imagination is required to realize under what conditions "*p* implies *q*." Possibly more than a coincidence is responsible for the well-known fact that those writers who have been most insistent in identifying mathematics with logic are usually found in that larger group of men who have failed to create. To the intrusion of extra-logical influences in mathematics is presumably due the marked differences in the presentations of the same subject by different mathematicians. The trained student may usually detect very easily by the form of the presentation whether a demonstration is due, say, to Riemann or to Weierstrass. A similar situation arises of course in poetry and in music; but there it is so familiar that it scarcely warrants mention.

We may summarize this discussion by saying that mathematics involves two steps: firstly, the application of rules; secondly, *the ability to combine the rules*. The second step, however, cannot itself be expressed by rules.

Poincaré, discussing the various qualifications required by a creative mathematician, writes:

"Others will commit to memory the details one after another; they will be able to understand mathematics and sometimes to apply it, but they will be unable to create. Still others will have, in a more or less marked degree, that special form of intuition of which I have just spoken, and then they will not only be in a position to understand mathematics, even though their memory be none too good, but they may also become creators and seek to invent, with more or less success, according to the degree of development of their intuition.

"What indeed is mathematical invention? It does not consist in setting up new combinations with mathematical beings that are already known. For anyone could do this, and the combinations that might thus be formed would be infinite in number and the majority of them without the slightest interest. To invent consists precisely in constructing not useless combinations, but useful ones, and these form an infinitely small minority. To invent is to discern, to select.

"Sterile combinations will not even suggest themselves to the mind of the inventor. In the field of his consciousness will appear only those combinations that are really useful, some few of which he will reject but which nevertheless have something in common with the useful ones." *

---

*The quotations from Poincaré in this chapter are translations of passages appearing in his four popular books: *La Science et l'Hypothèse*, *La Valeur de la Science*, *Science et Methode*, and *Dernières Pensées*.

Some people will contend that those characteristics of a mathematical demonstration that we have been stressing are irrelevant to the nature of mathematics itself, and that a confusion is being made between the qualifications of the creator and the subject matter he is manipulating. This contention may be defended, but the point is that if the man who is a mere logician cannot develop into a creative mathematician, we must conclude that the methods of logic alone are insufficient to construct mathematics. Under these conditions how can the subject matter of mathematics be exhausted by logic alone? If the bare rules that are applied in mathematics are to be viewed as constituting the subject matter, regardless of whether or not we be capable of applying these rules and obtaining significant results, then according to the same criterion we should be justified in saying that music is nothing but the rules of counterpoint and harmony. We may grant, however, that a study of the qualifications of the creative mathematician is not the only way of obtaining an understanding of the nature of mathematics. More direct means of approach can be resorted to. These we shall now consider, and in the course of the chapter we shall summarize the conclusions to which the most competent authorities have been led.

The creative mathematicians of international fame who have investigated the foundations of mathematics are, in the historical sequence of their writings, Poincaré, Peano, Hilbert, Brouwer, and Weyl. To these we should add Borel, Hadamard, Lebesgue, and Baire, all of whom have expressed valuable opinions. Many of the great masters are not included in this list, Volterra and Picard for example. This is because their discoveries have been confined to the more advanced regions, and, so far as we know, they have expressed no definite opinions on the foundations. The fact is that mathematics offers a vast field of unsolved problems and unexplored territory; the foundations do not constitute the sole sphere of interesting research.

We are particularly fortunate to find Poincaré and Hilbert among those who have discussed the foundations; for Poincaré is universally recognized as one of the greatest mathematical minds of all time, and Hilbert is one of the greatest among the living. Yet in spite of their unquestioned competence, Poincaré and Hilbert are at loggerheads; and the same is true of all the other masters mentioned: no two of them agree on all points. This situation creates a mathematical scandal, for we have been taught to view mathematics as a science in which a rigorous demonstration should always put an end to any dispute. However, we must remember that the foundations of mathematics, and mathematics proper, are not the same thing. To stress the difference, Hilbert refers to the former as "metamathematics," a very appropriate name.

If we neglect minor differences of opinion, we may class the leading thinkers into two groups: the "formalists" and the "intuitionists." Between these two schools, mathematical thought is divided. The appellation intuitionist is likely to confuse the layman. He may be tempted to suppose that the mathematicians of the intuitionist school propose to rely on "hunches" in mathematics instead of appealing to rigorous mathematical demonstrations. It is scarcely necessary to say that no such absurdity is intended. As a matter of fact quite the reverse is true: the intuitionist is a rigorist. We shall resume the thread of this discussion when we examine the points of disagreement between the two schools. For the present we shall confine our attention to the results which appear to have been accepted unanimously. One of these is that mathematics and logic are not identical. This was always Poincaré contention, as the following passage shows:

"If all the propositions which mathematicians enunciate can be deducted one from the other by the rules of formal logic, why is not mathematics reduced to an immense tautology? The syllogism can teach us nothing that is essentially new, and, if everything is to spring from the principle of identity, everything should be capable of being reduced to it. Shall we then admit that the theorems which fill so many volumes are nothing but devious ways of saying that A is A?"

Let us now consider an expression of opinion from the opposing school. Hilbert writes:

"Arithmetic is indeed designated as a part of logic, and it is customary in founding arithmetic to presuppose the traditional fundamental principles of logic. But, on attentive consideration, we become aware that, in the usual exposition of the laws of logic, certain fundamental concepts of arithmetic are already utilized, for example the concept of aggregate and, in part, also the concept of number. We thus fall into circularity, and, therefore, to avoid paradoxes a simultaneous development of the laws of logic and arithmetic is requisite."

Thus, the formalist views logic and arithmetic as complementary, and not as identical.

Another point on which agreement has been arrived at (though only in part) concerns the impossibility of defining number, integers in particular. We say that the agreement is only partial because, though both schools agree that a direct definition of number in terms of logical concepts is impossible, the formalists claim that an indirect definition, based on a postulate system, is feasible. The following passage of Poincaré illustrates the attitude of the intuitionists:

"The definitions that have been suggested for 'number' are numerous and varied. I cannot even attempt to give the names of their authors. We need not be surprised at the plethora of definitions suggested. For if one of these definitions were satisfying, no further one would be required. If each new philosopher who has been interested in this matter has found it necessary to invent some new definition, it is because he was dissatisfied with those of his predecessors, and the reason he was dissatisfied was that he regarded the earlier definitions as circular."

Similar opinions have been expressed by the other intuitionists. The attitude of the formalists is much the same. Hilbert, for instance, discusses Frege's definition of integers* and also several others; but he rejects them all as unsatisfactory.

Though both schools of mathematicians recognize that a direct definition of number cannot be given, the formalists contend that an indirect definition (by the method of axiomatics) is possible. Peano's celebrated postulate system, which yields an indirect definition of integers, is a case in point. The intuitionists deny even this more modest claim. Hadamard, who occupies an intermediary position between the more extreme members of the two schools, states in his preface to a book by Gonseth on the foundations of mathematics:

"The notion of integer appears to Poincaré as a notion which is inherent in the mind and without which it would be impossible to think. Others have since implicitly upheld the contrary opinion, for they have tried to define integers. May I be permitted to be of Poincaré's opinion? And when those who state their definitions proceed to number their postulates, I am not so sure that the circularity which thus appears is purely on the surface. This criticism, however, does not prevent me from regarding the axiomatics of the integer as an unquestionable and enormous progress, provided we do not expect more than it, or than anything else, can yield."

Axiomatics—The new science of axiomatics, developed chiefly by Peano and by Hilbert of the formalist school, constitutes the one bright spot in these obscure controversies on the nature of mathematical reasoning. Although axiomatics was conceived for the purpose of clarifying the foundations, it has proved itself of use in genuine mathematics. The germ of axiomatics can be traced to the system of axioms and postulates with which Euclid prefaced his geometrical discoveries. The ancient

---

* Frege defines the number "two," for instance, as the class of all couples.

Egyptians and the earlier Greeks presumably discovered many of the geometrical theorems (such as those of Thales and Pythagoras) by direct measurement with rods, no rigorous demonstrations being given. But Euclid introduced rigor by deducing his geometrical theorems from clearly stated assumptions. These initial assumptions were embodied in his axioms and postulates.

Today we know that the assumptions explicitly stated by Euclid were insufficient to serve as a basis for Euclidean geometry. Yet Euclid constructed his geometry. We are thus faced with a paradox, for if the assumptions were insufficient, how could a geometry be constructed? The answer is clear when we find that Euclid, in addition to the assumptions which he stated explicitly, utilized others unconsciously in the course of his demonstrations. For example, in one of his theorems Euclid seeks to establish the equality of two triangles. To do so, he imagines that one of the triangles is placed on the other, and he then shows that the two triangles must necessarily coincide and hence be equal. Yet none of the axioms or postulates explicitly stated by Euclid entitled him to operate in this way; and in proceeding as he did he was unconsciously assuming that his triangles could be displaced without changing in shape and size, an assumption to which he was undoubtedly guided by the behavior of the rigid bodies observed in his daily activities. Incidentally, we have here a first indication of the empirical origin of Euclid's assumptions. We shall revert to this matter later.

Summarizing Euclid's idea, we see that it consists in stating a certain body of assumptions and then in deriving from these assumptions their necessary consequences by logical means. Euclid's execution of this idea was defective because, as we have said, he inadvertently omitted to state all the assumptions which he subsequently utilized in his demonstrations. Nevertheless the idea in itself has since been proved valid, and it forms the essence of the new science of "axiomatics."

A point of nomenclature must here be mentioned. Euclid made a distinction (which was none too clear) between axioms and postulates. Today this distinction is no longer recognized, and all the assumptions on which a doctrine is founded receive the same name. Hilbert uses the word "axiom" whereas English-speaking writers prefer the word "postulate." We shall adopt the latter.

The method of axiomatics was first applied with rigor by Peano, when he gave a system of postulates serving to define the integers of arithmetic. Hilbert pursued the same method for Euclidean geometry. He introduced 21 postulates which defined certain relations among points, straight lines, and planes. The aggregate of these postulates is called "Hilbert's postulate system." Hilbert then showed that, by applying the logical principles to his postulate system, he could deduce all the theorems of Euclidean geometry without having recourse to any assumption not explicitly stated in the postulate system. As may be gathered, Hilbert's postulates comprise Euclid's axioms and postulates (though in a modified form) and also those assumptions which Euclid made use of unconsciously in his demonstrations.

An example of one of Euclid's tacit assumptions, which Hilbert states explicitly in the form of a postulate, is Hilbert's postulate of continuity, also called the postulate of Archimedes. The significance of this postulate is as follows: If A and B are two points on a straight line, and if from A we mark off towards B equal intervals one after the other along the line, we shall eventually reach points beyond the point B. The postulate of Archimedes becomes intuitive when we use a rigid rod to mark off the equal intervals. And since Euclid was guided unconsciously by the properties of rigid bodies, we may well understand why he took this postulate for granted and did not state it explicitly. We have here a further example of the empirical origin of Euclidean geometry.

Other typical postulates of Hilbert, taken at random, are listed below:

(1) "Two distinct points, A and B, always determine a straight line."

(2) "Three distinct points, A, B, and C, that are not on a straight line always determine a plane."

(3) "Of three points on a straight line, one and only one is between the other two."

(4) "If a segment AB is congruent (equal) to a segment $A'B'$ and to a segment $A''B''$, then $A'B'$ is congruent to $A''B''$." *

These samples suffice for our purpose.

Hilbert's postulates in their aggregate express the relations which the points, straight lines, and angles of Euclidean geometry are assumed to satisfy. But points, straight lines, planes, and angles are not the only entities which satisfy these relations; and with a little ingenuity many others may be found. A particularly simple illustration can be given in the case of the Hilbert postulates which apply to plane geometry; these differ but slightly from the former postulates which correspond to space geometry. The illustration follows:

By a point in the plane we shall mean any point of the plane, with the exception of one particular point O chosen arbitrarily. By a straight line in the plane we shall mean any circle passing through the point O.

---

* Hilbert, Foundations of Geometry, Chicago.

The distance between two points, A and B, on one of our new straight lines (now circles) is defined by

$$\cot O\omega B - \cot O\omega A,$$

where $O\omega B$ and $O\omega A$ are the angles (in the ordinary sense) obtained by joining O, A, and B to the centre ω of the circle. Finally, the angle

Fig. 20

between two straight lines (now circles passing through O) is defined by the angle (in the ordinary sense) made by the two circles. It can then be verified that all the relations satisfied by the points and straight lines of plane Euclidean geometry will also be satisfied by our new "points" and "straight lines" (and vice versa). We may therefore view plane Euclidean geometry and the model just defined as two different models, or so-called "concrete representations," both of which satisfy Hilbert's postulates for plane Euclidean geometry.

At first sight, it seems absurd to call the circle traced in the figure a straight line. However, our hesitation to do so brings out precisely one of the aspects that Hilbert is stressing, for Hilbert requests us to refrain from assigning any *a priori* properties to the points and straight lines that he mentions in his postulates. Points and straight lines are to be viewed as representing mere undefined terms, which we may conveniently refer to as a's and b's. Under these conditions Hilbert's postulates, say for Euclidean geometry in space, may be rewritten by replacing everywhere in the postulates the terms point, line, and plane by the symbols a, b, and c. If in these postulates, which now express relations connecting the a's, b's, and c's, we replace these a's, b's, and c's by the words points, lines, and planes, we obtain Euclidean geometry; and if we replace them by other appropriate words representing other entities whose mutual relations happen to be the same as those of the points, straight lines, and planes of Euclidean geometry, a new model satisfying the same postulates is obtained. Superficially, the original Euclidean model may differ considerably in appearance from the new one; yet, beneath the surface, both models have the same form,* or logical structure, since in both of them the entities which are substituted for Hilbert's a's, b's, and c's exhibit the same relationships, expressed by Hilbert's postulates.

Hilbert gives an important example of a group of elements which exhibit the same relationships as do points, straight lines, and planes. The new elements are represented by triads of numbers and by algebraic

---

* Technically, "are isomorphous."

expressions connecting these numbers. He was led to this example when, in place of points, straight lines, and planes, he took their Cartesian representations in terms of numbers x, y, z. The fact that the new elements, here numerical, satisfy the Hilbertian postulates merely proves the equivalence of straightforward geometry and of the Cartesian method of analytical geometry. In other words, it proves the equivalence, as to logical structure, of the geometric continuum and of the arithmetic continuum. From considerations such as the foregoing, we see that there is no particular reason to associate Hilbert's system of postulates with the familiar model of Euclidean geometry rather than with any other model having the same logical form. What is characteristic of a postulate system is the form, the network of relations, and not the subject matter, or content; the subject matter is contingent.

These conclusions of Hilbert's have a familiar ring, for long before Hilbert constructed his postulate system mathematicians realized that the concern of mathematics lies in relationships, and not in subject matter. An example is afforded by Laplace's equation. Laplace's celebrated equation expresses a relationship connecting the derivatives of a function of three variables x, y, z. The relationship happens to be realized in the physical world by the Newtonian potential distributed through space; and it was in this connection that Laplace came upon the equation. But as Fourier subsequently showed, exactly the same equation, or relationship, holds for the distribution of the temperature in a solid when a condition of heat equilibrium is attained. It also holds for the velocity potential in the case of an incompressible fluid in a permanent state of flow. The equation itself and its peculiarities transcend the particular verifier which happens to fit it. We have here an additional illustration of the fact that mathematics deals with relations, and not with content. And so we may wonder why such prominence has been given to Hilbert's conclusion: namely, that from his postulate system we can construct Euclidean geometry even when we have no idea of what is meant by a point, a straight line, and a plane.

To obtain the answer to this question we must recall that Euclid deduced his theorems by utilizing, in addition to his axioms and postulates, certain tacit assumptions which were suggested by the properties of rigid bodies. If, then, we were to base our judgment solely on Euclid's demonstrations, we might suppose that Euclidean geometry could not be constructed without a prior knowledge of its subject matter, *i.e.*, the points, lines, and planes with which the study of rigid bodies makes us familiar. According to this view, geometry in contradistinction to other mathematical doctrines would appear to be inseparable from its content. The significance of Hilbert's investigations has been to show that this belief would be erroneous, and that, by reasoning on a's, b's, and c's devoid of all content but satisfying well-defined relations, we could obtain Euclidean geometry. To Hilbert therefore goes the credit of having shown that in geometry, as in other parts of mathematics, the specific content is of no particular importance.

Einstein, commenting on Hilbert's discoveries, contrasts the older and the newer views by saying:

"The older interpretation:—Every one knows what a straight line is, and what a point is . . .
"The more modern interpretation:—Geometry treats of entities which are denoted by the words straight line, point, etc. These entities do not take for granted any knowledge or intuition whatever, but they presuppose only the validity of the axioms (postulates), such as the one stated above, which are to be taken in a purely formal sense, *i.e.*, as void of all content of intuition and experience. . . ." *

The Mathematical Utility of Postulate Systems—Axiomatics has been of use not only in clarifying obscure points in the foundations, but also in leading to constructive discoveries. We may readily understand how this situation has arisen. Take Hilbert's postulate system for example. As it stands, it defines indirectly the model of Euclidean geometry or any equivalent model (in the sense previously stated). Suppose now we withdraw one of the postulates and replace it by a different one. We thus obtain a new postulate system, for which the former Euclidean model is no longer valid. But we may find that the new postulate system is verified by some new geometry, or model; and in this event the method of axiomatics will have entailed the discovery of some novel geometrical doctrine. For instance, Hilbert showed that when the postulate of parallels is replaced by an appropriately selected one, the new postulate system is verified by Lobatchewski's geometry. Since, however, Lobatchewski's geometry was already known to Hilbert, nothing new was discovered thereby. But a totally new kind of geometry was secured when Hilbert denied the postulate of Archimedes. Thus, by juggling the postulates, we may obtain geometries, and more generally models, which might never have been thought of. It is in its capacity to furnish new doctrines that axiomatics reveals its constructive side.

A point that may seem strange is that axiomatics should have come to the fore as a constructive branch of mathematics. In this connection we must remember that Riemann's great contribution to the develop-

---

* Einstein. Sidelights on Relativity. E. P. Dutton and Co., 1920; p. 29.

ment of non-Euclidean geometry consisted precisely in his supplanting the narrow axiomatic method of dissecting postulates (as practised by Bolyai and Lobatchewski) by the more powerful methods of analysis. In this way Riemann generalized the various possible types of non-Euclidean geometry; from these generalizations the theory of relativity draws its material. Weyl's new geometry is also an example of a doctrine discovered by following Riemann's method rather than the postulate one. Thanks to Hilbert, however, the axiomatic method is also seen to have its advantages.

The possibility of replacing Hilbert's postulates by others and of obtaining verifiers for the new postulate systems has clarified the significance of these postulates. As late as the early nineteenth century, it was permissible to suppose that Euclid's axioms, postulates, and tacit assumptions, though suggested by experience, were imposed by the requirements of the human mind, *i.e.*, were *a priori* in the Kantian sense. Had this view been justified, no geometrical doctrine which rejected the Euclidean postulates and assumptions could have been conceived. But when Lobatchewski and Riemann rejected some of Euclid's postulates and succeeded nevertheless in constructing geometries, mathematicians realized that the rejected postulates, at any rate, could not be *a priori*. The status of the other postulates and of the tacit assumptions, all of which were retained in the new geometries, was, however, undisclosed; they could still be regarded as *a priori*. Finally, when Hilbert, after grouping all Euclid's postulates and tacit assumptions into one class, modified his postulates one by one and then showed that doctrines verifying the new postulate systems could be constructed, mathematicians were constrained to reject the Kantian thesis not merely for some of Euclid's postulates, but for all of them and also for all his tacit assumptions. Thus, as a result of Hilbert's investigations, there was no longer a place for the *a priori* in the concepts that are at the base of Euclidean geometry. From a philosophic point of view, this is probably the most interesting contribution of axiomatics. We shall see, however, that the exclusion of all *a priori* notions from the postulate systems does not necessarily imply that no *a priori* notions enter into the construction of a geometrical doctrine. The fact is that the postulates are not the only elements that must be examined; we must also consider the rules to which these postulates are submitted.

The Compatibility of Postulate Systems—Thus far we have dealt only with the bright side of axiomatics. We must now mention its limitations. One of the aims of mathematicians has been to furnish so-called *existence theorems*. The purpose of such theorems is to ensure that the mathematical solution we are seeking really exists, for to search for something that does not exist would be a waste of time. But before making more precise the meaning of mathematical "existence," let us examine a non-mathematical illustration.

The metaphysician Meinong is reported to have maintained that since we can say truly "The round square does not exist," there must be such an object as the round square, although it must be a non-existent object. Mr. Russell, who mentions this example, confesses himself puzzled and comments on it by saying:

"The present writer was at first not exempt from this kind of reasoning, but discovered in 1905 how to escape from it by means of the theory of 'descriptions,' from which it appears that the round square is not mentioned when we say, 'The round square does not exist.' "

We shall leave Mr. Russell with his theory of "descriptions," presumably to be found in his "Principia." What we shall be concerned with here is the different meanings that may be attributed to the same word. In ordinary conversation, the two verbs "to exist" and "to be" are regarded as synonymous. For instance, we say that "a unicorn does not exist," or that "there is no such animal as a unicorn." Here both verbs imply that no unicorns can be found, *i.e.*, that they have no empirical existence. In mathematical discussion, the verbs "to exist" and "to be" are still regarded as synonymous, but they now imply "absence of any logical contradiction." According to this understanding we are equally well justified in saying: "The round square does not exist," or "There is no such thing as a round square." The two propositions are equivalent.

Now, if we revert to Meinong's puzzle, we find that "to exist" and "to be" are no longer regarded as synonymous; the former verb is taken to imply "absence of contradiction," whereas the latter verb is applied to anything that may play the part of the subject in a proposition. There is therefore no puzzle to Meinong's example when due attention is paid to the meaning of words determined implicitly by the context. The puzzle arises only if the reader is careless enough to overlook the fact that the verbs "to exist" and "to be" are not credited by Meinong with the same meaning. If Meinong's example is taken seriously, it denotes, in our opinion, an inability to think clearly. If viewed as a kind of a pun, it is of the same type as the following one: "Where does the light go when it goes out." Here also the verb "to go" is not used in the same sense in the two cases. The moral is that whatever meaning we may attribute

to a word and to its synonyms, we must be consistent and avoid introducing surreptitiously different meanings, at least in the same sentence.
Symbolism has sometimes been suggested as a safeguard against confusions of this sort; but symbols merely replace words, and if we are
unable to differentiate the various meanings of the same word, we shall
be no better off with symbols.

Let us pass to a more important subject. When a mathematician tells
us that a certain problem has a solution, or that a solution exists, he means
that no logical contradiction prevents us from assuming the existence of
the solution. An existence theorem is a rigorous proof that the solution
exists, even though we may be unable to discover it.

One of the most celebrated problems in mathematics is an existence
theorem. It is known as Dirichlet's problem, and over it even the great
Riemann stumbled. Briefly, the problem is to determine whether or not
there will always exist a solution of Laplace's equation, which satisfies
certain restrictions and certain boundary conditions. In other words
does the aggregate of relations expressed by Laplace's equation and by
the restrictive conditions form a consistent whole. Riemann, in a celebrated demonstration, thought he had established the existence of a
solution and hence the consistency of the aggregate of relations. But his
demonstration involved two unwarranted assumptions. One was pointed
out by Weierstrass and another has been uncovered more recently by
Hadamard. A physical interpretation of Dirichlet's problem was indicated in the last chapter in connection with the temperature distribution
throughout a solid.

Let us return to Hilbert's postulates. When Hilbert constructed his
postulate system, one of his aims was to establish the consistency of
Euclidean geometry, i.e., to prove that, however far the consequences of
the Euclidean theorems were extended, no contradiction would ever arise.
Now we have seen that Euclidean geometry furnishes a model which satisfies Hilbert's postulates. Hence we conclude that the consistency of
Euclidean geometry will be established if we can show that the relations
embodied in Hilbert's postulate system are compatible. If the postulates
are not compatible, then Euclidean geometry and also all the other models
which verify the postulate system will be inconsistent and will have no
greater claim to mathematical existence than have Meinong's round
squares. (The problem of proving the compatibility of Hilbert's postulates is thus seen to have very much the same significance as has Dirichlet's
existence theorem.) We therefore request the axiomaticism to furnish us
with the required existence theorem for Hilbert's postulates.

Now, quite generally the compatibility of postulates could be tested if the postulate system yielded only a finite number of consequences, for in this case we might pursue the deductions to their end and ascertain whether or not a contradiction occurred. If no contradiction occurred, our postulates would be compatible. We may observe, however, that this method of establishing the compatibility of a system of postulates would be unsatisfactory, since it would constitute an empirical testing and not a general demonstration. Besides, in the case of Hilbert's postulates the course just outlined is impossible because the consequences deducible from the postulates are infinite in number and so cannot be exhausted. Hilbert circumvents this difficulty by reversing the problem. He notes that his system of postulates will be proved consistent if we can establish the existence of a model verifying the postulate system, the word *existence* here implying the absence of any inner inconsistency. He then maintains that the model based on numbers * satisfies this requirement; in other words, he accepts the inner consistency of the arithmetical continuum. Such is the argument advanced by Hilbert in support of the consistency of his postulate system. But the trouble is that we are by no means certain of the consistency of the arithmetical continuum, *i.e.*, of the number system. Indeed, Brouwer and Weyl question it seriously. The net result is that only by an act of faith can we accept the compatibility of Hilbert's postulates † and of any of the models that are thought to verify them. Logic alone is helpless. The significance of Hadamard's reservations on the value of axiomatics, quoted on a previous page, now becomes apparent.

All mathematicians have acknowledged the debt their science owes to Hilbert; and aside from the difficulty we have just discussed, disagreements occur only on minor details. Some of these we wish now to examine. Suppose we start with a postulate system such as Hilbert's. We then deduce the various theorems of Euclidean geometry by purely logical means, without being guided by any extra-logical intuition of what constitutes a point, a straight line, and a plane. These elements may be regarded as vague a's, b's, and c's, serving solely as substrata to the relationships expressed in the postulates. Hence it would seem that no extra-logical elements are present in our mathematical reasoning. We

---

* See page 197.

† The impossibility of establishing the consistency of a postulate system entails the impossiblity of proving that the postulates are independent. For, to prove that two postulates A and B are independent, we must show that the denial of A still leaves us with a consistent postulate system.

must note, however, that these arguments have no bearing on the conclusions we arrived at previously; namely, that mathematics does not consist solely in the rules; that it also embodies the science, or art, of combining these rules; and that this science cannot itself be compressed into a set of rules.

But even if we dismiss these objections, we may still inquire: Is it true that Euclidean geometry can be obtained merely by applying the rules of logic to the Hilbert postulates? Poincaré answers the question in the negative. He observes that in some of Hilbert's demonstrations the "principle of mathematical induction" is utilized, and he maintains that this principle is the expression of some extra-logical intuition of the human mind. Poincaré therefore concludes that geometry cannot be derived from a set of postulates by purely logical means. The validity of Poincaré's conclusion obviously depends on the precise status of the principle of mathematical induction. This principle is of constant use in mathematics; it dates back at least to Euclid, who applied it in his proof of the unending sequence of prime numbers. The principle cannot be a rule of logic, for we may quite well construct a mathematics in which the principle is denied. Let us, then, inquire whether the principle is a mere postulate, which we are free to accept or to reject. Hilbert himself does not regard it as such, since he does not mention it in his list of postulates. Peano, on the other hand, accepts it as a postulate, stating it explicitly as one of his postulates for the definition of integers. Peano is then able to prove the consistency of his postulate system. Poincaré agrees with Peano in the contention that a set of postulates must be proved consistent before the postulate system can be credited with any real significance, but he claims that Peano's proof of this consistency is circular and hence worthless. The fact is that Peano utilizes the principle of induction in a dual capacity: first as a postulate, and then as a rule which he applies when he seeks to establish the compatibility of his postulates. Circularity, according to Poincaré, can be avoided only if we remove the principle of induction from Peano's postulate system and class it with the rules to which the postulates may be submitted. But then, since the principle of induction is not a principle of logic, Poincaré is led to the conclusion we mentioned previously: the principle of induction expresses some extra-logical intuition of the human mind, and hence logic alone is insufficient to construct a mathematics.

The uncertain status of the principle of mathematical induction is, however, not the only argument that can be brought against the identification of mathematics and logic. As we have seen, the methods of axiomatics require that we start from some postulate system and that we then deduce its logical consequences; axiomatics, as such, throws no light on the motives which prompt us to adopt one postulate system rather than another. But the trouble is that the task of the mathematician does not begin when the postulate system is given; it starts far earlier, in the construction of the system itself. No one, we may presume, imagines that Hilbert's 21 postulates suddenly fell into his lap, or that they were revealed to him through some inner vision, or were adopted by caprice. Hilbert's postulate system, like every other such system, was constructed according to the same routine. The method consists in starting from some previously established doctrine in the consistency of which we have at least a moral certitude. We then pick the doctrine to pieces, examine every step in the demonstrations, and state explicitly all the assumptions utilized. These assumptions are then listed and our postulate system is obtained; care is usually given to retain only a minimum of assumptions and to discard those that are redundant. It was by following this course that Hilbert constructed his postulate system. Euclidean geometry was before his eyes and all he was required to do was to dissect it, analyze every step taken by Euclid, and fill in the omissions.

The reason why constructors of postulate systems usually start from some accepted doctrine which seems consistent is that the majority of postulate systems involve an infinite number of consequences, and postulate systems of this sort cannot be proved consistent. Of course, once a postulate system is constructed, mathematical fancy may come into play; we may change some of the postulates and obtain thereby new doctrines of which we might never have thought. Nevertheless for the reasons stated, the initial postulate system, whence the others are derived, is usually moulded on some previously recognized doctrine.

These conclusions have important philosophical consequences, for we see that, in our attempts to analyze the nature of mathematics, we cannot merely start from a given postulate system and deduce its logical consequences. We must seek the origin of the postulate system, and we are thereby referred to the preestablished doctrine on which the postulate system was moulded. In the case of the Hilbert system, we are referred to Euclidean geometry. No modern mathematician would deny the empirical origin of Euclidean geometry. Indeed the very words "geometry," "cubit," "foot," show conclusively how geometry originated, and so does the importance the Greeks attributed to geometrical constructions that can be performed with the straight edge and compass. An examination of Hilbert's postulates also leads to the same conclusion. The empirical origin of the ancient's belief in the postulate of Archimedes has already been mentioned; let us give another illustration. Consider the following postulate of Hilbert. In ordinary language, it asserts:

> If *A*, *B*, and *C* are three points in a plane and are not in a straight line, and if a straight line intersecting the segment *AB* is drawn in the plane, this straight line must intersect the segment *AC* or the segment *BC*.

Fig. 21

The postulate appears so obvious that we cannot imagine the possibility of its being denied. Yet, as Hilbert's investigations have shown, the postulate may quite well be rejected. Hence its implicit acceptance by the ancients must be attributed to the fact that it is verified empirically when lines are traced in a plane. The chronological priority of Euclidean over non-Euclidean geometry has not been touched upon in the previous discussion, but we may justify it on empirical grounds. Thus the savage and the child naturally assume that stones and sticks retain the same volumes or lengths when displaced. Consequently, when primitive people sought to divide their fields into equal parts, they invariably utilized rods or ropes in their measurements. Now all measurements which the geometricians antedating Euclid could have performed would have shown that, within the limits of experimental error, similar triangles could exist. It would therefore have been gratuitous for Euclid to have constructed a geometry in which this elementary result of measurement was denied. These considerations suggest that even had non-Euclidean geometry been imagined by the ancients, it would in all probability have been discarded as absurd.

The work of Peano and of Hilbert made it at one time the fashion to seek systems of postulates for every known mathematical theory. The construction of these postulate systems requires ingenuity and is in many respects analogous to the construction of a complicated cross-word puzzle. Unless the mathematical theory which we propose to set into postulate form pertains to an advanced branch of mathematics, the construction of a postulate system makes but little demands on our mathematical knowledge. So many different postulate systems have been devised, frequently for the same doctrine, that it would be impracticable to mention the names of all their authors. But what is important is not the construction of one more postulate system; it is the fact first demonstrated by Peano and by Hilbert that postulate systems may be constructed.

We may summarize the major contributions of axiomatics in the following statements:

1. Axiomatics has been of inestimable value in mathematics, both from an analytical and from a constructive standpoint.

2. It has shown, more clearly than ever before, that mathematics consists in systems of relations and that the specific content is unimportant.

3. It has shown that logic, by itself, cannot establish the consistency of mathematics.

4. It has shown that, in a study of the nature of mathematics we cannot be satisfied with axiomatics alone but must go behind the postulates and discover their origin. In this way we are led to trace the historical development of a doctrine; and here psychological considerations cannot be ignored.

The foregoing conclusions appear to express the consensus of opinion of the leading authorities; no serious disagreements occur thus far.

Intuitionists and Formalists—In our discussion of the foundations, we have restricted our attention as much as possible to points on which only minor differences of opinion separate the various masters. But there are many other aspects of the subject on which no agreement has been arrived at or presumably ever will be. To elucidate the matter, an examination of the tendencies of the two opposing schools is necessary.

We mentioned that the appellation "intuitionist" does not refer to a mathematician who wishes to substitute hunches for rigorous demonstrations; but that, quite the reverse, an intuitionist is a "rigorist." His principal characteristic is that he rejects as invalid or insufficient some of the definitions and proofs which the formalist accepts. He claims that if a conclusion is established by means of what he regards as an unsatisfying proof, and that if this conclusion must nevertheless be accepted for mathematics to be possible, then the best course is to recognize that the conclusion cannot be established by logical means and admit that it is forced upon us by intuition. It is in this sense that the word "intuitionist" is used. The intuitionist accuses the formalist of circularity and of lack of rigor. The formalist would presumably retort by telling the intuitionist that he is narrow-minded, old-fashioned, and that the ideal of rigor to which he wishes to restrict mathematical proofs is gratuitous and certainly unattainable.

The net result of these divergences is that the formalists accept the proofs of the intuitionists, whereas, the latter reject many of the proofs of the former. For this reason the intuitionists are on the attacking side, the formalists on the defensive. A few quotations will clarify these points. Poincaré discusses a proof proffered by Zermelo, one of the formalists. In this proof Zermelo sets out to show that the continuum (*e.g.*, all the points of space) can be transformed into a well-ordered set. The latter technical expression means that the points may be placed one after another in such a way that each point has a definite successor. In connection with this demonstration, Poincaré imagines a typical argument between a formalist and an intuitionist, whom he refers to under the equivalent names of "Cantorian" and "Pragmatist." He points out that a Cantorian will be won over by the rigor, whether real or apparent, of the demonstration. And then follows the discussion:

"*Pragmatist*—You say that you can transform the space continuum into a well-ordered aggregate. All right then, do so.
Cantorian—It would require too much time.
Pragmatist—Then show us at least how a man having the necessary time at his disposal and sufficient patience could effect the transformation.
Cantorian—This cannot be done because the number of operations to be effected is infinite; it is even greater than Aleph-zero.*
Pragmatist—Can you at least express in a finite number of words the rule you would have to follow to secure the ordering?
Cantorian—No, I cannot.
"And the Pragmatist will conclude that the theorem is meaningless, or incorrect, or in any event unproved."

The following passage, due to Weyl, is characteristic of the same rigoristic attitude:

"We must learn a new modesty. We have stormed the heavens but have succeeded only in building fog upon fog, a mist which will not support anybody who earnestly desires to stand upon it."

As may be imagined, it is more especially in problems involving the infinite (whether the infinitely great or the infinitely small) that the disputes arise. Cantor's theory of infinite aggregates was thus the starting point of the controversies. When Cantor published his epochal papers, many mathematicians refused to follow him. But, as Poincaré remarks, owing to the great beauty of Cantor's discoveries the majority of mathematicians were won over, even though they felt they were treading on dangerous ground. Then, in 1897, Burali Forti obtained a result which flatly contradicted one of Cantor's theorems. An inner inconsistency was thus revealed in the Cantorian theory, and a general cry of "I told you so" was raised by those who had ever felt but lukewarm towards the

---

* Aleph-zero represents infinity as exemplified in the infinite number of integers.

new ideas. Since Burali Forti's initial discovery, a number of other antinomies, all of them of the same general pattern, were unearthed in rapid succession by Russell, Richard, König, and a host of other logicians. Today, this search for antinomies has about the same interest as searching for asteroids in astronomy. The important point is that such antinomies are inherent in Cantor's theory of aggregates.

Desperate attempts to save all that was possible of Cantor's theories were made by his supporters, the formalists. And it is in this field of mathematics that the controversies have been most acute: the formalists on one side, the intuitionists on the other.

Thus we have Poincaré writing: "I am here speaking of genuine mathematics, not of Cantorism."

Turning to Hilbert, we read:

"From the paradise created for us by Cantor, no one will drive us out."

With the passing of Poincaré, the formalists lost their most powerful opponent, and for a time discussion subsided. The attack was renewed at a later date by Brouwer and by Weyl, and the sphere of the controversy was extended to cover practically the whole field of mathematics. This may be seen in the following passage from Weyl's writings:

"The antinomies of the theory of aggregates are usually viewed as mere skirmishes, which relate only to the extreme frontiers of the mathematical empire, and which in no wise threaten the safety and the security of the empire itself.

"The explanations that are given for these disturbances in competent circles (with a view of contesting them or of weakening their significance) do not appear to be dictated by any solid conviction. Rather are they attempts at autosuggestion, not altogether sincere, such as are met with in politics and in philosophy. But in point of fact, an attentive and sincere examination of the matter can only lead to the conviction that these irregularities in the frontier regions of mathematics must be interpreted as symptoms. In this way we may bring to light the hidden evil, concealed by the seemingly perfect interplay of the mechanism in the central regions; this hidden evil is the inconsistency and lack of solidity of the foundations on which the whole empire is built. I know of only two attempts which go to the root of the evil. One is that of Brouwer. . . ."

The second attempt to which Weyl then refers is the one he makes himself in his book: "Das Kontinuum."

In view of these attacks of Brouwer and Weyl, Hilbert returned to the defense, temporarily interrupting his researches in the advanced parts. In a final effort he sought to establish the perfect consistency of mathematics. But the intuitionists were neither convinced nor silenced, and the controversy rages on. According to Weyl, the concept of irrational number must be sacrificed or at least profoundly modified. Brouwer also contends that in dealing with infinite collections the principles of Aristotelian logic cannot be applied as they stand, and that the principle of Excluded Middle in particular must be discarded.

Assuming for the sake of argument that this view (which is not shared by the more moderate intuitionists) is valid, let us see where it will lead us. We recall that, in Hilbert's postulates, a distinction is made between the postulates themselves and the rules of logic, which are applied to them. The postulates may be varied according to the doctrine we wish to construct; but the laws of logic applied to all postulate systems are the same. If we follow Brouwer and decide that the law of excluded middle applies to finite, but not to infinite, collections, we are in effect treating this logical principle as a postulate. Many will feel that this course is the more satisfactory one to adopt, for it forces us to the conclusion that the laws of logic (or, at least, the principle of excluded middle) are not *a priori* forms of the understanding but arise from our commonplace experience with finite collections. Small wonder then that when we deal with infinite collections we fumble, since we have nothing in the world of our experience to guide us.

In favor of Brouwer's stand is the fact that results which are valid when we deal with finite quantities often cease to be true when infinite magnitudes and aggregates are considered. For instance, the properties of the sum of a finite number of terms cannot be extended uncritically to infinite series. Similarly the properties of finite numbers differ from those of infinite ones. In view of these examples and of many others which could be given, what right have we to assert that the passage to the infinite cannot possibly affect the validity of the commonly accepted rules of logic?

Against Brouwer's stand there is the objection that the antinomies do not only arise in connection with infinite aggregates. In this connection we have the well-known example of Epiminedes, who told his listeners that all Cretans were liars. Since, however, Epiminedes was himself a Cretan, the precise significance of his statement constituted a puzzle for the ancients: the logical principle of excluded middle seemed to be invalid. As another example of the same kind, Mr. Russell mentions the amusing case of a barber who shaves everybody in the town except those who shave themselves. And he then asks us whether the barber shaves himself. Whatever answer we give appears to be wrong. Here also the principle of excluded middle cannot be applied; and let us observe that in neither of the two preceding examples are infinite aggregates involved. The intuitionists dismiss this objection to Brouwer's views by claiming with Poincaré that, for finite aggregates, the antinomies are mere toys. To quote Poincaré:

"The antinomies to which certain logicians have been led arose because they were unable to avoid circularity in some of their arguments. This has occurred when they considered finite collections, but it has happened more often when they professed to argue on infinite collections. In the former case they might easily have avoided the trap into which they fell; more exactly, they themselves set the trap into which it pleased them to fall, and indeed they were very careful not to fall outside the trap. In short, in this case the antinomies are mere toys. Very different are those that are generated by the concept of infinity; it often happens that here we fall without intending to, and even when we are forewarned, we cannot feel safe."

In our opinion, the most illuminating method of making headway in this labyrinth of assertions and counter-assertions is not to defend the thesis of any particular school, but to examine in an unbiased spirit typical demonstrations of the formalists, and to concentrate on the points where disagreements start. In this way at least we shall be in a position to understand clearly the points of disagreement. Poincaré followed this method in his last paper on the subject; and as far as we know he is the only writer who has done so.

In a chapter called "Les mathématiques et la logique" which appears in "Dernières Pensées," Poincaré gives a wealth of illustrations, one of which we have already referred to in connection with Zermelo's proof. Here is another. The formalist sees no objection in defining X in the following way:

(a) X (object to be defined) has such and such a relation with *all* the members of the genus G.

(b) X is a member of G.

Poincaré then notes that, according to the intuitionists, the foregoing definition of X is circular and therefore unacceptable. Poincaré proceeds to elucidate the objections of the intuitionists by pointing out that, in their opinion, definitions of the previous type can never be proved consistent, and that as a result it is impossible to say whether X has a mathematical existence or is a "round square." He notes that the antinomies usually issue from these kinds of definition.

Another controversy between the intuitionists and formalists arises from definitions which cannot be expressed by a finite number of words; these the intuitionists reject as unfit for mathematical use. A particular example will clarify the nature of the point at issue. Consider the infinite series

$$1 + \frac{1}{2} + \frac{1}{4} + \frac{1}{8} + \frac{1}{16} + \ldots \ldots$$

This series contains an infinite number of terms, which it would of course be impossible to write out since the terms are inexhaustible in number. Nevertheless the intuitionists accept a series of this sort as being susceptible of mathematical study. Their contention is that the law of formation of the series can be expressed by a finite number of words, *i.e.*, by the statement that each term has one-half the numerical value of the one immediately preceding it. But the intuitionists refuse to consider a series whose law of formation could be expressed only by an infinite number of words. Note that the distinction is *theoretical* and not *practical,* for the intuitionists would accept a series defined by means of, say, a quadrillion words; and from the practical standpoint, a definition containing a quadrillion words is just as impossible to state as one containing an infinity. But whereas a quadrillion words may be uttered *in theory,* the intuitionists claim that an infinite number cannot, for the simple reason that infinity cannot be exhausted. The result is that we can never complete the definition, even in theory, and for this reason we cannot determine whether a definition involving an infinite number of words has any meaning.

As may be gathered from these illustrations, the root of the disagreement lies beyond logic and mathematics in obscure psychological differences pertaining to the various kinds of minds. And so it would seem that a formalist will remain a formalist from the cradle to the grave, and that the same holds true for an intuitionist. There is not much hope of the one ever convincing the other, so that nothing is to be gained by perpetuating endless controversies. Yet in a sense these controversies have not been entirely futile, for they have convinced mathematicians that agreement is impossible—and this in itself is a form of agreement.

Poincaré ends his last paper on these fruitless controversies with a note of resignation. He writes:

"At all times there have been opposing tendencies in philosophy, and it would not seem that these differences are nearing any settlement. The reason is presumably that men have different minds, and that these minds cannot be changed. There is therefore no hope of expecting any agreement between the Pragmatists and the Cantorians. Men do not agree because they do not speak the same language and because some languages can never be learned.

"And yet they usually agree on mathematical questions; but this is precisely thanks to what I have called verifications. These latter are the judges of last appeal, and to their verdicts everyone submits. But where these verifications are lacking, mathematicians are as helpless as mere philosophers. When it is a matter of deciding whether a theorem which has no means of being verified may nevertheless have a meaning, who can act as judge, since a verification is impossible? The sole recourse would be to force one's adversary into a contradiction. But this has been tried without success.

"Many antinomies have been recorded, and yet disagreement has persisted; no one has been convinced. The fact is that we can always evade a contradiction by introducing a reservation: I mean a *distinguo*."

From the standpoint of the theoretical physicist, the study of the nature of mathematics would have been of interest had it led to definite conclusions; for the whole structure of theoretical physics is based on a mathematical scheme. And unless mathematics be reliable, what hope can we have of applying it with success to the physical world? In some respects this difficulty is mitigated in current theoretical physics because mathematics is here applied only for the establishment of probabilities; and pseudo-accuracy is less objectionable in the calculation of probabilities than it was in the days when rigorous laws were thought to exist. It might even be said that modern physics is witnessing the same crisis that we have been discussing in mathematics: the quantum theorists occupy the position of the intuitionists while Einstein and Planck occupy that of the formalists.

Thus the quantum theorists reject rigorous determinism for microscopic processes, though they grant its validity on the more commonplace levels of ordinary physical experience. Note the resemblance of this attitude to that of Brouwer, who rejects the law of excluded middle for infinite aggregates, while granting its validity for the finite aggregates with which our ordinary experience makes us familiar. Also, recall the insistence of the quantum theorists on accepting only those notions and results which, in theory at least, are susceptible of being observed. Is not this attitude very similar to that of the intuitionists, who insist on definitions that can be expressed by means of a finite number of words? The following quotation from Heisenberg, exhibiting his philosophy in physics, is very similar to one we might credit to an intuitionist in mathematics. He writes:

"In this connection, one should particularly remember that the human language permits the construction of sentences which do not involve any consequences and which therefore have no content at all—in spite of the fact that these sentences produce some kind of picture in our imagination;

c.g., the statement that besides our world there exists another world, with which any connection is impossible in principle, does not lead to any experimental consequence, but does produce a kind of picture in the mind. Obviously such a statement can neither be proved nor disproved. One should be especially careful in using the words 'reality,' 'actually,' etc., since these words very often lead to statements of the type just mentioned." *

The ideas expressed in this passage are strongly reminiscent of Poincaré's objections to Zermelo's proof.

---

* Heisenberg. The Physical Principles of the Quantum Theory; University of Chicago, 1930; p. 15.

CHAPTER XVII

ANALYTICAL MECHANICS

MECHANICS is divided into two parts: Statics, which deals with the equilibrium of forces; and Dynamics, the object of which is the study of the motions of bodies under the action of forces. Archimedes mastered the elements of statics, but dynamics was totally unknown to the ancients; for what little they thought they knew, as can be judged from the writings of Aristotle, was grossly erroneous. Not until Galileo and Newton, were the foundations of dynamics established.

Statics and dynamics, though forming distinct chapters in mechanics, may nevertheless be fused into one doctrine. D'Alembert's principle brings about this fusion; it shows that a problem of dynamics can be viewed as a problem of equilibrium, and hence as a problem of statics, when peculiar fictitious forces called the forces of inertia are taken into account. For instance, a planet describing a circular orbit around the sun can be viewed as in equilibrium under the equal and opposite gravitational and centrifugal forces.

As may be inferred from Chapter XV, mechanics does not restrict its attention to rigid bodies and to point-masses. It deals also with the motions of continuous deformable material media. In Chapter IX we mentioned the reasons why the mechanics of point-masses was the first to be developed. The mechanics of extended rigid bodies followed in the course of the eighteenth century, chiefly owing to the contributions of Euler. Finally, the mechanics of continuous deformable media was investigated in the early part of the nineteenth century.

Galilean Frames—When we wish to study the motion of a particle or of a dynamical system, we measure the various displacements and motions in some frame of reference. The frames of reference usually selected in dynamics are the so-called inertial, or Galilean, frames. From a phenomenological standpoint, these frames exhibit no rotation and no acceleration of any sort with respect to the star-system. The reason Galilean frames play so prominent a part in dynamics is that the laws of dynamics assume their simplest form when they are referred to such frames. The laws of dynamics are then expressed in the form established by Galileo and Newton.

Fields of Force—Consider a particle in empty space. We refer positions and velocities to a Galilean frame. If the particle remains at rest in the frame or moves with constant speed along a straight line, we agree to say that no force is acting on the particle. This statement expresses the law of inertia. If, on the other hand, the motion of the particle is not a uniform motion along a straight line, *i.e.*, if the motion is accelerated, we assume that a force is acting on the particle. The direction of the force coincides at any instant with the direction of the acceleration, and the magnitude of the force is proportional to that of the acceleration. In the majority of cases, the magnitude and direction of the force acting on a particle depend on the particle's position. For instance, the force of attraction exerted by the sun on a planet varies with the position of the planet. Let us observe that the gravitational force is disclosed only through the behavior of the planet; nevertheless, we may reason as though a force were still in existence at each point of space around the sun even in the absence of any planet. We are thus led to conceive of a region of space at each point of which a force is present. The aggregate of such forces is called a *field of force*. In particular, the field of force surrounding the sun is called the sun's gravitational field.

The magnitude of the force at the various points of a given field is proportional to the mass of the particle on which the force is acting.* There is no sense, therefore, in attributing any definite magnitude to the force at a given point until the mass of the particle has been specified. So as to remove all vagueness on this score, we shall suppose that the particle considered has unit mass.

A field of force may be represented by an aggregate of arrows (the feathered extremity of an arrow being present at each point of space). The length and the direction of each arrow will indicate the magnitude and direction of the force at the point where the feathered end of the arrow is situated. The gravitational field surrounding the sun is then represented by an infinite aggregate of arrows, all of which point towards the centre of the sun and decrease in length as they are further removed from this centre.

A field of force is called "permanent" if the forces at the various points do not change, either in magnitude or in direction, as time passes. Fields of this type are the only ones we shall consider.

---

* This is so in mechanics. In electrostatics, however, the force is proportional to the electric charge of the particle; and in electromagnetics the situation is more complicated.

Work and Potential—Consider a permanent field of force and in it a particle. We constrain the particle to move from a point A to a neighboring point $A + dA$. During this displacement a force acts on the particle. The *work* developed by the force over this tiny displacement is measured by the product of the length of the displacement dA and the projection of the force at A onto the displacement. If the projection of the force lies in the direction of the displacement, the work performed by the force is said to be positive; if the projection lies in the opposite direction, the work is negative. Suppose now we consider a finite displacement along a given path from a point A to a point B. The total work of the force over the entire path may be obtained by summing, or integrating, the elementary amounts of work corresponding to the consecutive tiny displacements.

Here two situations may arise. In the general case of a permanent field of force, when the particle is made to pass from any given point A to any given point B, the work expended (whether positive or negative) will depend on the path followed. But in many important cases the work from A to B depends solely on these terminal points and is the same regardless of the path that is described. When this situation occurs, the field of force is said to be derived from a *potential*.* For reasons that will appear presently, the field is also called "conservative" in this case.

Potential and Kinetic Energy—The important concept of potential energy in a field of force arises only if the field is derived from a potential, *i.e.*, is conservative. Consider a particle in a field of this type, and two arbitrary points A and B in the field. The relationship between the potential energy of the particle at A and at B is furnished by the following definition:

Potential energy at B minus potential energy at A = work (positive or negative) expended by the field of force when the particle is made to pass from B to A.

We see that the potential energy at B will exceed the potential energy at A if the work from B to A is positive; the opposite will be true if the work is negative. Incidentally we may note that, for the difference in the potential energies at A and at B to be completely determined by the points A and B, the work must be the same regardless of the path

* The field of gravitational force developed by a massive body affords an example of a field which is derived from a potential (the Newtonian potential). The magnetic field surrounding an electric current is a field having no well-defined potential.

followed. Hence it is only in conservative fields that the concept of potential energy has a clear meaning. Let us also observe that our definition is restricted to differences in the value of the potential energy; absolute values elude us. However, if we agree to assign a definite value to the potential energy at some selected point, the potential energy is determined without ambiguity at all other points. For instance, if we agree that the potential energy at the point A is zero, the potential energy at any arbitrary point B is given by the work (positive or negative) expended by the force when the particle is moved from B to A.

As a concrete illustration let us consider the field of gravitational force developed by the earth. This field is permanent and is conservative. Suppose a stone is situated at a point B above the earth's surface. We shall assume that the point A, at which the potential energy is taken as zero, is any point on the ground. The potential energy of the stone at B is then defined by the work expended by the force of gravity when the stone falls to the ground; this work is positive, and so the potential energy of the stone at B is positive. Since the natural motion of the stone, when released from a position of rest, is towards the ground, the natural motion starts in the direction defined by a decrease in the potential energy. This rule is general: Whenever a particle is released from a position *of rest* in a conservative field, it will always start moving towards regions of decreasing potential energy. An immediate consequence of this rule is that, in the particular case where the initial position of rest is one of minimum potential energy, the particle remains motionless; the particle is then said to be in a position of equilibrium. Thus a stone that has fallen to the ground has the smallest potential energy possible and hence remains at rest.

Kinetic energy is energy due to motion. It is defined by $\dfrac{mv^2}{2}$, where m is the mass of the particle and v its velocity in any direction.

In a permanent and conservative field of force, the sum of the potential and kinetic energies of a particle remains constant during the particle's motion. We express this fact by saying that the total mechanical energy of the particle is conserved. The conservation of energy would not hold in a non-permanent field; nor would it be realized in a permanent field which is not derived from a potential (*i.e.*, which is not conservative). It is because the conservation of energy holds only in the case of permanent fields derived from potentials that such fields are called conservative.

These summary indications must be amplified, however. We have said that, if a particle is acted upon by a permanent and conservative field of force, the total energy of the particle remains constant. But this statement is valid only when the presence of the particle causes no disturbance on the field. In a theoretical discussion, we may assume, for the purpose of argument, that the foregoing condition is satisfied; but in all rigor we have no right to make this assumption, for we know that the particle must necessarily disturb the field. The fact is that, in practice, a field is generated by some physical system; and, according to the law of action and reaction, the particle reacts on this system and thereby modifies the field.

We shall clarify the matter by considering the case of a stone falling towards the earth.* The earth develops a permanent and conservative field, and the stone is moving under the action of the field. We first treat the problem rigorously. According to the law of action and reaction, the attraction of the earth on the stone is accompanied by an equal and opposite attraction of the stone on the earth. As a result, while the stone is falling towards the earth, the earth is rising to meet the stone. Now the earth, during its accelerated motion towards the stone, carries its field with it; and so the distribution of this field in a Galilean frame is not permanent. Consequently, the total energy of the falling stone is not conserved. On the other hand, if the stone were to exert no reaction on the earth, and hence if the earth's field were permanent, conservation of energy would hold for the stone: while the stone was falling, its potential energy would be decreasing and its kinetic energy increasing at an equal rate, so that the sum of the two kinds of energy would remain constant. Of course, the assumption that the stone exerts no reaction on the earth is invalid. Nevertheless, when we are satisfied with approximations, we may disregard the stone's reaction in view of the small mass of the stone as contrasted with that of the earth. We must note, however, that the approximation just considered would be manifestly untenable if, in place of a small stone, we were dealing with a large mass, such as the moon. As we shall see presently, the conservation of energy will hold rigorously in the example of the falling stone provided we consider not the energy of the stone alone, but the total energy of the system formed of all the bodies involved, i.e., the earth as well as the stone. But first we must discuss mechanical systems.

Mechanical Systems—By a mechanical system we mean any aggregate of two or more bodies which interact. In an extended sense we sometimes speak of a single body as constituting a system.

An isolated system is a system the various parts of which may interact but must not influence, or be influenced by, bodies outside the

---

* Friction and air resistance are disregarded in all our illustrations.

system. A stone thrown in the air does not constitute an isolated system, for it is subjected to the gravitational attraction of the earth: the earth acts on the stone, and the stone acts on the earth. Let us, then, consider the system formed of the earth and stone jointly. Is this system isolated? It is a better approximation to an isolated system than was the earth alone, but it is not a truly isolated system, for the stone and the earth are acted upon by the planets and sun and they react on these bodies. To obtain an isolated system, we must therefore include in our system the sun and all the planets. We are then dealing with the solar system. But the solar system in turn is not truly isolated, since the planets and sun are attracted by the various stars and nebulae. Hence, only by considering the entire universe can we attain a rigorously isolated system. In many cases, however, a sufficient approximation is secured when we view the earth and stone, or better still the solar system, as forming an isolated system.

Let us now examine the meaning of a conservative system. Consider a system composed of different bodies which exert forces on one another. These forces are called "internal forces." At a given instant the bodies of the system occupy certain positions, the aggregate of which defines a configuration of the system. Consider two different configurations A and B, and let us constrain the system to pass from one of these configurations to the other. We may suppose that we can displace the various bodies with our hands, removing them from their positions in the configuration B to their positions in the configuration A. The displacements will be resisted or aided by the internal forces, and as a result the internal forces will perform work (negative or positive). Now in the general case of an arbitrary system, two different situations can occur. In certain systems the sum total of the work performed by the internal forces, during the passage from an initial to a final configuration, is completely determined by these terminal configurations and is independent of the particular sequence of intermediary configurations through which the system is made to pass. Systems of this kind are said to be *conservative*. As we shall see presently, this appellation is justified by the fact that the total internal energy of such systems is conserved. In other systems the work of the internal forces depends on the sequence of intermediary configurations and is not determined by the terminal configurations. These systems are called non-conservative, because conservation of energy does not apply to them.

As an example of a conservative system we may mention a system formed of gravitating masses (*e.g.*, the solar system); its internal forces are represented by the mutual gravitational attractions. One of the essential requirements for a system to be conservative is that changes in configuration be unaccompanied by friction.

The concept of potential energy, which we examined in the case of a particle in a conservative field of force, arises afresh for conservative systems. In a conservative system if we select some arbitrary configuration A to represent the configuration of zero potential energy, the internal potential energy of the system for any configuration B is measured by the work (positive or negative) developed by the internal forces when we pass from the configuration B to the configuration A.* In dealing with the solar system, we usually agree that the configuration A of zero potential energy is realized when all the planets are at infinite distances; i.e., when the solar system is dispersed. We shall adopt this convention. Consider then any configuration in which the planets are not scattered at infinity. The work developed by the internal forces (i.e., mutual gravitational forces), when we assume that the system is constrained to pass from the configuration of interest B to the configuration A at infinity, is a negative work. Consequently, the potential energy of the system in the configuration B will always be negative. It should be observed, however, that the sign and value of the potential energy depend on our choice of the configuration of zero potential energy, and that this configuration may be selected at pleasure.

In any event, if the planets be released from arbitrary positions at rest, they will proceed to move in the directions required by the internal gravitational forces. These forces will thus perform positive work at the start, and, as a result, any configuration immediately following the original one will be a configuration of lesser potential energy. We have here a general rule: An isolated and conservative mechanical system released from a position of rest will tend to move toward configurations of lesser potential energy. It is essential to stipulate that we start from a position *of rest,* for otherwise, the change might quite well be towards configurations of higher potential energy.† We also see that if the initial configuration of rest corresponds to a minimum for the potential

---

* We are considering here the ''internal potential energy'' of the system, i.e., the potential energy that the system would have if it were isolated. Of course, if the conservative system were situated in a conservative field of force, it would have an additional potential energy due to the action of the field.

† For instance, the system represented by the earth and by a pendulum swinging freely under the earth's attraction (friction ignored) forms an isolated and conservative system. If we release the pendulum from a position of rest, it will start to fall, and the potential energy will thus start to decrease. But suppose we communicate an initial upward velocity to the pendulum. The pendulum will now start to rise, and its potential energy will increase.

energy, the system will not evolve but will remain at rest. Such a configuration is one of equilibrium. In the solar system this situation would be realized if all the planets (viewed as so many billiard balls) were lying on the surface of the sun.

The total kinetic energy of a mechanical system is easily defined; it is simply the sum of the kinetic energies of the various masses. The total energy at any given instant is the sum of the potential and of the kinetic energies at this instant.

The Principles of Conservation—If a system is isolated and conservative, the sum total of the potential and kinetic energies will always retain the same value: the total mechanical energy is conserved. For instance, the solar system may be viewed, with good approximation, as isolated and conservative, and so although its kinetic and its potential energy vary separately the sum of the two energies remains constant.

Let us now consider the case of a conservative system which is submitted to the action of external forces; such a system is, of course, not isolated. If the external forces are performing positive work on the system against the internal forces, the total internal energy of the system will be increased, and the increase in this energy will be equal to the work performed by the external forces. Conversely, if our conservative system is performing work on the surroundings, the work will be accomplished at the expense of the internal energy of the system. If, then, we take the configuration of minimum potential energy to define the configuration of zero potential energy, we see that the total internal energy of the conservative system will represent the maximum amount of work that the system can perform on the surroundings. In the particular case where the system is in a state of rest, its total energy reduces to its internal potential energy, so that in this case it is the internal potential energy which measures the total work that the system can perform.

Our discussion of mechanical systems enables us to clarify certain points we left in suspense on a previous page. We mentioned that the total energy of a stone falling towards the earth would be rigorously conserved if the stone exerted no reaction on the earth; but we mentioned that since the stone reacts, the conservation of the stone's energy is not rigorous. Let us, then, examine the approximation made when we assume that this energy is conserved. The stone and earth jointly form a system which we may regard as isolated and also as conservative (if we neglect the friction of the atmosphere). The total energy of this system is therefore conserved. This total energy is represented by the sum of the kinetic energies of the stone and of the earth, to which we must add the potential energy of the earth-stone system. Now the potential energy of the earth-stone system is the same as the potential energy we ascribed to the stone alone when the problem was not treated rigorously. Consequently, if we were to assume that the energy of the stone were conserved, our assumption would be correct only insofar as we could neglect the kinetic energy of the earth in its fall towards the stone. In practice, owing to the large mass of the earth, its velocity of fall towards the stone is negligible, and so also is its kinetic energy. For this reason, if absolute rigor is not demanded, we are justified in saying that the energy of the stone is conserved.

The foregoing considerations show that the conservation of mechanical energy holds rigorously only in an isolated and conservative system. We have already observed that no system is rigorously isolated. But in addition no system is rigorously conservative, for friction and inelastic deformations are inevitable. Indeed in many cases the systems with which we are concerned are not even approximately conservative. Thus, if two inelastic balls (assumed isolated from all outside influences) enter into collision, the total energy (here kinetic) is decreased considerably by the collision. Only if the balls are ideally elastic will conservation hold.

The restricted conditions under which conservation of energy holds in mechanics precluded the earlier physicists from viewing this form of conservation as the expression of a general principle. Only later, in the nineteenth century, when the concept of energy was considerably enlarged, was the Principle of Conservation of Energy accepted. Since, however, the systems with which we deal in mechanics are usually of a kind in which the energy is conserved, we may speak of a mechanical principal of conservation of energy without fear of being misunderstood.

There are two other principles of conservation in mechanics; they are called the Principles of Conservation of Momentum and of Angular Momentum. These two principles are of wider application than the principle of conservation of energy because they apply to all isolated systems whether conservative or non-conservative. For example in the illustration of the impact of two non-elastic balls, given above, though the energy of the system is not conserved, the total momentum and the total angular momentum are both conserved.

Let us first examine the principle of conservation of momentum. The momentum of a mass m moving with a velocity v in a given direction is defined by mv. The momentum, like the velocity, is a directed quantity which may be represented by an arrow. The total momentum of a system is obtained by compounding, according to the rule of the parallelogram, the individual momenta of the various masses of the system. The principle of conservation of momentum then states that the total momentum of an isolated system is constant in magnitude and in direction. In contradistinction to the principle of conservation of mechanical energy, the principle of conservation of momentum does not require the isolated system to be conservative.

We may apply this principle to a particular case. Suppose that a billiard ball is moving on a billiard table; the ball has a definite momentum. Each time the ball rebounds from a cushion its momentum changes in magnitude and in direction, and eventually as a result of friction the ball comes to rest. At first sight it would seem that the principle of conservation of momentum is at fault; but this is not so, for the ball by itself does not represent an isolated system. An isolated system will be obtained, however, if we regard the ball, the table, and the earth as forming a single system. The principle then asserts that the total momentum of all three bodies is conserved. This implies that each time the ball rebounds and its momentum changes in direction, the momenta of the table and of the earth change in reverse manner. In other words, as the ball moves forwards and backwards, the earth must move backwards and forwards. We may understand how this situation arises by noting that when the ball rebounds from the cushion it communicates a certain motion to the cushion and thereby to the billiard table. Since the latter is fixed to the ground, it drags the earth along, so that the earth's momentum is changed in consequence.

The principle of conservation of momentum may also be illustrated in the fall of a stone towards the earth. Thus the principle enables us to assert that, during the reciprocal fall of the stone towards the earth and of the earth towards the stone, the centre of gravity of the isolated earth-stone system will remain fixed in a Galilean frame. The principle thus informs us on the displacement the earth will sustain when the stone reaches a given elevation.

Finally, let us consider the principle of conservation of angular (or rotational) momentum. A top set spinning has angular momentum. The angular momentum vanishes when the top ceases to rotate and topples over. The principle of conservation appears to be at fault. But this is not so, for, in order to apply the principle correctly, we must apply it to an isolated system, i.e., to the earth-top system. The total angular momentum of this system remains constant, because any variation in the rotational speed of the top generates a rotation of the earth in the opposite direction.

Applications of the Principles of Conservation—When we are dealing with a mechanical system for which the principle of conservation of mechanical energy may be regarded as satisfied, we may often utilize this principle to simplify our calculations and obtain more rapid answers to our problem. For instance, let us suppose that we wish to compute the velocity with which a mass m, released at rest from spatial infinity, will fall onto the earth's surface (the attractions of all other celestial bodies are ignored). We may assume, as a first approximation, that our planet remains motionless during the stone's fall. Our assumption implies that the stone exerts no reaction on the earth, so that the field developed by the earth is permanent and is regarded as given. The problem can be solved by integrating the equations of dynamics; but we may proceed more simply by utilizing the principle of conservation. Thus, if the aforementioned approximation is made and all frictional resistances are disregarded, we know that the total energy of the stone will remain constant during the fall. Let us consider this total energy. The potential energy of the stone at any given point may be assigned at pleasure. Let us, then, agree that, at an infinite distance, the potential energy has the value zero. The kinetic energy of the stone at the instant of its release from a position of rest is obviously zero. Consequently the total energy, which is the sum of the potential and of the kinetic energies, also has the value zero. According to the principle of conservation of energy, this same zero value will be retained by the total energy during the fall;[*] and so the kinetic and the potential energies must be equal and opposite at each instant during the fall. In particular must this be true when the stone strikes the earth's surface and hence when it is situated at a distance R from the earth's centre (where R is the radius of the earth). If we call M and m the masses of the earth and stone respectively, the potential energy of the stone on reaching the earth's surface is known immediately to be $-k\dfrac{Mm}{R}$, where k is the gravitational constant. Therefore conservation of energy requires that the kinetic energy $\dfrac{mv^2}{2}$ of the stone should have the value $k\dfrac{Mm}{R}$. From this relation, in which everything is known except the velocity v of the stone, this velocity v may be deduced—and our problem is solved. We might also proceed rigorously by taking into consideration the simultaneous fall of the earth towards the stone. Here, however, it would be necessary to

---

[*] A zero total energy does not imply the absence of energy, for any other value could be assigned to the total energy if we changed in an appropriate way the zero-point of the potential energy.

take into account not only the conservation of energy, but also the conservation of momentum. In any case the application of the general principles of conservation leads to a rapid solution of the problem.

The effect of the tides on the motion of the moon affords a good example of a problem which is clarified by the principle of conservation of angular momentum. The earth and moon, considered jointly, may be assumed to form an approximately isolated system. The total angular momentum of this system is obtained by compounding the angular momentum of the earth, due to its rotation, with the angular momentum of the moon, due to its revolution around the earth.* Now the frictional effect of the tides causes a slowing down in the earth's rotation and thereby decreases its angular momentum. Some compensatory increase in angular momentum must appear somewhere; otherwise the principle of angular momentum would be violated. The required increase exhibits itself in an increased angular momentum of the moon. Our satellite must therefore start moving more rapidly on its circular orbit. But this more rapid motion will generate an increase in the centrifugal force which tends to tear the moon away from the earth; and as a result the moon will recede from the earth. Thus a mere application of the general principle allows us to anticipate without calculation that the moon must be receding from our planet owing to the frictional effect of the tides.

A characteristic feature of the foregoing general principles of conservation is that they yield at least partial information on the behavior of a mechanical system, without compelling us to go through tedious calculations and to analyze the various processes step by step. The fact is that the principles of conservation are examples of so-called "first integrals" of the motion, and that the knowledge of first integrals advances the solution of dynamical problems. This point will be explained on a later page.

General Characteristics of the Laws of Dynamics—We mentioned in Chapter XV that if a point-mass is acted upon by a force directed along a straight line, and that if we call x the position of the point-mass at any instant, then the ordinary differential equation

$$(1) \qquad\qquad mx'' = F(x)$$

(where $F(x)$ denotes the force at the point x) expresses the mechanical law. More generally, if the point-mass is moving in space, it will have

---

* There is also the angular momentum of the moon, due to its rotation on its axis; but this angular momentum is so small that it may be disregarded.

three variable coordinates x, y, z; and in place of the single differential equation (1) we shall have the three equations

$$(2) \quad \begin{cases} mx'' = F_x(x, y, z) \\ my'' = F_y(x, y, z) \\ mz'' = F_z(x, y, z) . \end{cases}$$

In these equations, $F_x(x, y, z)$, $F_y(x, y, z)$, $F_z(x, y, z)$ denote the values, at a point x, y, z, of the projections of the known force on the three coordinate axes Ox, Oy, and Oz. We need not go into further detail, for we shall only be concerned with general conclusions.

The law of motion of the point-mass is given by the three equations (2); but in order to know exactly how the body will move, we must integrate these equations and obtain their general solutions. The three general solutions will furnish general expressions, which will tell us how the three coordinates x, y, z of the point-mass vary with time and hence how the body may move. In the present case, since our three differential equations (2) are of the second order each, six ($3 \times 2 = 6$) arbitrary constants will be present in the general solutions. Until we have attributed definite values to these constants, no definite motion is determined. It can be shown that the six arbitrary constants may be taken to represent the three initial coordinates of the body and the three components (or projections on the coordinate axes) of the initial velocity. In simpler language, this means that the motion of the particle under the action of a given force is completely determined by its initial position and by its initial velocity (the latter being specified both in magnitude and in direction). As for the initial instant, it may be chosen arbitrarily. What is called the *state* of the body at any instant is defined by the position and velocity at the instant considered. We may therefore compress our previous findings by saying that the state of a point-mass at any instant, moving under a given force, is completely determined by the initial state.

These conclusions are general and apply equally well to any dynamical system, *e.g.*, to the solar system formed of n point-masses moving under their mutual gravitational attractions. Thus, the motions of the various bodies of the solar system are unambiguously determined by their initial spatial distribution and by their initial velocities. The Greeks were ignorant of the laws of dynamics and believed that the velocity of a body, and not its acceleration, was proportional to the force. If this contention of the ancients were correct, the differential equations of mechanics would be of the first order and not of the second. As a result, the position and motion, at any instant, of a body moving under a given force would be dependent on its initial position alone, and not on its initial velocity. The initial velocity would be automatically determined by the force acting on the body, and to specify it would be redundant.

The laws of mechanics and their consequences, the principles of conservation, are valid only when positions and motions are referred to particular frames of reference, usually called inertial, or, as Einstein prefers, "Galilean."* It is the existence of these privileged frames that is primarily responsible for our difficulty in avoiding the introduction of an absolute space serving to differentiate a Galilean frame from a non-Galilean, or accelerated, one. The general theory of relativity attempts to obviate absolute space by expressing the laws of mechanics in a form so general as to apply to all frames of reference. To a certain extent, however, the privileged nature of Galilean frames still subsists, for it is in these frames that the relativistic laws assume their most simple expression. Thus, the theory of relativity, like classical mechanics, confronts us with the problem of understanding why Galilean frames should be distinguishable from frames of other kinds. The theory of relativity attributes this situation to the structure of space-time; but, in so doing, it is compelled to ascribe a physical significance to this structure and thereby to confer on space-time many of the attributes of an absolute continuum. Only after Einstein had formulated his hypothesis of the cylindrical universe was it possible to reconcile the theory of relativity with the doctrine of the relativity of motion.

There is, however, another way of justifying the relativity of motion. It was discussed by Poincaré long before the advent of the theory of relativity; and though today Poincaré's analysis has but a historical interest, we shall examine it because it illustrates some of the characteristic features of differential equations. Poincaré's argument may be understood from the following considerations.

From the standpoint of the mathematician, absolute space seems to impose itself in classical mechanics because the laws of dynamics are valid only when measurements are referred to Galilean frames. Let us observe, however, that laws which are valid only under special conditions may usually be regarded as restricted formulations of more general laws that are valid under a greater variety of conditions. We are thus led to inquire whether the laws of mechanics may not be so generalized that they will be expressed independently of any frame of reference.

---

* See page 214.

To take a specific example, let us consider a system, such as the solar system, formed of N point-masses moving under their mutual gravitational attractions. If we follow the usual procedure, we shall take as variables the coordinates of the planets with respect to the Galilean frame attached to the centre of mass of the system. The laws controlling the motions will then be expressed by a system of second-order differential equations connecting the coordinates and their derivatives. Let us call this system of equations the equations (A).

In order now to formulate the dynamical laws in a manner consistent with the relativity of motion, we must rid the equations (A) of their dependence on the frame of reference. Hence we must rid these equations of the coordinates of the point-masses, replacing the coordinates by the mutual distances between the various point-masses. There is no difficulty in securing this objective, for the coordinates and the mutual distances are connected by simple mathematical relationships.

The elimination of the coordinates from the equations (A) may be made to yield a system of second-order differential equations in which the mutual distances between the point-masses are the variables. The new equations will be called the equations (B). Since the mutual distances are determined independently of any frame of reference, we might suppose that the new dynamical equations would satisfy the requirements we have imposed. But such is not the case, for we have omitted to mention that the equations (B) contain in their expression a constant of undetermined value; hence the equations are indeterminate till the value of the constant is specified. If, at this stage, we choose to disregard the hypothesis of absolute space, we have no means of deciding on the value of the constant. On the other hand, when we accept absolute space, the equations may be so written that the constant measures the constant angular momentum of our mechanical system with respect to the Galilean frame. The constant is thus connected with a rotation in absolute space; and the equations (B) like the equations (A) are thus seen to incorporate the notion of absolute space. Obviously, therefore, in spite of the elimination of the coordinates, the equations .(B) do not have a form consistent with the relativity of motion.

The situation is clarified further when we seek to deduce the actual motion of the system from the equations (B). In order to predict this motion, we must ascertain how the mutual distances vary in the course of time. The required information is obtained when the equations (B) are integrated and when the initial conditions are specified. The equations being of the second order, the initial conditions are represented by the initial values of the mutual distances and by the initial rates of change of these mutual distances. But in addition, we must be informed of the value of the constant which is present in the equations (B). Owing to the aforementioned significance of this constant, we see that the hypothesis of absolute space is still retained in the formulation of the mechanical laws.

Let us then modify the form of the equations (B) so as to remove all reference to absolute space. To secure this result, we must rid the equations (B) of the constant. Now the elimination of the constant may be secured if we first differentiate the equations (B). As a result of this differentiation and of the subsequent elimination, we obtain a set of differential equations involving solely the mutual distances and their derivatives. We shall call these mechanical equations the equations (C). One of the new equations is of the *third* order. This fact implies that the equations, when integrated, will define the actual motion provided we specify not only the initial values and the initial rates of change of the mutual distances, but also the initial rate of change of the rate of change of one of the mutual distances. Let us observe that the last item of information takes the place of the value ascribed to the constant of the equations (B), so that all reference to absolute space is now avoided. In fine, the equations (C) express the mechanical laws in a form compatible with the relativity of motion. Our revised dynamical laws (C) are more general than the former ones. The greater generality of the revised laws is illustrated by the fact that the new equations comprise a third-order differential equation, whereas the original laws (A) were expressed by second-order differential equations; and we have mentioned elsewhere that the higher the order of a differential equation, the greater the number of possibilities it includes.

It is of interest to contrast the second-order differential equations (A) (which are connected with the notion of absolute space) with the third-order equations (C) (which dispense with this notion). Either one of these systems of equations, when integrated, determines the motion which follows from prescribed initial conditions. But the equations (A), owing to their reduced order, are the simpler. Poincaré concludes that the introduction of Galilean frames, and hence of absolute space, is a mere artifice which enables the mathematician to express the dynamical laws in the form of second-order differential equations (A), instead of leaving them in the more complicated though more general form (C). According to Poincaré absolute space, though meaningless in theory, fulfills a useful office in practice.

We have mentioned Poincaré's analysis on account of its philosophic interest. In the remaining part of this chapter, however, we shall follow the usual procedure and refer all motions to Galilean frames.

Degrees of Freedom—The number of degrees of freedom of a dynamical system is defined by the number of separate magnitudes which must be specified in order to determine the exact position and configuration of the system. Thus the position of a point-mass is specified in a given frame when we have stated its three coordinates x, y, z; a point-mass has therefore three degrees of freedom. Two separate point-masses will form a system of six $(2 \times 3 = 6)$ degrees of freedom; and a system of N point-masses will have $3N$ degrees of freedom. Passing now to a rigid body of finite dimensions, we note that its position and orientation are not fully determined by the position of one of its points alone, e.g., by the centre of mass, for the body may be rotated in many ways about this fixed point. To determine the position of the solid, we must determine three other magnitudes; they may be chosen in various ways. Euler defines them by three angles. At all events whatever choice is made, the net result is that in addition to the three first magnitudes which determine the position of a given point of the solid, three additional ones are required, making six in all. We say therefore that a finite rigid body has six degrees of freedom.

Lagrange Coordinates—Whereas the position of a point-mass is determined by three lengths x, y, z, the position of a rigid body may, as explained, require the specification of angles. Now angles and lengths, though constituting different kinds of magnitudes, play the same rôle in the determination of the position and orientation of a rigid body. In view of their sameness of function, Lagrange suggested designating both lengths and angles by the same letter q. Thus in the case of the rigid body, the six magnitudes, whether lengths or angles, which serve to define its position are designated by

(3) $q_1, q_2, q_3, q_4, q_5, q_6.$

These magnitudes are called Lagrange's generalized coordinates. When the solid moves, the value of at least one of the six coordinates q changes; hence these coordinates are functions of time. We may also consider their time-rates of change, i.e., their first derivatives:

(4) $q_1', q_2', q_3', q_4', q_5', q_6'.$

It can be shown that if the values of the six magnitudes (3) and of the six magnitudes (4) are specified at any instant of time, the instantaneous position, orientation, and motion of the solid are determined at the instant considered. In short, the values of the twelve magnitudes (3) and (4), at any instant, define the state of the body at that instant. Thanks to the introduction of his generalized coordinates, Lagrange was able to devise a new form for the equations of dynamics; he obtained what are known as the "Lagrange equations." We shall omit to consider these equations, because a more convenient form of the equations of dynamics, a form due to Hamilton, will be mentioned presently.

The Coordinates q and the Momenta p—We have seen that Lagrange defined the state of a dynamical system at any given instant by means of coordinates q and of their first derivatives q'. An alternative means of expression, introduced originally by Poisson but which owes its success to Hamilton, consists in replacing the derivatives q' by new magnitudes p. To each coordinate q corresponds a p, and the mechanical significance of a p depends on that of the corresponding q.

For instance, let us revert to the example of the rigid body. The position and orientation of the body in a given Galilean frame are determined when the numerical values of the six coordinates q (one for each degree of freedom) are specified. We have seen that three of these coordinates q, *e.g.*,

$$q_1, q_2, \text{ and } q_3,$$

may be taken to represent the ordinary Cartesian coordinates x, y, z of the centre of gravity of the solid. The remaining three coordinates q,

$$q_4, q_5, \text{ and } q_6,$$

may then represent three angles which will serve to complete the specification of the solid's position. With the three first coordinates q, we associate three magnitudes p, *i.e.*,

$$p_1, p_2, \text{ and } p_3,$$

the mechanical meanings of which are given by the components of the momentum of the solid with respect to the three coordinate axes Ox, Oy, Oz. More precisely, if M is the total mass of the solid and v_x, v_y, and v_z are the components (along the axes) of the velocity of the centre of gravity, then p_1, p_2, and p_3 will be the momenta

$$Mv_x, Mv_y, Mv_z.$$

We now pass to the three other coordinates q, which define angles. With them will be associated respectively three magnitudes p, *i.e.*,

$$p_4, \ p_5, \ \text{and} \ p_6.$$

These latter p's represent the angular momenta of the rigid body in its spinning motion. We shall not go into further detail. What is important to understand is that the magnitudes p represent momenta or angular momenta. For this reason they are often called *generalized momenta*, just as the coordinates q are referred to as *generalized coordinates*.

When the body moves, its position, orientation, momenta, and angular momenta, vary with time, so that the coordinates q and momenta p are functions of time. If at any instant the q's and p's are specified, the state of the system is thereby defined at the instant considered. In this respect the q's and p's of Hamilton fulfill exactly the same office as the q's and q''s of Lagrange. When we recall that the evolution of a dynamical system moving under given forces is determined by the initial state, we may say: The entire history of a dynamical system moving under given forces is determined by the initial values of the q's and p's.

Obviously, whether we choose the Lagrange or the Hamilton method of presentation, the final results will always be the same. But we must remember that the mathematical difficulties to be overcome, even in some of the seemingly elementary mechanical problems, are often so great that every kind of simplification must be resorted to. And it so happens that the equations yielded by the q's and p's are more symmetric and hence easier to manipulate.

Hamilton's Equations—Suppose we are dealing with a conservative mechanical system having n degrees of freedom; the system itself may be acted upon by a permanent and conservative field of force. As we explained previously, a permanent field of force which acts on a system or on a body is necessarily affected by the reaction of this system on the system generating the field. But we saw that in many cases the reaction may be disregarded owing to its small relative importance. We shall assume in the present chapter that this simplification is always permissible.

We have already mentioned that the differential equations which regulate the motion may be expressed in various forms. Among these is the form introduced by Hamilton; it has the advantage of conferring greater symmetry on the equations. In Hamilton's method, the state of a mechanical system of n degrees of freedom is determined at any instant by the values at this instant of the n coordinates q, which fix the configuration of the system, and by the n corresponding momenta p. We shall assume, as always, that measurements are referred to a Galilean frame. The coordinates q may be chosen in many different ways. Thus, if our system is represented by a single point-mass moving in space, we may select for its three coordinates q the three Cartesian coordinates x, y, z which define the position of the point-mass; but we may also utilize spherical coordinates or cylindrical ones. Inasmuch as the three momenta p associated with the point-mass are related to the three coordinates q, any change in our choice of the q's will automatically entail a corresponding change in the p's. These remarks may be extended to the case of more complicated mechanical systems. In dealing with a system of n degrees of freedom, we shall suppose that a definite choice of the n coordinates q has been made.

Now we have assumed that the mechanical system is conservative and that the field of force which may be acting upon it is also conservative. Under these conditions the system has, in addition to its kinetic energy, a well-defined potential energy at each instant. The sum of the two kinds of energy gives the total energy of the system; and from the general principles of mechanics we know that the total energy will remain constant during the actual motion. The kinetic energy of the system is determined at any instant by the instantaneous values of the n coordinates q and of the n momenta p. The potential energy is determined by the n coordinates q alone. Collecting these results, we see that the total energy of a conservative system of n degrees of freedom is expressed by an appropriate combination of the n coordinates q and of the n momenta p. This expression of the total energy is called the "Hamiltonian function"; it is denoted by $H(q, p)$. There is no difficulty in constructing the Hamiltonian function for a given dynamical system. The precise form of this function depends, however, on the particular mechanical system considered, on the field of force to which the system is submitted, and also on our choice of the n coordinates q. This latter point is of importance, for the difficulty of solving a dynamical problem depends on the form of the Hamiltonian function. An appropriate change in our selection of the coordinates q may therefore facilitate considerably the solution of a problem. In particular, as will be explained later, if we can choose n coordinates q which will cause the Hamiltonian function to depend solely on the n corresponding p's, the dynamical problem is solved immediately. Let us also observe that, for a given choice of coordinates, the form of the Hamiltonian function will depend only on the nature of the dynamical system and on the field of force. To this extent the form of the Hamiltonian function is characteristic of the problem.

Having explained the significance of the Hamiltonian function, we now pass to Hamilton's equations. For a conservative system of n degrees of freedom, Hamilton showed that the motion is controlled by a system of $2n$ differential equations of the simple kind. These are:

(5)

$$
\begin{cases}
q_1' = \dfrac{\partial H}{\partial p_1} \; ; \; p_1' = -\dfrac{\partial H}{\partial q_1} \\[2ex]
q_2' = \dfrac{\partial H}{\partial p_2} \; ; \; p_2' = -\dfrac{\partial H}{\partial q_2} \\[2ex]
\;\; \vdots \qquad\quad \vdots \qquad \vdots \qquad\quad \vdots \\[2ex]
q_n' = \dfrac{\partial H}{\partial p_n} \; ; \; p_n' = -\dfrac{\partial H}{\partial q_n}.
\end{cases}
$$

The right-hand members of these equations, are known functions, for they are the various partial derivatives of the Hamiltonian function, which is assumed known. Hamilton's equations cannot always be written in the form just given; complications arise in some cases.* But in all the problems we shall consider in this chapter, the equations of the motion will be expressed by Hamilton equations having the simple form (5).

Hamilton's equations are of paramount importance in dynamics; they dominate the whole subject. We shall also find them of incessant use in the quantum theory. In statics likewise, Hamilton's equations may arise, but with a different physical significance attached to the variables. Thus Marcolongo has shown that the equations of equilibrium of a thread may be thrown into the Hamiltonian form. In pure mathematics, notably in the theory of contact transformations, equations of the same form as Hamilton's are found; and this fact was utilized by Hamilton and Jacobi in the methods they devised for the integration of Hamilton's equations in dynamics. Incidentally, it was the connection between con-

---

* The form given in the text for Hamilton's equations ceases to be valid when the field of force, though permanent, has no potential. Complications also arise if the constraints that may be imposed on the system are of a type which Hertz has called "non-holonomous." A situation of this sort is illustrated when a hoop is rolling on the ground: the constraints imposed by the earth's surface on the motion of the hoop are non-holonomous. We may also mention that Hamilton's equations of the simpler type (5) are valid in certain cases even when the field of force and the constraints imposed on the system vary with time.

tact transformations and Hamilton's equations that led Hamilton to his conception of wave mechanics.* This same wave mechanics was rediscovered by de Broglie by different methods and adapted to the requirements of the quantum theory.

The Integration of Hamilton's Equations—The Hamilton equations for a particular mechanical system constitute the differential equations which the motion of this system must satisfy. But only in a very implicit form do differential equations contain the information we are seeking; and, unless we can integrate these equations and obtain their solutions, the nature of the system's motion remains hidden. For this reason, the main problem in dynamics is to integrate the Hamilton equations that arise in connection with one problem or another. Let us first make clear what is meant by integrating Hamilton's equations. The equations (5), which refer to a system of n degrees of freedom, constitute $2n$ first-order differential equations; they express relationships that must be satisfied by the $2n$ functions of time, the q's and the p's. To integrate these equations means to determine the $2n$ functions of time $q_1(t), q_2(t)$ . . . $q_n(t), p_1(t), p_2(t)$ . . . $p_n(t)$ that satisfy the equations. Since there are $2n$ differential equations of the first order, the $2n$ solution functions of type $q(t)$ and $p(t)$ will contain among them, in addition to the variable t, $2n$ arbitrary constants. These constants may conveniently be taken to define the initial values of the q's and p's and hence the initial state of the system. The $2n$ solution functions $q(t)$ and $p(t)$ of Hamilton's equations will then define the state of the system at any arbitrary instant in terms of the initial state. The motion of the system is thus completely determined. In the majority of cases direct integration is impossible, and methods of approximation are resorted to, *e.g.*, in the Three-Body Problem.

The Method of Jacobi and Hamilton—Hamilton proved that if his equations could be integrated, a function sometimes known as the "Action Function" of the dynamical system, could be obtained. Jacobi reversed the proof and showed that if the action function could be obtained, Hamilton's equations could be integrated. Now this action function is a special kind of solution of a certain partial differential equation, called the "Hamilton-Jacobi equation." Consequently, Hamilton's equations can be integrated if we can solve this partial differential equation. Our original problem is thus transformed into another one. In

---

* See Chapter **XX**.

some cases the new problem is the easier, and when this is so, the method of Jacobi and Hamilton is of considerable advantage.

A second method due to Jacobi was suggested by the fact that, if in Hamilton's equations the Hamiltonian function $H(q,p)$ contains only the p's, the integration is immediate. This situation will always arise if we choose the coordinates q suitably. But the difficulty is to discover a means of securing appropriate q's. Jacobi's method for obtaining a correct choice of the q's will be explained more fully in Chapter XX. For the present, we shall merely state Jacobi's result. He proved that an appropriate choice of the q's for any given problem would be secured if we could integrate the partial differential equation mentioned in the first method. Jacobi's two methods are thus seen to be practically equivalent since both entail the integration of the same partial differential equation.

First Integrals—We have assumed that the systems we are dealing with are conservative, and that they are isolated or else that they are acted upon by conservative fields. Under these conditions, the total mechanical energy of the system retains the same constant value when the system evolves. Now the total energy is expressed by an algebraic combination of the q's and p's, namely, by the Hamiltonian function $H(q,p)$. The q's and usually the p's both vary in value as the system evolves; yet in spite of these variations, the algebraic combination $H(q,p)$, which defines the total energy, must necessarily remain constant in value. We may interpret this constancy of $H(q,p)$ by assuming that the variations of the q's and p's in the expression of the Hamiltonian function compensate one another. Expressions which are built up from the q's and p's of a dynamical system and which, like the Hamiltonian function, retain their initial values when the system evolves are called *first integrals of the motion*. The Hamiltonian function itself is one of the first integrals.

It can be shown that, for a system of n degrees of freedom, there exist $2n$ first integrals which are independent * of one another. One at least of these first integrals contains the time t explicitly. If $2n$ independent first integrals can be found for a system of n degrees of freedom, Hamilton's equations are integrated immediately, and the motion of the system is thereby disclosed. For this reason, mathematicians busied themselves with devising methods for obtaining first integrals.

---

* Two first integrals are said to be independent if the constancy of the one is not a necessary consequence of the constancy of the other.

Let us examine the physical significance of some of the first integrals. Consider a double-star system, in which the two stars are viewed as two point-masses. We have here an isolated and conservative system having six degrees of freedom (three for each point-mass), and hence there are twelve independent first integrals. One of these, as we know, is the Hamiltonian function and it represents the constant total mechanical energy of the system; it is called the *energy integral*. But we also know from the fundamental laws of mechanics that the total momentum of an isolated system is conserved. The total momentum of our binary system is therefore constant both in magnitude and in direction. The conservation of both magnitude and direction implies that the three components of the total momentum, along the three axes Ox, Oy, Oz of a Galilean frame, are individually conserved. Each of these components is given by a combination of the q's and p's (or of the p's alone); and since the values of these components remain constant during the motion, we obtain three additional first integrals. The conservation of the total momentum also yields three other first integrals, which we need not discuss. Thus far, then, seven first integrals have been found.

Three others can be obtained from the principle of angular momentum. This principle states that the total angular momentum of an isolated system remains constant (both in magnitude and direction) during the motion. Consequently, the three components of the total angular momentum are themselves constant and hence furnish three first integrals. Thus ten first integrals can be obtained.

A point of historical interest is that Kepler's first law (according to which the rate of increase of the area swept by the line joining the sun to a planet is constant) merely expresses the constancy of the total angular momentum of the system sun-planet, and hence the existence of the three first integrals just mentioned. Thus, the constancy which Kepler detected in the behavior of a planet, and which he viewed as a marvelous manifestation of harmony in Nature, reveals itself today as a mere consequence of the laws of dynamics.*

Let us revert to the problem of the Two Bodies. We have said that this problem will be solved if twelve independent first integrals can be

---

* In point of fact Kepler's law is incorrect, because it is not the angular momentum of the system sun-planet which remains constant during the motion, but the angular momentum of the entire solar system. The angular-momentum vector of the entire system is perpendicular to the so-called invariable plane of Laplace. Fortunately for Newton, who utilized Kepler's laws, Kepler's error has negligible consequences because the interactions of the planets are extremely weak when contrasted with the attraction of the sun.

found. One of these, however (the one that contains time explicitly), may always be discarded when we are dealing with a conservative system —and our present system is conservative. Thus eleven first integrals must be found. Ten of these have already been obtained, so that only one integral is still missing. However, Jacobi's method of the "last multiplier," which we shall discuss in the next paragraph, allows us to dispense with the last of the integrals. Hence the problem of the Two Bodies can be solved by the method of first integrals.

Jacobi's Method of the Last Multiplier—We have said that, in a mechanical system of n degrees of freedom, there are $2n$ independent first integrals of Hamilton's equations, one of which can be dispensed with in the case of conservative systems. This leaves $2n - 1$ independent first integrals to be found. Thanks to a method discovered by Jacobi, we may dispense with one of these $2n - 1$ first integrals provided the $2n - 2$ remaining ones have been secured. Thus let us suppose that $2n - 2$ integrals have been found, so that only one more integral is required. Jacobi showed that the problem of obtaining the last integral could be replaced by that of obtaining a function which he called a *multiplier* of Hamilton's equations. A multiplier is defined by any solution of a certain partial differential equation (the equation of the multiplier), so that Jacobi's method consists in discovering a solution of this equation. Now in the general case, a partial differential equation is not easily solved. We might therefore suppose that the discovery of a multiplier would be as difficult as the discovery of the missing first integral. However, owing to the symmetric form of Hamilton's equations, a multiplier for these equations is available immediately; for any constant, *e.g.*, the number 1, is a multiplier. As a result Jacobi's method may be applied whenever $2n - 2$ of the first integrals are known.*

The method of the last multiplier is a generalization of the method of the integrating factor, introduced a century earlier by Euler. The multiplier itself is of considerable importance in pure mathematics and in many physical problems. In the kinetic theory of gases the multiplier is seen to measure the density of a probability. We shall revert to this point in Chapter XXII.

---

*The name "Last Multiplier" is given to Jacobi's theorem because the actual solving of the problem requires that we discover a function called a last multiplier. The latter is easily obtained, however, from the multiplier itself and from the $2n - 2$ known first integrals.

Poisson's Method—For the purpose of obtaining first integrals, Poisson devised a method which yields a new first integral when two are already known. Associating then the new integral with one of the former two, we obtain a fourth integral, and so on. We might suppose that, by proceeding in this way, all the first integrals would eventually be secured. Unfortunately Poisson's method is not always successful. In some cases, after a certain number of integrals has been secured, we obtain a mere constant, after which no additional integrals are forthcoming. In other cases, though our succession of integrals does not come to a stop, we still do not obtain all the different integrals of the problem; because after a certain number of different integrals has been obtained, these same integrals recur over and over again and no new ones appear.

These various methods of Jacobi and of Poisson exhaust the list of the general devices for securing the first integrals of Hamilton's equations of dynamics.

Special Problems in Dynamics—Among the famous problems of dynamics are those concerning the motion of a rigid body of arbitrary shape, one of whose points is fixed; the body is supposed to be moving under no forces, or else under the force of gravity. We shall see that, if no force is acting, the problem can be solved. On the other hand, if the body is submitted to the force of gravity, the problem is insoluble in the general case,* though a solution can be obtained in certain particular cases. Let us make clear, however, that when we speak of a dynamical problem as being insoluble, we usually mean that it cannot be solved by *quadratures*, *i.e.*, by means of the known elementary functions or the indefinite integrals of such functions. This does not imply that the problem cannot be solved by means of successive approximations.

Let us revert to our problem of the rigid body. The motion is controlled by Hamilton's equations, and the problem consists in obtaining the first integrals. In the present case, since one of the points of the body is fixed, the position of the body may be defined by means of three angles. The system has therefore three degrees of freedom, and so $2n - 1$ (with $n = 3$), *i.e.*, five first integrals, must be found.

---

* By the general case we mean not only that the shape of the solid and the position of the fixed point are arbitrary, but also that the initial motion we confer on the body is arbitrary; the body may receive a twisting motion as well as any other. For instance, if we merely release the body without velocity from any given position, the solution is simple: the body oscillates about the fixed point as does a pendulum. But this is not the general case, for the initial conditions are very special.

Now it so happens that one of the three q's, which represent the three angles (defining the position of the solid), does not enter into the expression of the total energy and hence into the Hamiltonian function. A mere inspection of Hamilton's equations then shows that the p which corresponds to the missing q must be constant during the motion. Hence this p furnishes one of the five first integrals. Only four others need be found. It can also be shown that one more first integral may be dispensed with. The net result is that we must find three first integrals. One of these is the integral of energy (for the total energy of the solid remains constant). Two additional first integrals must be found. But if one of these is obtained, then thanks to Jacobi's method of the last multiplier the other may be dispensed with. Consequently, all we require is one first integral which is independent of those already mentioned. If this integral can be found, the problem is solved. As we have just said, the problem in the general case is insoluble: in other words the missing first integral cannot be obtained. This impossibility does not appear to be due to accidental circumstances; deeper reasons are involved.

Mathematicians, however, have obtained the missing first integral in three particular cases. In the first case, the solid is assumed to be of arbitrary shape; the fixed point is arbitrary; but no force is acting. Alternatively, we may suppose that the solid has weight but that the fixed point is at the centre of gravity. In this event, the force, acting as it does upon the fixed centre of gravity, cannot affect the motion; the situation is thus the same as if there were no force. Euler solved the problem; and a century later an elegant geometrical interpretation of the motion was furnished by Poinsot; whence the name "Poinsot motion" given to the complicated motion that occurs.* Incidentally, when a stone of aribtrary shape is thrown in the air and its centre of gravity describes a parabola, the turning and twisting of the stone about this centre is exactly the same as if the centre were fixed. Consequently, the twisting and turning motion of the stone about its centre of gravity

---

* Poinsot's representation of the motion requires that we explain what is meant by "an ellipsoid of inertia." A rigid body which is free to move about a fixed point has well-defined inertial properties relative to this point. The inertial properties may be illustrated geometrically by means of an appropriate ellipsoid; it is called the "ellipsoid of inertia" of the body with respect to the fixed point. Suppose, then, our rigid body is fixed at a point P. Poinsot considers the corresponding ellipsoid, whose centre is fixed at P, and shows that the rigid body will move as if it were attached to the ellipsoid, and the ellipsoid were to roll on a fixed plane situated at a well-defined distance from the fixed centre P. The angular velocity of the rolling motion at any instant can also be expressed geometrically.

illustrates a Poinsot motion. The mathematical representation of this type of motion involves elliptic functions.

A second case of integrability is illustrated by the symmetrical spinning top. The solid is symmetrical around an axis, and the fixed point is situated on this axis (*e.g.*, the point where the top touches the ground). The force of gravity is assumed to be acting. Owing to the restrictions of symmetry and to the special position assigned to the fixed point, this problem, like the former one, is but a very special case of the general problem. It was solved by Lagrange and by Poisson. It also involves elliptic functions. The motion of a spinning top is fairly well known from toys. We know that in spite of its weight the top does not topple over so long as it is spinning. We also know that the upper extremity of the axis of spin describes a circle round the vertical. In point of fact the motion is more complicated, for in the general case the tip of the axis describes a wavy curve or one with loops. The vibrations of the top's axis corresponding to the irregular shape of this curve are called *nutations*. It is these nutations which, in view of their rapidity, give rise to the humming sound emitted by the top. The nutations cease when the top is spinning vertically; the humming sound ceases, and the top is said to have gone to sleep.

For more than a century after the discoveries of Euler and Lagrange, no further special cases of integrability were found: the missing first integral could not be discovered. But Mme. Kowalewski * in the latter part of the last century integrated the equations in another special case. Poincaré mentions that still other cases of integrability were thought to have been established by her, but she died soon after, and the notes she left were insufficient to reconstruct her proofs.

All these motions we have been considering come under the heading of gyroscopic motions; they are often exceedingly baffling to common sense. Only because of our long association with spinning tops does the behavior of the top cease to astonish us. Yet the man who would see a top spinning for the first time would regard its motion as paradoxical. He would inquire: Why is it that the top does not fall; how can the mere spin generate a force pulling the extremity of the top upwards, thereby counteracting the force of gravity? Elementary, but less familiar, experiments with gyroscopes would convince us that their behavior is most unexpected and would show that our calm acceptance of the spinning top as a matter of no particular curiosity is due to force of habit. Gyro-

---

* Mme. Kowalewski, a pupil of Weierstrass, is the only woman to have attained a position of preëminent importance in mathematics.

scopic motions are of considerable importance in engineering. It is claimed that many aviation accidents have been caused by curious gyroscopic motions which the pilot was unable to anticipate or to remedy. In astronomy, gyroscopic motions are responsible for the precession of the equinoxes, a phenomenon first discovered by Hipparchus. In Bohr's atom they also play a considerable part.

The Problem of Three Bodies—The Problem of Three Bodies is the most famous of all the problems of dynamics. Whittaker estimates that, since the year 1750, over 800 memoirs, many of them bearing the names of the greatest mathematicians, have been published on the subject. The problem consists in determining the motions of three bodies (viewed as point-masses) attracting one another according to the law of gravitation. In the most general form of the problem, no restriction is placed on the masses or on the initial conditions represented by the initial positions and velocities. Special cases of the problem arise when restrictions are imposed on the initial conditions or on the masses, *e.g.*, when one of the masses is assumed to be infinitesimal.

The problem of Two Bodies, which we discussed previously, is extremely simple and was solved by Newton. But the situation is very different when three bodies are involved. Let us examine the problem of the Three Bodies from the standpoint of first integrals. There are three bodies and therefore nine degrees of freedom. Consequently, there will be eighteen $(2 \times 9 = 18)$ independent first integrals. Of these, one, as usual, may always be dispensed with, and therefore only seventeen first integrals need be found. Ten first integrals are given immediately by the principles of energy, of momentum, and of angular momentum. (We explained on page 237 how the corresponding integrals arise.) Hence seven additional independent first integrals must still be found. Could we but obtain six of these, we could dispense with the last, thanks to the device of Jacobi's last multiplier. The problem was solved by Lagrange in a special case, but in the general case the missing first integrals could not be found. The suspicion grew among mathematicians that the missing integrals could not be expressed by means of any known functions; and that, thought they existed in theory, they would be useless to the mathematician. At this stage Bruns proved that the missing first integrals could not be algebraic (as were the integrals already known). Bruns's theorem was followed in 1889 by a memoir of Poincaré in which it was proved that the missing first integrals could not even be uniform functions. Poincaré's paper put a stop to further attempts to solve the problem of Three Bodies by the classical methods,* and mathematicians realized that they would have to be content with methods of approximation.

The impossibility of solving the problem of Three Bodies by the classical methods refers, however, only to the general problem. Indeed we have mentioned that particular solutions were furnished by Lagrange. But before examining Lagrange's findings, we must explain what is meant by "periodic solutions," because Lagrange's solutions are of this kind.

Quite generally, regardless of the number of bodies involved, a solution is called periodic if at regular intervals of time the bodies resume the same positions in a Galilean frame attached to the centre of mass of the system. A less restricted definition is often given. Thus a solution is sometimes called periodic if the relative configuration of the system defined by the bodies repeats itself at regular intervals, while the orientation of the configuration in space may rotate uniformly and need not be fixed.

Metaphysicians, among others, have upheld the thesis that if the bodies or atoms of the universe are assumed to be finite in number, a periodic return to the initial configuration is a logical necessity. This belief is at the basis of the idea of recurring cycles. However, the most elementary mathematical knowledge proves that the metaphysicians' argument is utter nonsense. Inasmuch as questions of periodicity occur frequently in the quantum theory, we shall give a simple illustration showing that even in the simple case of two points, or atoms, a periodic return is only a slim possibility. To see this, consider two perpendicular segments OA and OB, and two points P and P' moving back and forth with uniform motion along the horizontal and the vertical segment, respectively. We shall suppose that the points leave O simultaneously, and that P executes a com-

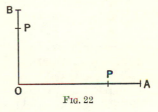

Fig. 22

plete vibration (from O to A and back to O) every second. Let n represent the number of vibrations performed every second by P' along the vertical segment. We now inquire whether the two oscillating points which leave O at the same initial instant will ever find themselves together again at O.

---

* Poincaré took as variables the elements of the orbits of the bodies. Levi Civita has shown that if different variables are used, other uniform integrals may exist.

If n is a rational number, the answer to our question is in the affirmative. For instance, let $n = \frac{7}{8}$. In this event, P' executes $\frac{7}{8}$'ths of a vibration every second and will therefore have performed exactly 7 vibrations at the end of 8 seconds. During this time, P has vibrated 8 times and hence will be back at O simultaneously with P'. From then on, at regular intervals of 8 seconds, the system of the two points will resume exactly the same configuration, and periodicity will hold. In short, the motion will be periodic whenever the number n is an integer or a rational number. But suppose n is irrational, e.g., $\sqrt{2}$ or π. In this case the points will never again be found at O simultaneously.* The probability that the motion of the two points will be periodic is thus the same as the probability that a positive number taken at random will be an integer or a rational number. This probability is zero, because the class of irrational numbers is infinitely more extensive than the class of rational numbers. We must note, however, that a zero probability does not mean an impossibility; for it is not impossible that a number chosen at random should happen to be rational. A zero probability must be interpreted therefore as implying an extremely small probability rather than an impossibility. From this discussion we conclude that the probability of our system of two points undergoing a periodic motion is extremely small. Since rigorous periodicity is such an exceptional occurrence, even when only two points are considered, we may well imagine how insignificant is the probability of a periodic motion when a large number of points or bodies is involved.

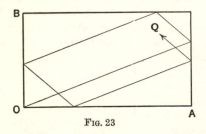

Fig. 23

A simple geometric representation of the motion of our two points P and P' may be given. Calling x and y the distances from O of the two points respectively, we have but to consider a point Q defined by x and y as Cartesian coordinates. If the motions of the points P and P' are uniform, the graph traced by Q will be a sequence of straight segments, all of which are contained in a rectangle, as shown in the figure. We thus obtain a broken line originating from O. If the motion is

---

* In the event of n being irrational, the two points cannot ever find themselves simultaneously at O after the initial instant; nevertheless, in the course of time and on an infinite number of different occasions, they will come as close as we choose to realizing this coincidence. The proof results from a theorem of the theory of numbers.

periodic, the line will eventually double back on itself and return to the point O. But if the motion is aperiodic, our broken line will never again pass through O; it will never close. The path of Q is the path that would be traced by a billiard ball shot from one of the corners of the table. (We of course assume no friction so that the billiard ball will go on rolling for ever.) Our previous analysis shows that only under very exceptional circumstances will the ball return to the corner point whence it started, and will the motion repeat itself.

We may now revert to the astronomical problems. In the problem of Two Bodies, periodic solutions arise when the two bodies describe ellipses one around the other. In the problem of Three Bodies, it is by no means certain *a priori* that periodic solutions can exist. Lagrange, however, on investigating this matter, established an important class of periodic solutions. He proved that if the three bodies are thrown simultaneously with appropriate velocities in appropriate directions, they will describe similar ellipses in the same time. Thus at regular intervals the three bodies will be back at their initial positions. More restricted cases uncovered by Lagrange are those of the so-called equilateral-triangle, and straight-line, solutions. In the former, if suitable initial motions are assumed, the bodies may be started from the three summits of an equilateral triangle, and they will then move as though attached to the triangle, which is itself rotating in its plane about the centre of mass. The straight-line solution may arise when the three bodies are thrown from positions on the same straight line. If the initial motions are appropriate, the bodies will move as though fixed to the straight line, which itself is pivoting in its plane about the centre of mass.

The solar system, in view of the large number of bodies it contains, does not furnish illustrations of the Three-Body problem. In some cases, however, it is permissible, as an approximation, to single out three of the bodies belonging to the solar system and view them as existing alone. This possibility occurs when, owing to their larger masses or smaller mutual distances, the three bodies of interest exert interactions that exceed considerably in importance the perturbing actions of the remaining masses. The system formed by the sun, the Earth, and the moon is an apt illustration. If the conditions conducive to Lagrange's straight-line solution were satisfied by these three masses, one-half of the Earth might always be illuminated by the sun and the other half by the full moon. Since this situation does not happen to occur in the solar system, Laplace remarked caustically:

"If the moon has been given to the Earth by Providence to illuminate the night, as some have maintained, the end sought has been only imperfectly achieved."

As for Lagrange's equilateral-triangle solution, no evidence of its realization was found in the solar system until recently. But in 1906 a certain asteroid, presently called Achilles, was discovered, and its motion taken in conjunction with that of the large planet Jupiter and of the sun was found to illustrate Lagrange's solution: the asteroid, Jupiter, and the sun move as though fixed to the corners of a revolving equilateral triangle. Subsequently, other asteroids of this family were found: they are called the Trojan asteroids.

Since Lagrange's discoveries, other periodic solutions have been investigated by Hill, Poincaré, and Darwin. But these all necessitate restrictions on the masses of the bodies. Thus Poincaré has shown that if two of the masses are small in comparison with the third, there will be an infinite number of periodic solutions. Periodic solutions illustrate highly exceptional cases which arise only for very special initial positions and motions. They would therefore seem to be more in the nature of curiosities. But their theoretical importance in the problem of Three Bodies was set forth clearly by Poincaré in his monumental memoirs. To quote Poincaré:

"Besides, what makes these periodic solutions so valuable is that they constitute, so to speak, the only opening through which we may penetrate into a domain hitherto regarded as inaccessible."

The fact is that when a periodic solution is known, many other solutions may be obtained by changing slightly the initial conditions. For instance, let us start with a known periodic solution; the three bodies describe closed orbits, and also revert simultaneously to their initial positions over recurrent equal intervals of time. Suppose now we vary slightly the initial conditions by imposing slightly different initial positions and velocities on the three bodies. It may happen that the new orbits branch away very rapidly from the periodic ones; in this event the periodic solution is called *unstable*. But it may also happen that the new orbits never depart very far from the periodic ones. The periodic solution is then called *stable*.

Poincaré established the existence of two novel classes of solutions, which he called asymptotic and doubly asymptotic respectively. But first let us understand the meaning of the word asymptotic. Consider two curves. If, when we follow one of the curves, we find that it tends to coincide with the other curve, we say that the two curves are asymptotic to each other. If this coinciding of the two curves is also realized when we follow the first curve in the opposite direction, we say that the two curves are doubly asymptotic. Imagine, then, two bodies describing two asymptotic orbits. If, as time passes, the motions of the two bodies tend to coincide, the motions along the two orbits are said to be asymptotic to each other. Similarly if the orbits are doubly asymptotic and the motions tend to coincide in the distant future and also in the remote past, the motions along the orbits are called doubly asymptotic.

Let us, then, consider a periodic solution of the problem of Three Bodies. Poincaré proved that solutions exist in which the motions are asymptotic and doubly asymptotic with respect to the periodic motions. These solutions constitute the asymptotic and doubly asymptotic solutions to which we referred in the previous paragraph. Asymptotic and doubly asymptotic solutions never arise in the problem of Two Bodies. However, when Einstein's more rigorous law of gravitation is taken in place of Newton's, these solutions occur, even in the case of two bodies.

For the study of stability conditions, Poincaré introduced mathematical expressions which he called *integral invariants*. We shall encounter these integral invariants in the kinetic theory of gases and in the quantum theory. Briefly, they are dynamical magnitudes which remain constant during the motion, and in this respect they are analogous to first integrals. They are distinct, however, from first integrals in that the constancy they express refers not to one single motion, but to a whole class of motions which differ only little in their initial conditions.

Developments in Series—The periodic solutions of the Three-Body problem define very special motions of the three bodies. But when we seek to obtain the most general motions, the problem becomes exceedingly difficult. Mathematicians therefore devised indirect methods of approach. Typical of these is the so-called "method of perturbations." A rough sketch of the method will suffice for our purpose. Suppose we wish to determine the motions of three bodies, say, the sun, Jupiter, and the Earth. For simplicity we assume that the attractions of the other planets may be ignored and that the sun is so massive that its motion may be disregarded. In the method of perturbations we first neglect to consider the mutual gravitational attraction of Jupiter and of the Earth. The motion is then easily determined: Jupiter and the Earth describe ellipses independently of each other around the sun. Using this simplified motion as a starting point, we next take into account the mutual gravitational attraction. The former simple motion is now disturbed, or perturbed, so that the actual motion may be regarded as due to a perturbation impressed upon the simplified motion. But instead of seeking to compute the effect of the perturbation at one performance, we proceed by successive approximations. The actual motion is therefore expressed by the limit of an infinite series, and as each successive term of the series is calculated, our results approximate more and more closely to the actual motion. The method of perturbations thus reveals itself as a method of successive approximations.

Now in Chapter XIV we mentioned that unless a series converges, it defines nothing. Hence the method of perturbations will be illusory if the series we obtain should happen to diverge. At first sight, we might suppose that this danger need not be feared. We might contend that in our previous example Jupiter and the Earth must move in some definite way and that, as a result, the mathematical series which represent the motion cannot diverge and be meaningless. This argument, however, carries no weight, for the divergence of a series is not necessarily due to the meaningless nature of the problem; it may be due to the unsuitability of the particular type of series selected. Many of the series investigated for the purpose of solving problems in celestial mechanics were found to diverge, but, owing to the remark just made, mathematicians did not give up hope; they merely sought to modify the form of the series so as to ensure convergence.

The history of these series of celestial mechanics extends from the days of Euler to the present day. In the latter part of the last century the so-called Lindstedt series were constructed. The problem of deciding whether a series converges is often extremely difficult; and in the case of the Lindstedt series no rigorous proof of convergence was forthcoming, though the consensus of opinion was that these series converged. Had this convergence been established, the problem of Three Bodies would have been solved formally by the method of perturbations. But Poincaré proved that the Lindstedt series do not converge. Poincaré showed, however, that a kind of semiconvergence holds. More precisely, if we break off the series at a certain point, the truncated series that remains gives a highly accurate description of the motion, provided we content ourselves with the not-too-distant future. The Lindstedt series are thus not entirely useless, but they cannot inform us of the motion over extremely long periods of time. The general theory of series which are similar to Lindstedt's was further elaborated by Poincaré; he called such series "asymptotic expansions."

Other methods of development in series can also be applied to the problem of Three Bodies. In any dynamical problem involving, say, the motions of point-masses, the coordinates of the masses may be expressed by the limits of Taylor series. The values of these limits change with time and thereby reveal the motion. But Taylor series do not always converge for all values of the variable (here time); singular points may be present * in the complex plane, and in such cases the series converge only within a limited range of time. To determine the motion of the system at instants beyond the range of convergence of the series, recourse must be had to analytic continuation. Repeated analytic continuation would thus furnish the motion of the system throughout time. Unfortunately, the method of analytic continuation is only of value in a theoretical discussion; it cannot be applied in practice. Hence the problem of Three Bodies, though it was solved in theory by this method, still remained unsolved in practice.

Poincaré circumvented this difficulty by transforming the Taylor series (which define the motion) into other power series that converged for all values of time. He was compelled, however, to impose certain restrictions on the motion; namely no collisions were to arise, and the bodies were to remain at finite distances. Shortly after Poincaré's death, Sundman in 1912 removed one of these restrictions. He showed that provided all three bodies do not enter into collision simultaneously, Taylor series which converge for all values of time may be constructed. If, then, we exclude the case of triple collision, the problem of Three Bodies may be said to be solved. It must be borne in mind, however, that in order to be informed of the exact positions of the three bodies throughout time, it is necessary to calculate the limits of the Taylor series; and this is an extremely tedious task. That the present solution of the Three-Body Problem is not as yet satisfying to mathematicians may be gathered from the following remarks of Whittaker:

"The series discussed in the preceding articles are all open to the objection that they give no evident indication of the nature of the motion of the system after the lapse of a great interval of time: they also throw no light on the number and character of the distinct types of motion which are possible in the problem: and the actual execution of the processes described is attended with great difficulties."

The Problem of n-Bodies—The problem of n-Bodies, where n is any integer greater than three, is, in effect, the problem of the solar system. All the difficulties we have met in the problem of Three Bodies are here aggravated. Mathematicians have confined themselves, in the main, to investigating the general stability of the solar system. The word stability is, however, used with different meanings. Obviously, if we could show that the solar system must resume exactly the same configuration at recurrent equal intervals of time, the motion would be periodic and therefore would endure forever: stability would be assured. But we know

---

* See Chapter XIV.

that the motion is not periodic, so that this situation need not be considered.

A more general definition of stability is due to Lagrange. The solar system would be stable in the Lagrangian sense if, in the course of time and on an infinite number of different occasions, it were to assume configurations differing by as little as we might choose from the initial one. Furthermore, between times, the planets would have to remain within restricted régions and not move to extremely great distances. Lagrange and Laplace showed that this kind of stability holds for the solar system. The significance of their results must be made clear, however. Lagrange and Laplace pursued their calculations only to a certain degree of approximation, neglecting terms which appeared to be of minor importance. But over a protracted period of time, the influence of these neglected terms may become highly important and vitiate our conclusions. Consequently, the foregoing calculations cannot give any definite assurance that the solar system is truly stable, and that it will never disperse, or that some of the planets will never fall into the sun.

Another conception of stability, still less restricted than that of Lagrange, was given by Poisson. It differs from Lagrange's in that the planets may now move to extremely great distances. Poisson, pursuing his calculations to a higher degree of approximation than did Lagrange, showed that the solar system is stable in this more extended sense. But inasmuch as Poisson, like Lagrange and Laplace, was compelled to resort to approximations, his results are subject to the same misgivings that we mentioned previously.

Poincaré, by introducing new methods based on the use of integral invariants, proved that in a large number of cases the Poisson form of stability is rigorously assured. Unfortunately, these cases do not cover the general problem of the solar system.

The Dynamics of Fluids—The type of dynamics we have discussed deals with particles and with rigid bodies. But the laws of dynamics may also be applied to continuous fluids. Euler and Lagrange established the dynamical equations for continuous fluids: they are called the equations of hydrodynamics. In the following paragraphs we shall mention briefly some of the more important results that have been obtained in connection with the configurations of equilibrium for rotating liquids.

An incompressible fluid, *e.g.*, a liquid, is supposed to be in rotation; the various parts of the liquid attract one another according to the law of gravitation. The problem is to determine the shape or shapes which the rotating mass may assume under the combined action of the centrifugal force and of the gravitational attractions. Because of the continuity of the liquid, the equations expressing the problem are not those of the simpler kind which occurred in the dynamics of point-masses and of rigid bodies, but are partial differential equations, more difficult to integrate. Their solutions determine the shapes, or configurations, which the rotating liquid may adopt, *i.e.*, the configurations of equilibrium.

If the liquid is not in rotation, there will be no centrifugal force, and the shape of the liquid will be spherical. But if there is any trace of rotation, the rigorously spherical form is impossible. Maclaurin, in 1742, obtained solutions of the problem of the rotating liquid. He established thereby the existence of a whole class of spheroidal configurations and showed that, according to the velocity of rotation of the fluid mass, one or the other of these configurations would be realized. The Maclaurin spheroids become progressively flattened as the speed of rotation of the liquid increases. An illustration of a Maclaurin spheroid is afforded by the shape of our planet. For a long time it was thought that all other possible configurations of the liquid mass would necessarily be surfaces of revolution about the axis of rotation. But one hundred years later Jacobi dispelled this belief. He obtained solutions which determined the existence of a whole class of ellipsoids with unequal axes; some of these have the form of a cigar. A third family of solutions was furnished by Poincaré: these solutions yield the so-called pear-shaped configurations. Some of the configurations of this family have the appearance of an elongated body with a waist in the middle.

The interest of Poincaré's investigations consists not so much in the disclosure of a new family of configurations as in the discovery of a general law which expresses the connections between the various families. But before examining Poincaré's results we must explain the difference between stable and unstable configurations. A configuration is said to be stable if a small disturbance causes the liquid to vibrate like a jelly and does not bring about a permanent disruption. The vibrations set up are gradually damped owing to the viscosity of the liquid, and the liquid resumes its original shape. A configuration is called unstable if the slightest disturbance destroys it permanently.

Suppose, then, we start with a non-rotating liquid mass. The shape of the liquid is necessarily spherical. We now set the mass into rotation. A Maclaurin spheroid is formed, and it will become progressively flattened as the speed of rotation is increased. We shall say that the configuration of the fluid is proceeding along the Maclaurin series. Now the earlier spheroids of this series are stable; but when the flattening has increased to a certain point, the spheroidal configuration becomes unstable and, from then on, all subsequent spheroids are likewise unstable. We might therefore suppose that an increase in the speed of rotation would cause the liquid to adopt the successive unstable spheroidal configurations, with the result that the slightest disturbance would bring about a disruption of the liquid. Poincaré proved, however, that an entirely different sequence of events was to be expected. He showed that the last of the stable configurations of the Maclaurin series also constitutes one of the stable configurations of the Jacobi ellipsoids; and he further proved that the Jacobi ellipsoids immediately following this configuration are stable. The net result is that when the shape of the liquid reaches the last of the stable Maclaurin spheroids, it may proceed to further stable configurations by abandoning the Maclaurin series and following the Jacobi series. From this account, the last of the stable Maclaurin spheroids is seen to represent a configuration at which a bifurcation occurs from one series into another. It is therefore convenient to call a configuration of this sort a configuration of bifurcation.

We are now in a position to state the general law established by Poincaré. It may be expressed as follows: When we proceed along the stable members of a series and reach the last of these stable configurations, this last configuration is one of bifurcation, acting as a link between two different series of configurations.

Let us follow the evolution of the liquid mass after it has passed into the Jacobi series from the configuration of bifurcation. As the speed of rotation is increased the shape of the liquid mass proceeds along the Jacobi series of increasingly elongated ellipsoids. The earlier ellipsoids encountered are stable, but when the speed of rotation is sufficiently high and the ellipsoid is correspondingly elongated, instability sets in; and thenceforth all the following ellipsoids of the Jacobi family are unstable. Poincaré's law indicates that the last of the stable Jacobi ellipsoids must constitute a configuration of bifurcation. Poincaré showed that the bifurcation here connects the Jacobi series with the Poincaré series of pear-shaped figures. If the Poincaré figures were stable, we should expect the configuration of the liquid to proceed along the Poincaré series. However, the problem of deciding whether the Poincaré figures are stable is one of great difficulty. Poincaré himself, Liapounoff, Darwin, and Jeans have investigated the subject. The latest results have shown that these configurations are unstable. Presumably therefore the liquid, after passing through the second configuration of bifurcation, cannot be in a stable condition, so that the slightest disturbance should disrupt it. Possibly, the effect of a disturbance will be to split the liquid into two separate masses gravitating one around the other.

As may well be imagined, the foregoing theoretical investigations are of considerable interest in cosmogony. For example, the partitioning of a rotating liquid star into two gravitating bodies may have given rise to double-star systems. The formation of the moon at the expense of the Earth may likewise be accounted for in this way, the tidal actions subsequently causing the moon to recede from our planet.

The study of the configurations of rotating incompressible fluids was extended by mathematicians to gases, which of course are highly compressible. The figures of equilibrium differ from those assumed by liquids. Laplace initiated these investigations when he developed his nebular hypothesis. Since his time many important results have been established, and they too are of wide application in cosmogony.

Conclusions—The problems of analytical mechanics were studied for the purpose of elucidating various physical phenomena. But in the transition from the physical to the mathematical, many simplifications were introduced by the mathematicians. In the problem of n-Bodies we view the bodies as point-masses surrounded by radially symmetric gravitational fields. Yet this assumption does not correspond to reality, for we know that the planets are more or less irregularly shaped, and it can be shown that a body which is not a point-mass or a sphere (composed of homogeneous concentric layers) generates a field of force that is no longer radially symmetric. Thus a spheroid does not exert, at an outside point, a force inversely proportional to the square of the distance; nor is the force the same in all directions.

This fact was seized upon at one time by the opponents of the general theory of relativity. They noted that, if the sun is treated as a spheroid or if we assume the presence of matter around the sun, the attraction exerted on a planet will not be correctly given by a force varying inversely to the square of the distance from the planet to the centre of the sun. Under these conditions a planet should describe not an ellipse, but a rosette-like orbit entailing an advance of the perihelion. Inasmuch as this rosette motion is precisely the type of motion which is exhibited by the planet Mercury, it was argued that Newton's law would suffice to account for the motion of Mercury and that there was no necessity of following Einstein in revising the law of gravitation. We mention this example to show that when we treat the planets and the sun as point-masses, we are simplifying the physical problem actually given in Nature.

Similar considerations apply to the tidal actions which the various planets and satellites exert on one another. The tidal actions are not even considered in the problem of n-Bodies, and yet they affect results. For instance, the tidal action of the moon on the Earth is causing our satellite to recede and is thereby modifying the configuration of the entire solar system. We may also mention the pressure of light, the presence of meteorites and of clouds of cosmic dust throughout the solar system; none of these influences is taken into consideration. Finally, we have argued throughout as though the solar system were isolated, and yet we know that each planet is attracted by every star and nebula in the universe.

We are thus placed in a dilemma. If we introduce simplifications, our model is not true to Nature, and if we do not introduce them, even the simplest of problems (let alone the n-Body problem) will be insoluble in practice. Fortunately in the solar system and in many other physical systems, the various influences at play are of such unequal importance that we may safely disregard most of them and yet secure fairly accurate previsons (provided our simplified problems can be solved).

This discussion brings out an important point. Our knowledge of celestial mechanics issues from the discoveries of Kepler, Galileo, and Newton. Kepler was able to formulate mathematically simple laws which were utilized by Newton. But simple laws were found by Kepler only because the mutual gravitational actions of the planets are relatively small, and because the mass of the sun happens to be far superior to the masses of the planets. Hence we can afford to treat each planet separately as an individual point of infinitesimal mass moving around a fixed sun. As we may well imagine, the situation would have been much more complicated had there been two suns, such as exist in the double-star systems. The motion of a planet would then have been that of an infinitesimal body moving under the attraction of two finite masses; a particular case of the Three-Body problem. Under such conditions, Kepler could never have obtained simple laws for the planetary motions; and it is doubtful whether celestial mechanics, as a mathematical science, could ever have arisen.

We must conclude therefore that the rise of celestial mechanics has been due to the fortunate circumstances we have just mentioned. We must remember, however, that influences which may be negligible over relatively small intervals of time may contribute cumulative effects over long intervals. For this reason the problem of n-Bodies, even if it could be solved and even if Einstein's more rigorous law of gravitation were applied, would still be utterly useless in revealing the ultimate fate of the solar system. Nevertheless no one can maintain that the labors of the mathematicians in investigating the stability of the solar system have been vain. By devising new methods of approach, mathematicians have created new instruments and concepts, and these have proved of service in many other problems.

This rapid survey of analytical mechanics clearly demonstrates the artificial nature of the distinction which is often made between pure and applied mathematics. According to some writers, since the subject matter of analytical mechanics is physical, the mathematical doctrine should be regarded as of the applied variety. But in point of fact, the physical interpretation of the equations whose solutions yield answers to the mechanical problems plays no part in the discovery of these solutions. The solutions are obtained (when possible) by purely mathematical means, the physical significance of the equations being lost sight of completely. The fundamentally mathematical nature of the problems encountered in mechanics explains why those who have contributed most to the development of this science are to be found among the pure mathematicians.

CHAPTER XVIII

MINIMAL PRINCIPLES AND PRINCIPLES OF ACTION

HERO, in the days of Greek antiquity, stated that when a ray of light issuing from a point A reaches a point B after being reflected against a mirror, it invariably follows the shortest path. Hero's statement is not quite accurate, for the path is not necessarily the shortest path; it may also be the longest.

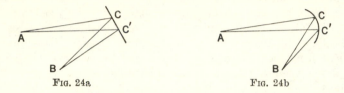

FIG. 24a   FIG. 24b

To illustrate the various situations that can arise, we shall first suppose that the light issues from a point A and is reflected against a plane mirror (Figure 24a). The path actually followed is the path ACB. Hero's statement is correct in this case, for the actual path ACB is indeed shorter than any other imagined path, such as $AC'B$. But suppose now that in place of the plane mirror we take a concave one (Figure 24b). In this case the actual path ACB may sometimes be longer than any neighboring path, such as $AC''B$. From these examples we infer that the characteristic of an actual path is that it is a *minimum or a maximum*; in other words, it is an *extremum*. This characteristic holds, however, only if we compare the actual path with infinitely near paths. For instance, let us suppose that a point source A emits rays of light which are reflected against one or another of different mirrors and then converge to a point B. In Figure 25 three plane mirrors, C, D, and E, are represented. The figure shows that ACB, ADB, and AEB will all three be actual paths for the rays. The mirrors may be placed in such

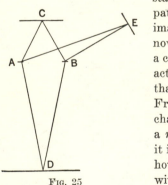

FIG. 25

a way that the paths are unequal in length; one of the paths is then neither the longest nor the shortest (the path *AEB* in the figure). But even so, each one of these actual paths will be shorter than any one of the infinitely near paths. We express this result by saying that all three of the actual paths are *relative*, or *local*, minima. The same general conclusions would hold if we were to take concave mirrors; but the actual paths might now be relative maxima instead of relative minima.*

The various particular situations examined enable us to refine Hero's original statement. We must say: The paths of the rays of light are "relative extrema" or, if we prefer, are "stationary paths."

We have dwelt at some length on these refinements because Hero's law is typical of a whole class of laws, or principles, that were subsequently discovered in the most diverse domains of physical science. The common characteristic of these principles is to assert that a physical system evolves from an initial to a final state in such a way that an appropriately selected magnitude (depending on the nature of the system) will exhibit the property of an extremum. The earlier investigators, however, failed to realize that this magnitude could be a maximum as well as a minimum; and to this error was due the names "Least Time" and "Least Action" given by Fermat and Maupertuis to principles of the foregoing type which they discovered in optics and mechanics. Today, all such principles are referred to as "minimal principles" or "extremal principles" or "stationary principles."

Usually, a natural law which is expressed in the form of a minimal principle can also be expressed in other forms; and in many cases these other forms are more convenient. As an example in point we have but to mention Descartes' presentation of the law of reflection. Whereas Hero established this law in the form of a minimal principle, Descartes stated it in another, though equivalent way. According to Descartes, when a ray of light is reflected against a mirror, the angle of reflection is equal to the angle of incidence. Similarly the law of refraction, which, as we shall see presently, may be given the form of a minimal principle, was stated by Descartes in a non-minimal form. Descartes' law of refraction is expressed by the formula

$$\mu_1 \sin i = \mu_2 \sin r$$

---

* We might also consider a vertical mirror, the cross section of which would have the shape of an *S*. In certain cases the actual path of the ray would be neither a relative maximum nor a relative minimum; but it would still have a property which mathematicians call "stationary." We shall, however, ignore these complications.

where i and r are the angles of incidence and of refraction of a ray of light which passes from a medium of refractive index μ_1, to one of index μ_2. Obviously when the refractive indices of two media are known, Descartes' law enables us to predict the exact degree of bending of a ray of light which passes under given incidence from one medium into the other. The refractive index of a vacuum is taken as standard and is credited with the value 1; the other transparent media have indices greater than 1.

We have mentioned Descartes' law of refraction because its discovery enabled the physicists of the seventeenth century to state the laws of reflection and of refraction in the form of a single minimal principle, which coincides with Hero's principle in the particular case of reflection.

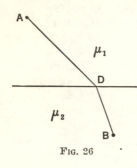

This minimal principle utilizes the notion of "optical path," and is sometimes called "the principal of the optical path."

The optical path of a ray of light, extending between two points A and B in a homogeneous medium of refractive index μ, is defined by μ times the spatial distance of the two points.

Suppose, then, that a ray of light passes from a point A, in a medium of refractive index μ_1, to a point B, in a medium of index μ_2. If $\mu_2 > \mu_1$, the path followed by the ray may be represented by the broken line ADB in the figure. The total optical path of the ray is then

Fig. 26

(1) $$\mu_1.\overline{AD} + \mu_2.\overline{DB}.$$

In the present example we have considered the optical path in connection with a single refraction, but the definition of the optical path may readily be extended to include any number of refractions and also reflections.

The law of the optical path, controlling reflections and refractions may be stated:

When a ray of light is emitted from a point A and after any number of reflections and refractions reaches a point B, its optical path will be a relative extremum—usually a minimum.

For example, in the phenomenon of refraction represented in the figure, the optical path (1) will be a minimum.

We may verify that in the case of the reflection of a ray of light against a mirror in a homogeneous medium, the law of the optical path coincides with Hero's law. Thus, if we call μ the constant refractive index of the medium, the optical path reduces to the spatial length of the path multiplied by the constant, μ; and obviously, when this optical path is a minimum (or an extremum) the same will be true for the spatial path, and vice versa.

The law of the optical path is an immediate consequence of Descartes' laws of reflection and refraction and is in no wise dependent on our adopting the corpuscular or the undulatory theory of light. But different forms may be given to the law according to the theory accepted. We first consider the undulatory theory. In the undulatory theory, the refractive index of a medium is defined by $\dfrac{c}{v}$, where c is the velocity of light *in vacuo* and v its velocity in the medium. (Incidentally this relation shows that the higher the refractive index, the smaller the velocity of light in the medium.) Let us call v_1 and v_2 the velocities of light in the media of refractive indices μ_1 and μ_2; the optical path (1) becomes

$$(2) \qquad c\left(\frac{\overline{AD}}{v_1} + \frac{\overline{DB}}{v_2}\right).$$

Now we know that the optical path (1), and hence (2), is an extremum. Since c is a mere constant, the extremal property will also hold for the expression enclosed in the bracket of (2). Let us examine the physical significance of the bracketed expression. The bracket contains the sum of two terms, each one of which is the ratio of a length to a velocity. Ratios of this sort measure intervals of time, and in the present case the ratios define respectively the times the light takes to pass from A to D in the first medium and from D to B in the second. Altogether then, the expression in the bracket of (2) defines the time required for a wave of light to advance from A to B over its path ADB. In short, if the undulatory theory of light is accepted, we have the following result:

A ray of light proceeding from a point A and reaching a point B, after any number of reflections and refractions, will always lie along the path for which the time of transit is a relative extremum (usually a minimum).

Fermat assumed that the extremum would always be a minimum, and so he called this principle *The Principle of Least Time*.

The law of the optical path is valid whether we accept the undulatory or the corpuscular theory of light; but Fermat's principle is valid only in the undulatory theory.   If we adopt the corpuscular theory, a minimal principle differing from Fermat's is obtained; it is called Maupertuis' Principle of Least Action.   Before examining Maupertuis' principle we must first define what is meant by the "Maupertuis" action.

Suppose a particle of mass m is moving with a variable velocity along some path, straight or curved, limited by the points A and B.   At each point P on this path the corpuscle has a velocity v and hence a momentum mv.   We suppose the path divided into tiny consecutive segments of length ds.   The Maupertuis action of the particle over the tiny stretch ds, about the point P, is then defined by the product $mvds$ of the particle's momentum at P and the tiny stretch ds.   The total Maupertuis action over the entire path AB is obtained by summing, or integrating, the elementary amounts of action which correspond to the successive tiny stretches.   Of course, if the velocity v of the particle happens to be constant during the motion, the Maupertuis action over the entire path is simply the constant momentum of the particle multiplied by the total length of the path.

Let us now revert to the corpuscular theory of light.   In this theory the refractive index μ of a homogeneous medium is measured by the ratio $\dfrac{v}{c}$ where v is the velocity of the corpuscle of light in the medium and c its velocity *in vacuo*.   From this expression we see that, in contradistinction to what occurs in the undulatory theory, the velocity of light is now greater in a refractive medium than it is *in vacuo*.   Suppose, then, we consider, as before, the path ADB of a ray of light which passes from a medium of refractive index μ_1 to one of index μ_2.   If we call v_1 and v_2 the velocities of the corpuscle of light in the two media, we have the relations

$$(4) \qquad \mu_1 = \frac{c}{v_1}, \quad \mu_2 = \frac{c}{v_2}.$$

Substituting these values for μ_1, and μ_2 in the expression (1) of the optical path, we obtain

$$(5) \qquad \frac{1}{c}\left(v_1 . \overline{AD} + v_2 . \overline{DB} \right).$$

According to the principle of the optical path, the expression (1), and hence (5), is an extremum.   The same extremum property will then hold for the expression (5) when it is multiplied by any constant, *e.g.*, by the constant mc, where m is the mass of a corpuscle of light.   But, as a result of this multiplication, the expression (5) becomes the Maupertuis action of the corpuscle of light along its path *ADB*. We conclude therefore that, in the corpuscular theory of light, the Maupertuis action of a light-corpuscle over its actual path is an extremum. This statement expresses Maupertuis' *Principle of Least Action*. In fine, according to whether the undulatory or the corpuscular hypothesis is adopted, the phenomenon of light-propagation is seen to be governed by the principle of Fermat or by that of Maupertuis.

Maupertuis originally discovered his principle through his investigations on Newton's corpuscular theory of light. The principle was soon extended by Euler and Lagrange to mechanics. In view of its importance we shall examine it in some detail. We first investigate what is meant by a "constrained path." Suppose a particle is moving in a field of force. In theoretical discussion it is often advantageous to suppose that the particle is constrained to describe some path differing from the one it would normally follow. For instance, let us assume that the particle is a smooth ball, and let us constrain the ball to move in a hollow tube the shape of which is selected arbitrarily. We shall suppose that the interior of the tube is frictionless. The ball will be guided by the tube and will thus be deflected from its normal path; but owing to the absence of friction no additional impediments will be placed on the normal motion. The forced path along which the ball is guided is called a *constrained path*.

The motion of a particle along a given constrained path in a given permanent field of force depends on the velocity with which the particle is initially thrown along the constrained path. If the field is conservative, the particle has a well-defined potential energy at each point. Consequently, if the velocity, and hence kinetic energy, with which the particle is thrown from a point *A* is stated, a definite total energy is associated with the particle; and this total energy will remain constant during the motion (whether actual or constrained). We conclude that the motion and the Maupertuis action of the particle along a given constrained path, between two prescribed points *A* and *B* in a given conservative field of force, are determined by the total energy of the particle.

In Maupertuis' principle we have to consider the actual path *AB* described by a particle in a conservative field of force, and also the constrained paths extending between the same terminal points *A* and *B* and differing slightly in their lay from the actual path. We compare the Maupertuis action along the actual path with the Maupertuis action for any one of the constrained paths which the particle might be made to describe with the same total energy as it has on the actual path. Maupertuis' principle for a free particle may then be stated thus:

> Along an actual path AB followed by a particle in a conservative field of force, the total Maupertuis action is a relative extremum. More precisely, it is an extremum when contrasted with the actions that would be developed by the particle if it moved with the *same total energy* along any one of the *neighboring* constrained paths extending between the same terminal points A and B.

The statement we have given of Maupertuis' principle is due to Lagrange. Maupertuis himself, who discovered a particular application of the principle in 1732, did not realize that the action could be other than a minimum, and this circumstance may account for the extravagant implications he attributed to his so-called "Principle of Least Action." In his "Essai de Cosmologie," he writes:

> "Here then is this principle, so wise, so worthy of the Supreme Being: Whenever any change takes place in Nature, the amount of action expended in this change is always the smallest possible."

Maupertuis' theological interpretation loses much of its plausibility when we recall that the action is only a relative extremum or, more precisely, is stationary. For we must now suppose that the Supreme Being is intent not on economizing, but more especially on going to extremes one way or the other; and even so, only in a relative way.

Nearly a century later Hamilton introduced another principle of "Least Action" or rather of "Stationary Action." It is more convenient than Maupertuis' and also more general, for it does not restrict us to conservative fields of force. We shall not present the principle in its full generality here. Consequently, we shall assume that the field of force is conservative, as a result of which the particle has a definite potential energy and a definite total energy.

To understand Hamilton's principle, we must first explain what is meant by the "Hamiltonian Action." Let us suppose that a particle is moving in a conservative field of force. At each instant the particle has a definite kinetic energy T and a definite potential energy V; hence a definite value is connected with the difference $T - V$ of the two energies at the instant considered. In an infinitesimal interval of time dt, the particle has moved over an infinitesimal element of its path. The Hamiltonian action during this time dt is then defined by the product $(T - V)\, dt$. We obtain the total Hamiltonian action expended during the motion of the particle from A to B along any path, whether actual or constrained, by integrating the elementary amounts of action. (Notice that the Maupertuis, and the Hamiltonian, actions are different mechanical concepts.)

Hamilton's principle states:

> If a free particle moving in a conservative field of force is thrown from a point A and reaches a point B, the path actually followed, when contrasted with all other neighboring constrained paths (extending between A and B) which would be covered *in the same interval of time*, will be such that, along it, the Hamiltonian action is an extremum or, if we prefer, is stationary.

We have already explained the meaning of a constrained path in connection with Maupertuis' principle. It has the same meaning here. We must also remember that the constrained paths to be considered are exclusively those which differ but little in lay from the actual path.

Another minimal principle in dynamics was formulated by Gauss. It is called the *Principle of Least Constraint*. According to this principle, a magnitude which Gauss calls "the constraint" is a minimum in any actual motion.

The application of the various minimal principles to dynamics leads to a solution of the dynamical problems; but this does not mean that the mathematical difficulties are any less considerable than they would be if more straightforward methods were applied. In the last analysis the minimal principles merely illustrate equivalent ways of expressing the same dynamical equations.

All the minimal, or extremum, principles have a characteristic in common. All of them state that, in certain classes of natural phenomena, the process of change is such that some appropriate physical magnitude will be an extremum (usually a minimum). The fact that the characteristic property is that of being an "extremum" rather than of being a "minimum," suffices in itself to deprive Maupertuis' theological opinion of much of its plausibility. But other reasons also show that Maupertuis' contention is untenable. First of all, the property which is to be an extremum varies from one treatment to another. Thus in dynamics, if we stipulate that the energy of the particle is the same along the constrained path and along the actual path, the magnitude which is an extremum along the actual path is the Maupertuis action. If, on the other hand, we stipulate that the times of transit are the same along the different paths, the Hamiltonian action is an extremum, whereas the Maupertuis action will usually no longer be so. No one definite magnitude has the extremum property in all cases; this property is therefore contingent.

Further insight is obtained when we examine the problem from the standpoint of the mathematician. Thus, when we say that Maupertuis' principle is valid in dynamics, we mean that the principle is a mere consequence of the equations of dynamics and is an equivalent expression of these equations, or laws. The presence of laws and principles in dynamics which are so different in appearance, and yet equivalent, merely exhibits the possibility of transforming mathematical expressions into forms that differ in appearance and yet are basically equivalent. Many examples of this sort of equivalence have been mentioned on various occasions. For instance, we have seen that functions may equivalently be expressed by different types of series. We must not be surprised, therefore, at different mathematical principles furnishing equivalent expressions of the same fundamental laws, or equations, of dynamics. These considerations prompt us to suspect that many other physical laws may be expressible in different equivalent forms, some of which may have the aspect of extremum principles. This surmise is correct. Thus the dynamical equations of the theory of relativity may be given the form of Maupertuis' and of Hamilton's principles; but of course the two actions in the relativity theory will not be quite the same as in classical dynamics. Einstein's gravitational equations, Maxwell's electromagnetic equations, and Schrödinger's wave equation may likewise be expressed by stationary, or extremum, principles.*

In short, practically all laws expressed by differential equations, whether these laws be actually realized in this universe or be mere fictions born of the mathematician's fancy, could be shown to entail the extremum property for some magnitude appropriately manufactured to suit the occasion. The existence of extremum principles in physics is thus no more characteristic of this universe of ours than of many other universes which we might imagine.

Nevertheless, although the possibility of expressing the same law in various ways cannot come as a surprise to the mathematician, the peculiar form of the stationary principles is somewhat disturbing because of the teleological scheme it seems to connote. Take Maupertuis' principle, for instance. According to it, a particle moving from A to B, with given total energy in a given field of force, will always follow the path along which the action from A to B is less (or greater) than it would be along all neighboring paths. But how can the particle know what precise path has this peculiar property until it has tested all the paths? And since the particle performs no tests, how does it happen to pick out the

---

* For the connection of Schrödinger's wave equation with an extremum principle, see page 707.

right path without hesitation? In this discussion we are arguing as though the particle had intelligence. We may, however, avoid this absurd assumption by saying: The particle does not know what it is doing, but it is guided by an influence which operates with a definite end in view. In other words, the future would appear to control the past.

To many thinkers this teleological aspect of the principles of action has appeared unsatisfying. But in point of fact the possibility of the future controlling the past is consistent with any thoroughgoing deterministic scheme. Thus, if we accept rigorous determinism, we must assume that an isolated physical system which is in a state A at one o'clock will necessarily be in the state B at two o'clock. Conversely if the system is in the state B at two o'clock, it must necessarily have been in the state A at one o'clock. Hence it is just as true to say that the future determines the past as to say that the past determines the future. This sort of teleology is an inevitable consequence of rigorous determinism. In our ordinary existence, however, the passage of time plays so important a part that when we speak of one event determining another, we usually mean that the first event antedates the second; and the words "cause" and "effect" are coined in consequence. But from the standpoint of theoretical discussion, the concept of "being determined by" is static and has no particular reference to any time flow. For these reasons we may speak of the present being determined by the future, though we should not say that the future is the cause and the present the effect.

The point we wish to stress is that the possibility of introducing teleological interpretations is inherent in any deterministic scheme. Let us clarify this point further in the particular case of dynamics.

We have seen that if we disregard the ordinary dynamical laws, as given by Newton, and base our arguments on either one of the principles of action, we may say:

1. The path and motion of a corpuscle in a conservative field are determined by the terminal points, by the energy (or the time of transit), and by the condition that the Maupertuis action (or the Hamiltonian action) should be an extremum.

In this presentation the teleological feature is obvious.

Next let us suppose that the ordinary laws of dynamics are utilized. In this event we may say:

2. The path and motion of a particle in a conservative field are determined by the initial point and the initial velocity.

In this presentation the teleological feature is obviated.

But even when we take into account the ordinary laws of dynamics, a teleological presentation may still be given. Thus we may say:

3. The path and motion of the corpuscle are determined by the terminal points and by the time of transit.

This last teleological presentation is the same as the presentation (1) except that we are now utilizing the ordinary laws of dynamics in place of Hamilton's principle. We conclude that even if the principles of action had never been discovered, the science of dynamics would still have been amenable to teleological interpretations.

The question of deciding which of the three presentations (1), (2), or (3) is to be preferred remains a matter of choice, for in the last analysis all three presentations are equivalent.

Extension of the Principles of Action to Systems—In our presentation we have assumed for simplicity that we were considering the motion of a particle. However, the presentation may be generalized so as to apply to mechanical systems. Consider a system of n point-masses moving under their mutual gravitational attractions. The system has $3n$ degrees of freedom, exhibited by the $3n$ Cartesian coordinates of the n masses. The coordinates may be written:

$$x_1, y_1, z_1, x_2, y_2, z_2, \ldots x_n, y_n, z_n.$$

Each point-mass is submitted to a force which is the resultant of all the forces exerted on the point-mass by the other point-masses. The three components of the force acting on the first point-mass are, say, X_1, Y_1, Z_1. Similarly, for the other $n-1$ point-masses. Thus, there will be $3n$ components of force altogether. The problem is to determine the motion of the system.

In dynamics it is often convenient to investigate a problem of this kind by introducing what is known as a "configuration space." We imagine a space of as many dimensions as our system has degrees of freedom ($3n$ in the present case). We then consider, in this $3n$-dimensional space, a point P which has for its $3n$ coordinates, at any instant, the $3n$ coordinates of the n point-masses. The $3n$-dimensional space is called a configuration space. As our point-masses move, their coordinates change and the point P moves in the configuration space; conversely, at any instant the position of P determines the positions of the n masses in ordinary space. This point P, which determines the configuration of the system, is called the configuration point. To clarify the situation, let us consider the problem of Three Bodies—three point-masses moving in ordinary space under their mutual gravitational attractions. If the motion of the system is periodic, the three point-masses will describe the same paths periodically, returning simultaneously to their initial positions; and the configuration point P, which here moves in a 9-dimensional configuration space, will describe periodically a closed curve. If, on the other hand, we take the general case, where the motion of the system is no longer periodic, the configuration point P will describe an open curve. It will never pass twice through the same point and be moving in the same direction both times.

The device of the configuration space enables us to apply the principles of action more simply. Thus let us suppose we are dealing with the problem of n-Bodies. The configuration point P in the $3n$-dimensional configuration space is defined as explained above. We now assume that this configuration point is itself a point-mass, say of unit mass. We imagine it to be acted upon by a force having, for its $3n$ components in the configuration space, the $3n$ components $X_1, Y_1, \ldots$ of the forces in ordinary space. It can then be shown that if a suitable metrics is imposed on the configuration space, the configuration point-mass P will move under the action of the force in a manner which will correspond completely to the motions of our n masses in ordinary space. To this point-mass P in the configuration space, we may apply the principle of action of Maupertuis or of Hamilton in exactly the same way as we did in ordinary space; the only difference is that our present configuration space has $3n$ dimensions. The motion of the point-mass P in the configuration space is thereby obtained, and from it we may derive the motions of the n particles in ordinary space.

In short, the essence of the method is to replace a more or less complicated dynamical system in ordinary space by a single point-mass moving in a hyperspace. The configuration space is utilized frequently in dynamics, and in wave mechanics it is appealed to constantly. In general, if we wish to preserve the correspondence between the motions in ordinary space and in the configuration space, we must assume that the latter space is non-Euclidean. The method of the configuration space in dynamics does not decrease the mathematical difficulties which beset us when we attempt to solve a problem; for, in the final analysis, the mathematics is the same whether we appeal to a point-mass moving in a configuration space or to a mechanical system evolving in ordinary space. The advantage of the configuration space is solely that it often permits of a more concise presentation of the motion.

The Mathematics Involved in the Principles of Action—The principles of action lead us to determine under what conditions a certain magnitude subjected to specified restrictions will be an extremum. Problems of this sort occur not only in physics but also in pure mathematics. Their study pertains to a branch of mathematics developed chiefly by Lagrange and by Weierstrass. It is called the "calculus of variations." Typical among these problems are the following:

What shape must an incompressible fluid assume if its total surface is to be a minimum? The answer is: a sphere (a result illustrated in the spherical form of raindrops).

What shape must a closed curve of prescribed length assume if the area it encloses is to be a maximum? The answer is: a circle.

What path must a point-mass moving under the action of gravity be constrained to follow in order that, on being released at rest from a point *A*, it should reach a given point *B* (at lower elevation) in the shortest possible time. The answer is: an arc of a cycloid.

The similarity between such problems and those raised by the principles of action is obvious.

EUCLID (about 300 B.C.)

ARCHIMEDES (287?-212 B.C.)

GALILEO (1564-1642)

SIR ISAAC NEWTON (1642-1727)

LEONARD EULER (1707-1783)

JOSEPH LOUIS LAGRANGE (1736-1813)

PIERRE SIMON LAPLACE (1749-1827)

AUGUSTIN CAUCHY (1789-1857)

WILLIAM ROWAN HAMILTON (1805-1865)

CARL G. J. JACOBI (1804-1851)

EVARISTE GALOIS (1811-1832)

CHARLES HERMITE (1822-1901)

MARIUS SOPHUS LIE (1842-1899)

HENRI POINCARE (1854-1912)

DAVID HILBERT (1862-1943)

CHAPTER XIX

THE UNDULATORY THEORY OF LIGHT

ORDINARILY, when we speak of a wave we are referring to a protuberance in motion. It is in this sense that we speak of a wave breaking on the beach. But in a theoretical discussion a wave of this sort would be called a wave crest, the name wave being given more particularly to the whole sequence of wave crests. So as to avoid confusion, we shall speak of "waves" or of a "train of waves" when we wish to stress that a succession of waves (in the ordinary sense) rather than a single wave crest is being considered.

The usual conception of waves is illustrated by the waves which travel over the surface of the ocean. But ocean waves furnish only a very special example, and other kinds of wave motions are encountered in physics. In acoustics we have sound waves; they are represented by the local compressions and rarefactions of the air or of elastic bodies. We may understand some of their peculiarities by supposing that a tuning fork is vibrating and emitting a musical note. If at a given instant we could determine the regions of the atmosphere that are compressed or rarefied, we should find that the regions of maximum compression and of maximum rarefaction alternate over the surfaces of concentric equidistant spheres having as common centre the position of the tuning fork. As time passes, these spherical surfaces expand at some uniform rate, remaining thereby at the same constant distances from one another. As a result, any point of space, situated in the region attained by the waves, will be the seat of alternating compressions and rarefactions.

In the present illustration we are dealing with so-called "spherical waves." This name is given because the mobile surfaces of compression and rarefaction (the wave fronts) are spherical. The velocity with which the wave fronts expand is called the *wave velocity*. In a sense these spherical sound waves are reminiscent of the circular water waves that are generated on the surface of a pond when a pebble is thrown into the water. But, aside from the fact that the sound waves occupy a volume of space, whereas the water waves are confined more or less to the surface of the water, other differences also distinguish the two kinds of waves. In particular, whereas the molecules of the air oscillate back and forth along the line of advance of the spherical waves, the water molecules rise and fall, or, more precisely, describe vertical circles.

Instead of spherical waves we may have "plane waves." In these the wave fronts are parallel planes advancing in the same direction with the same velocity. In other cases we may have waves whose wave fronts have irregular shapes and undergo deformations as they move.

The wave motions we have mentioned occur in matter, and they involve motions of the particles of matter. But we must not suppose that a wave motion entails a transport of matter. This point is made clear when we watch the waves which, under the action of the wind, may be advancing against the direction of flow of a river. Obviously in this example, the particles of water are following the direction of the stream, and not the opposite direction of the waves. The situation is further clarified when we observe the motion of a piece of cork floating on the sea. The cork merely rises and falls as the waves pass beneath it, and is not carried along by the advance of the waves. From these illustrations we see that a wave motion in matter does not entail a transfer of matter, but only a transfer of motion.

Wave motions are not restricted to matter; we may conceive of them in any continuous medium. Thus we may imagine color waves propagated through space. The regions of deep color correspond, let us suppose, to the regions of compression encountered in sound waves, whereas the regions of light color are the analogues of the regions of rarefaction. More generally, we may conceive of any property X being propagated in the form of waves. As an example, electromagnetic waves illustrate the wave propagation of the electric and magnetic intensities; and there is nothing material about them. Since, however, waves in matter are the most familiar, we shall examine the characteristic features of wave motions through material media.

Waves in matter are due to the vibrations of particles, and they may arise in any medium which is susceptible of undergoing some kind of deformation, and which, when deformed, tends to resume its initial state by the interplay of elastic forces. Such media are called "elastic." A discrimination of importance arises when we consider the direction in which the particles of matter vibrate during a wave propagation. In the main, we may differentiate between *longitudinal waves,* in which the particles of matter oscillate back and forth in the direction of the propagation, and *transverse waves,* in which the oscillations occur at right angles. For instance, longitudinal waves are generated in the air by a vibrating diapason (sound waves). Transverse waves are set up in an infinitely extended rope when we agitate one of its extremities with rhythmical motion; when the end of the rope is moved up and down, waves are propagated along the rope, and the vibrations are perpendicular to the direction of advance of the waves, *i.e.*, to the length of the rope.

The properties of the medium through which the waves are propagated determine the kind of waves that may be generated. A medium which, when compressed, tends to resist the compression and to expand again (*e.g.*, air) is a medium in which longitudinal waves may be produced. A medium which cannot be compressed because of its infinite rigidity, or a medium which opposes no resistance to compression, is always a medium in which longitudinal waves are impossible. Similarly, if the lateral distortion produced when we glide one layer of the medium over the other calls forth an elastic reaction, transverse waves are possible; and if the medium is infinitely rigid so that no such distortion is possible, or if the sliding of one layer over another produces no elastic counteracting force (a situation which occurs in fluids), transverse waves are impossible. In practice both kinds of waves may be generated in solid bodies, but the propagation velocities of the two kinds of waves usually differ. Thus an earthquake occurring at a point of the earth's surface may be felt as two distinct shocks at a distant point; the two kinds of waves have passed through the earth with different speeds.* According to these explanations only longitudinal waves should be possible in liquids, a conclusion which seems to be contradicted by the existence of water waves, whose oscillations are transverse. No contradiction is involved, however, because waves propagated over the surface of the water are not of the same type as waves transmitted in the interior substance of the liquid. In the case of the former, the molecules on a wave crest are pulled downwards by the force of gravity, and this force plays a part analogous to an elastic restoring force.

Standing Waves—A highly important form of wave motion is exhibited by the so-called *standing waves*. Such waves are generated when similar wave trains propagated along opposite directions are superposed. Thus, if a train of plane waves is advancing over the surface of the sea and is reflected by a wall, the incident and the reflected trains interfere, and we find that, along certain equidistant lines parallel to the wall, the agitation of the water is a maximum. Along intermediary lines (nodal lines) the water remains still. In the agitated regions the water merely rises and falls periodically; the waves do not advance, but stay where they are. To this peculiarity is due the name "standing waves." Standing waves may be of the longitudinal or of the transverse variety. In pipe organs and in wind instruments generally, the vibrations of the

---

* A third shock is also felt owing to the propagation of waves along the earth's exterior crust.

air are longitudinal, and longitudinal standing waves are produced. In elastic strings, the vibrations are transverse, and so if standing waves are generated they are necessarily of the transverse kind. For example, when a horizontal string whose extremities are fixed is set into vibration, the various points of the string rise and fall perpendicularly to the line of the string, but no wave motion appears to be propagated. We have here an illustration of transverse standing waves.

Polarization—Polarization is a condition which occurs in connection with transverse waves. In these waves the vibration is in a direction perpendicular to the direction of the propagation; and since an infinite number of such perpendicular directions exist, the precise direction of the vibration is arbitrary, and various situations may arise. No such complications can occur with longitudinal waves, for, with the latter, the direction of the vibration is the same as the direction of the propagation and is thus well determined. Now in the case of transverse waves, the vibration may very well change in direction from place to place at a given instant or from time to time at a given point. But when the direction of the vibration does not change and is everywhere the same, the wave is said to be *plane polarized.* Sometimes the vibration at a given point, instead of taking place back and forth along a straight line, executes a motion in a circle or in an ellipse situated in a plane perpendicular to the line of advance. In these cases we have circular or elliptic polarization.

Frequency, Wave Length, and Phase—Suppose a point P describes a circle with constant speed (Figure 27). We project the point P at each instant onto the vertical OB and thereby obtain a point Q which vibrates back and forth along the vertical diameter. The motion of the point Q is called a simple harmonic vibration; and the *phase*, or state, of the vibration at any instant is measured by the angle θ formed by OP and the horizontal OA. The phase thus increases from zero to 2π during a complete vibration. The *frequency* of the vibration is defined by the number of vibrations that are completed every second. The *amplitude* of the vibration is defined by the maximum elongation OB, *i.e.*, by the radius of the circle.

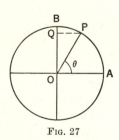

Fig. 27

Let us now suppose that a point Q is vibrating harmonically along a vertical and that this vibrational motion is propagated to the right. The particles to the right will progressively enter into vibration, and a wave will be formed. If the velocity V of the propagation is uniform, we shall obtain what is known as a sinusoidal wave motion. A snap-shot picture of the wave at any instant may then be illustrated by a sinusoidal curve drawn about a horizontal (see Figure 28). The elevation of the curve, above or below the horizontal, measures the elongation of the vibration at the point and at the instant considered. Points such as D are called the wave crests, whereas the points E correspond to the wave troughs. At the points D, the phase of the vibration is the same; similarly so at the points E. The phases at D and at E are opposite; they differ by π. More generally, the phase is the same at points, such as F, which are similarly situated on the successive folds of the curve. The distance between two consecutive points of this sort is called the *wave length*. Thus

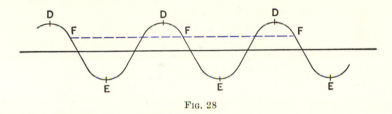

Fig. 28

the distance between any two successive points F, or between any two successive points D, measures the wave length.

Thus far we have restricted our attention to the appearance of the wave at a given instant; but we may also consider what occurs when time passes. In the course of time the wave motion proceeds to the right, and we may picture the motion by displacing the curve rigidly to the right with the uniform velocity V. This velocity is called the "wave velocity," or the "phase velocity."

An important relation connects the wave length λ, the frequency ν, and the velocity V. It is expressed by

$$(1) \qquad\qquad \lambda\nu = V.$$

This relation is easily justified. Thus we note that, at a fixed point P, ν wave crests pass every second and that each wave crest is separated from the next by the same distance λ. Consequently, the distance separating the wave crest at P from the wave crest which passes P one second later is $\lambda\nu$. On the other hand, this distance is also the distance over which the first wave crest has advanced during one second of time, and hence it also measures the velocity V of the waves. The relation (1) is thus demonstrated.

Interference phenomena occur when two or more wave trains are superposed. Thus let us assume that, superposed on the wave train described in Figure 28, there is a second wave train of the same frequency and amplitude. We shall suppose that the second wave train is advancing in the same direction with the same velocity, the wave crests of one train falling just above the wave troughs of the other. A mutual cancellation will then occur. The two similar wave trains are said to be of opposite phase. If the phases of the two similar trains are not exactly opposite, cancellation will not occur; and we shall have a resultant wave train. If the two similar wave trains are advancing in opposite directions, standing waves will be the result. In any particular case we have but to add the elevations at the various points along the two curves representing the two wave trains, and the resultant curve will define the resultant wave disturbance at the instant considered.

In the foregoing we have assumed that the wave motion at an instant was represented by a sinusoidal curve; this situation implies simple harmonic vibrations for the particles. More generally, the curve representing the wave may have a less regular form, but if the curve repeats at regular intervals, Fourier's theorem on Fourier series shows that it may be regarded as due to the superposition of a large number (or an infinite number) of sinusoidal curves. These sinusoidal curves will have appropriate amplitudes and frequencies and will be suitably disposed with respect to one another. In the theory of light, sinusoidal vibrations are identified with monochromatic radiations, so that a wave motion which is not strictly sinusoidal will correspond to the superposition of lights of different color. A similar situation holds for sound waves.

The waves with which we shall be concerned usually occur in space and not only along a line. We must therefore extend our former definitions to the case of waves in space. As an example, consider a train of plane sinusoidal waves propagated in a horizontal direction from left to right. So as to facilitate the discussion we shall suppose that we are dealing with sound waves propagated in the air. We mentioned some of the peculiarities of this propagation at the beginning of the chapter, and we said that the propagation consisted in a transfer of compressions and rarefactions of the air. Let us, then, examine the phenomenon more fully. If, at any instant, we could take a snapshot picture of the condition of the air, we should find that the compression was a maximum over certain equidistant planes set perpendicularly to the direction of the propagation, and the rarefaction a maximum over similarly disposed planes, each plane of maximum rarefaction being situated midway between two planes of maximum compression. Furthermore, between a plane of maximum compression C and the following plane of maximum rarefaction R, there would be other planes over which smaller compressions and rarefactions would occur; and, in particular, the plane midway between C and R would define a region in which the air was neither compressed nor rarefied at the instant considered. In the course of time all these planes would advance with the wave velocity in the direction of the propagation.

A graphical representation of our sinusoidal sound wave is afforded by the curve of Figure 28. Thus, if we consider planes set perpendicularly to the horizontal line, the planes passing through the wave crests will represent the planes of compression, those passing through the wave troughs the planes of rarefaction, and those passing through other points of the curve the intermediary planes to which we referred previously. The propagation of the sound wave is then represented when we displace the curve, as a unit, along the horizontal with the wave velocity.

When we discussed a sinusoidal propagation along a line, we mentioned that points similarly situated on the successive folds of the curve (e.g., points such as D or F in Figure 28) were points associated with the same phase. Similar conclusions may be extended to waves in space. For instance all planes of maximum compression define regions where the phase is the same; and so do all planes of maximum rarefaction. On the other hand the phases over these two kinds of planes differ by π.

Consider now any one of the planes we have just discussed, e.g., a plane of compression. This plane advances with the wave velocity, and while it moves, the phase over it remains unchanged. A plane of this sort is called a "wave front." Incidentally we note that the wave velocity is the velocity with which we must progress if we wish to remain in a region where the phase does not change. The wave velocity is thus what may be called the velocity of the phase. For this reason it is often named "the phase velocity."

Instead of considering a mobile wave front, let us restrict our attention to any one of the fixed planes with which the wave front coincides at some instant. We may suppose that at the initial instant this fixed plane represents a region of maximum compression; but since the region of compression moves on, our motionless plane will soon be the seat of a rarefaction, then of a compression again, and so on. The phase over our fixed plane thus varies rhythmically with time, though at any instant the phase does not differ from one point to another. Fixed surfaces of this sort are sometimes called "equiphase surfaces"; we shall refer to them by this name in the sequel. The foregoing analysis shows that the wave fronts coincide in turn with each one of the equiphase surfaces.

The frequency ν of the wave is expressed by the number of rhythmical variations (vibrations) of the phase that take place every second at any fixed point, or equivalently over an equiphase surface. The wave length λ is given by the distance between two consecutive wave fronts whose phases are the same. The relation (1) holds as before between the wave velocity V, the wave length, and the frequency.

In the particular example we have just discussed, the waves were plane and the equiphase surfaces and wave fronts were therefore plane also. But if we are dealing with spherical waves, e.g., sound waves emitted by a tuning fork, the equiphase surfaces and wave fronts will be concentric spheres, the former remaining motionless and the latter expanding with the wave velocity and coinciding in succession with the contiguous equiphase spheres. In the general case where the wave velocity differs from one point to another, the equiphase surfaces and wave fronts are distorted and no longer have simple geometrical shapes.

The conclusions we have been led to in the case of sound waves may be extended to wave motions generally. Of course, the peculiarities of the propagation will differ from one situation to another, and we shall not always be dealing with compressions and rarefactions. But the general significance of equiphase surfaces and wave fronts is the same in all cases. It will be noted that wave crests and wave troughs, which have a clear meaning when we are dealing with the waves of the ocean, cease to have any intuitive significance in connection with waves in space. Sometimes, however, it is advantageous to retain these names even when we are considering waves in space. The wave crests and troughs then refer to those wave fronts over which the disturbance has its greatest intensity. As for the distinction between crests and troughs it is expressed by a difference π in their respective phases.

The Wave Theory of Light—Huyghens, in the seventeenth century, proposed a wave theory of light. In support of this theory he pointed out that when two beams of light crossed each other, each beam proceeded as though the other were inexistent. He remarked that this was precisely what we should expect if light were a form of wave motion. The corpuscular theory of light, on the other hand, seemed to require that collisions should occur between the corpuscles of light and that lateral scattering of the light should ensue. But the phenomenon of double refraction, discovered by Huyghens, seemed to disprove his original theory of light. In double refraction a ray of monochromatic light penetrating into a crystal, such as iceland spar, is split into two different rays. The two rays are of the same color, and yet they have different properties, since they do not follow the same path and are not refracted in the same way when they emerge from the crystal. Huyghens, who assumed that his waves were longitudinal, was unable to account for any intrinsic difference between two rays of the same color. Newton, explained the differences between the two rays, by supposing that rays of light may exhibit a feature which can be expressed by the statement that they have different "sides." A situation of this sort, unaccountable with longitudinal waves, becomes more comprehensible if we assume a corpuscular theory of light; the corpuscles may then be viewed as exhibiting shape, and hence different "sides." Newton utilized this argument in support of his corpuscular theory of light.

One century later, Fresnel showed that if waves of light are transverse instead of longitudinal, double refraction is no argument against the wave theory. He pointed out that with transverse waves, different directions of vibration (i.e., different polarizations) could be realized for waves of the same color. The difference in the behavior of the two rays in the phenomenon of double refraction could then be attributed to a difference in the directions of their polarizations with respect to a direction characteristic of the crystal's structure. Experiments susceptible of detecting differences in polarization confirmed Fresnel's views, for the directions of vibration of the two rays, when they emerged from the crystal, were found to differ.

An important difficulty militating against any wave theory, whether of the longitudinal or of the transverse type, arose from the well-known fact that light seemed to travel along well-defined rays instead of scattering in all directions as sound waves were known to do. In other words, light cast clear-cut shadows, whereas sound waves turned corners. Such at least was the belief current in Huyghens's time. These difficulties were removed in the early nineteenth century by Young and especially by Fresnel. Fresnel showed that the wave theory was not incompatible with the propagation of light along rays; for, owing to the phenomenon of interference, wave crests might be superposed on wave troughs and extinction occur everywhere except along a certain path which would thereby constitute the ray. Kirchhoff subsequently refined Fresnel's mathematical proof and remedied some of its deficiencies. He did this by integrating the equation of wave-propagation, namely, d'Alembert's equation.

Fresnel also proved that the interferences which gave a ray-like aspect to the propagation of light would not occur if light were made to pass through a tiny hole whose dimensions were of the order of a wave length

of the light. In this event, as evidenced in Young's experiments, the beam
of light, after passing through the small hole, would spread out in all
directions like a wave. A further result of Fresnel's analysis was that
the interferences which cause light to proceed along rays should also be
prevented from occurring if a fine needle were placed in the path of a
beam of light; the width of the needle would, however, have to be of
the same order of magnitude as the wave length of the light. The light
would then suffer deflections on passing by the needle, and no shadow of
the needle would be formed. Fresnel verified this prediction.

The significance of Fresnel's conclusions as relating to wave motions
in general is that, according to conditions, waves may scatter or may
appear to be propagated along well-defined beams. But then, if we apply
these conclusions to sound waves, we must infer that they too, though
usually turning corners, may be propagated along beams. Let us examine the matter more closely. We can hear a conversation even though
a screen hides the speakers from our sight; obviously, the sound waves
pass around the obstacle and turn corners. According to Fresnel's theory,
this circumstance arises because the wave lengths of the sound waves are
great in comparison with the dimensions of the screen interposed. The
situation is thus much the same as occurred for light in Fresnel's experiment with the needle.

To secure the production of sound-beams, we should, if Fresnel's theory
is correct, make the sound waves pass through a slit whose width is far
greater than the wave length of the waves. This experiment is realized
when an explosion occurs in a mountain gorge leading to the open plain.
The width of the slit, represented by the distance between the mountain
walls which form the gorge, is considerable when contrasted with the
wave length of the sound; and the sound is projected from the gorge
along a straight beam. If we place ourselves in the path of the beam,
the sound is heard distinctly. If we are standing outside the beam,
practically no noise is heard : we are then in the sound-shadow.

Reverting to light, we are thus led to recognize two kinds of optics.
In the first, called "geometric optics," or "ray-optics," light is propagated along rays (straight or curved) ; it thus behaves much as beams
formed of corpuscles might be expected to do. In the second form of
optics, called "wave-optics," the wave nature of light becomes conspicuous. Of course the distinction we are here making is artificial. In all
truth, there is only one kind of optics, namely, wave-optics ;* as for

---

* In this chapter we are explaining the situation as it appeared to the physicists
of the nineteenth century, and we are not taking into consideration the more recent
discoveries in the quantum theory.

ray-optics, it is a mere ideal limiting case never rigorously realized in practice. The present situation is very similar to the one we have met with time and again. Under certain limiting conditions, a theory or a manner of interpreting things, though wrong in the last analysis, is nevertheless so nearly correct in its anticipations that we are often justified in accepting it, even when we recognize that it is a mere approximation. An example was mentioned in connection with classical mechanics: We have every reason to suppose that the theory of relativity is more nearly correct than classical mechanics; and yet, since under the limiting conditions of low velocities and weak gravitational fields, the predictions of the theory of relativity tend to become indistinguishable from those of the classical theory, we are justified in retaining classical mechanics in many practical applications. Likewise, when the energy values are high and the energy transitions are small, we may neglect the refinements of the quantum theory and base our deductions on the classical laws of mechanics and of radiation.

The relationship between wave-optics and the less rigorous ray-optics is of the same type. Whenever the irregularities or inhomogeneities of the medium in which light is transmitted are insignificant over extensions of the order of the wave length of the light, the conclusions derived from wave-optics tend to coincide with those obtained from the application of ray-optics. In such cases, ray-optics may advantageously be applied because of its greater simplicity. Obviously ray-optics is the kind of optics that we may adhere to on the commonplace level of experience; and to this circumstance is due the considerable progress registered by the science of optics even before the wave properties of light were recognized. The earlier telescopes and microscopes, for instance, were constructed on the assumption that ray-optics was valid. Specific illustrations of cases in which ray-optics may or may not be valid will now be given.

Suppose that monochromatic light is propagated through a medium which is inhomogeneous. We shall assume that the inhomogeneity is due to a variation in the refractive index from place to place. Let us then imagine the medium to be sectioned into equal tiny cubes. We shall suppose that the sides of these cubes are of the same order of magnitude as the wave length of the monochromatic light transmitted through the medium. If, within each one of these cubes, the refractive index varies imperceptibly from one point to another, the medium will be approximately homogeneous over extensions comparable with the dimensions of a wave length; and in this event ray-optics will be valid. In such a medium, monochromatic light is propagated along rays and does not scatter to the right and left, as it would if the foregoing conditions were not satisfied. The rays will of course be curved, owing to the continuous variation of the refractive index throughout the medium. A practical realization of an inhomogeneous medium, which is yet sufficiently homogeneous to secure the validity of ray-optics, is illustrated by the atmosphere. Its refractive index is not constant, yet this index varies sufficiently slowly from place to place to permit the passage of the sun's light along well-defined rays.

This illustration shows that the degree of homogeneity credited to a medium is not an absolute characteristic of the medium; it is relative to the wave length of the light considered. The shorter the wave length, the greater must be the homogeneity of the medium if the conditions of ray-optics are to be realized. Consequently, in a given medium, light of a certain wave length (*e.g.*, red light) may proceed along rays, whereas light of shorter wave length, such as blue light, may be scattered.

The inhomogeneities in the medium considered in the previous paragraphs were due to a variation in the refractive index from point to point. But inhomogeneities may also be caused by the interpositions of screens, or obstacles. Thus let a playing card be placed in the path of a wide beam of monochromatic light. The card will stop some of the light, while the remainder passes freely around the card. Commonplace experience shows that a sharp shadow will be formed. But more accurate observation shows that the contour of the shadow is not clearly defined, as it should be if ray-optics were valid. Instead, alternating dark and bright bands appear along the edge of the shadow. This aspect of the phenomenon proves that, strictly speaking, we are not under the limiting conditions of ray-optics—a fact which must be ascribed to the discontinuity, or lack of homogeneity, that the presence of the card introduces into the otherwise homogeneous medium. The discontinuity is determined by the edge of the card: on one side of the edge the light is blocked, whereas on the other it passes freely. Though abrupt, this discontinuity occurs only along a line, and for this reason attentive observation is required to detect the wave effects. But we may increase the importance of the heterogeneity, or discontinuity, by extending the region of space which it covers. This may be done if we interpose a screen in which a fine slit is made. The region of discontinuity now embraces the two edges of the slit; it has therefore a certain width and is no longer restricted to a single line. If the width of the slit is of the same approximate magnitude as a wave length of the light, the discontinuity of the medium becomes important over an extension of the order of a wave length; and, according to our general rule, the wave effects should become still more noticeable. As is proved by experiment, the light, on passing through the slit, spreads out in all directions like a wave. The experiment we have just suggested was precisely one of those performed by Young, and we now understand why ray-optics was found to be at fault in this case.

Similarly Fresnel's experiment, in which a fine needle is placed in the path of a beam of light, may easily be interpreted. The two sides of the needle create discontinuities in the medium, and if the width of the needle is of the order of a wave length of the incident light, the discontinuities become noticeable over an extension of this order of magnitude, and pronounced wave effects should arise. We no longer obtain a geometric shadow of the needle but, instead, a series of alternating bright and dark parallel bands. In particular, owing to an overlapping effect of the waves, a bright band may be found where the shadow would normally be expected. These illustrations give at least a general idea of the limiting conditions which separate more or less confusedly the domains of validity of wave-optics and of ray-optics.

All these phenomena in which light does not appear to proceed along well-defined rays are called "diffraction phenomena," and they illustrate the wave nature of light. Their importance was stressed by Young and by Fresnel, both of whom devised experiments to generate them. A point of historical interest is that Grimaldi, in the first half of the seventeenth century, had also observed these diffraction phenomena. In particular he had noted that a fine thread does not cast the shadow we should expect. But the importance of this fact was not appreciated at the time, and Grimaldi's experiment was not regarded as having any bearing on the possible wave nature of light.

Mathematical Rays —Let us consider an infinite number of parallel planes placed one in front of another. We shall suppose that these planes form a continuous sequence, so that, through any point arbitrarily chosen, one plane passes. In mathematical terminology our succession of planes is said to form a simply infinite family of parallel planes. Instead of plane surfaces, we may consider spherical ones, *e.g.*, the spherical surfaces which have the same centre O and all conceivable radii. In this case we shall have a simply infinite family of concentric spheres. The successive spherical surfaces do not have exactly the same shape, for, when we proceed from the centre O, the curvatures of the spheres decrease. As before, we shall suppose that if any point of space is selected, one of the spherical surfaces will pass through it. More generally, a simply infinite family of surfaces may contain surfaces of arbitrary shape;

but, as in our previous examples, through every point of space one and only one of the surfaces passes. In practice, however, we may suppose that the surfaces are limited laterally so that the volume of space occupied by them is itself limited.

In mathematics the name "orthogonal trajectories" is given to the curves which intersect at right angles all the surfaces of a given simply infinite family. Thus, in the case of the simply infinite family of parallel planes, the orthogonal trajectories are all straight lines perpendicular to the aggregate of the planes. In the case of the simply infinite family of concentric spheres, the orthogonal trajectories are all straight lines issuing from the common centre O of the spheres. In both examples the orthogonal trajectories are straight lines, but if the surfaces of the family have more complicated shapes, the orthogonal trajectories will usually be curves. Through each point of space (or of the region occupied by the surfaces), one and only one orthogonal trajectory always passes; * we recall that this feature is also characteristic of the surfaces of the infinite family.

With these mathematical preliminaries disposed of we may return to our waves of light. Suppose that a train of monochromatic waves of light is propagated through space. We shall first consider the case of a propagation through empty space. The equiphase surfaces (whatever their shapes) will define a simply infinite family of surfaces. The orthogonal trajectories of this simply infinite family of equiphase surfaces are called the *mathematical rays*. The justification for this name is easily understood. Let us suppose that a screen perforated by a small hole is placed in the path of the waves. A part of the wave motion will be stopped by the screen, while the remainder passes through the hole. Now if the conditions of ray-optics are satisfied, and hence if the dimensions of the hole are not too small, the waves which have passed through the hole will form a narrow beam (or a ray) of light. It can then be shown that this ray of light will coincide with the mathematical ray which passes through the hole. To this extent a mathematical ray and a physical ray of light coincide. But we must remember that whereas mathematical rays can always be defined, the optical rays come into existence and coincide with the mathematical rays only when the conditions of ray-optics are fulfilled.

Simple examples of these general conclusions are illustrated in the case of plane waves and of spherical waves. With spherical waves, for

---

* Exceptional points, such as the common centre of the concentric spheres, may lie on more than one of the orthogonal trajectories, but usually a single trajectory passes through a given point.

instance, the equiphase surfaces are concentric spheres of centre O, and the mathematical rays are all straight lines issuing from O. If a screen perforated by a small hole is placed in the path of the waves, a straight ray of light emanates from the hole, and this ray, when prolonged backwards, passes through the point-source of light situated at the centre O of the spherical waves. Thus, the ray of light and the mathematical ray coincide.

Propagation in a Refractive Medium—Waves of visible light are transmitted with less speed through transparent media than in vacuo. The higher the refractive index of the medium, the greater the retardation of the waves. If we call μ the refractive index of a medium at a point P, and V the velocity of the waves in the medium at this point, the relation between μ and V is given by

$$(2) \qquad\qquad V = \frac{c}{\mu},$$

where c is the velocity of light in vacuo. In practice the refractive index μ at a point P is not a well-defined magnitude; its value depends on the color, or frequency, of the waves transmitted. In glass it is greater for violet light than for red light. Consequently, waves of violet light will travel more slowly through glass than will waves of red light. When the refractive index of a medium varies with the frequency of the light, the medium is said to be *dispersive*. To refraction is due the bending of a ray of light when it passes from one medium into another of different refractive index. To dispersion is due the unequal bending of the variously colored radiations; it is this unequal bending that causes the decomposition of white light by a prism into its colored components. In view of the different actions of the same dispersive medium on the various radiations, we shall confine our attention to monochromatic radiations. By this means the complications due to dispersion may be disregarded; and for a given radiation a refractive index from point to point assumes determinate values characteristic of the medium at each point.[*]

We consider first a homogeneous refractive medium. The refractive index has here the same value throughout the medium (for a given radiation), and the mathematical rays will be straight lines as they are in vacuo. The wave length of the radiation, however, will not be the same as in vacuo: it will be shorter. The reason for the change in the wave length

---

[*] We are restricting our discussion to isotropic media. Anisotropic bodies, such as crystals, generate double refraction, and the refractive index at a point varies with the direction considered.

is explained when we note that the frequency ν of the radiation remains the same, whether the propagation occur in vacuo or in a refractive medium; and that the wave-velocity c in vacuo is greater than the wave velocity V in the medium. Consequently, we see, by equation (1), that the wave length λ in the medium

$$(3) \qquad\qquad \lambda = \frac{V}{\nu}$$

will be smaller than the wave length in vacuo $\left(\text{which is } \dfrac{c}{\nu}\right)$. Thus, the successive wave crests of a train of plane waves will be more closely packed in a refractive medium than in vacuo. Since the color of the radiation remains the same, we conclude that the frequency, and not the wave length, characterizes the color. The frequency, being independent of the medium in which the radiation is propagated, is thus in many respects a more fundamental characteristic than the wave length.

We now suppose that the refractive index μ, for a given radiation, varies continuously from place to place. As the medium is no longer homogeneous, the conditions of ray-optics are not necessarily fulfilled even when no obstacles, or screens with holes, are interposed. In order to secure the validity of ray-optics, we shall assume that the refractive index μ varies so little from place to place that its variations are insignificant over extensions of the order of a wave length of the transmitted light.

Let us, then, imagine that a train of plane monochromatic waves advancing in vacuo falls normally on the plane surface AB of our refractive medium. The surface AB, being perpendicular to the direction of advance of the plane waves, constitutes an equiphase surface. For simplicity, we shall assume that at the instant considered the equiphase surface AB is occupied by a wave crest. The wave crest now penetrates into the medium, and at each instant its position defines an equiphase surface within the medium. Owing to the variability of the medium's refractive index μ from place to place, we may suppose that this index varies over the area AB. In this event we see from (2) that the velocities of advance V of the different parts of the incident wave crest will not be the same. Hence the wave crest, which was originally situated over a plane AB, will be deformed as it advances, and the successive equiphase surfaces will be more or less distorted. In view of the permanency in the conditions of the propagation, the equiphase surfaces thus defined will remain equiphase surfaces throughout time. The mathematical rays in the medium are, as usual, the orthogonal trajectories of the equiphase surfaces, and, in view of the irregular shapes of these surfaces, the rays will usually be curves. Since we are assuming that the conditions of ray-optics hold, the rays of light in the medium will coincide with the curved mathematical rays—a fact that may be verified when we utilize the perforated screen, as was explained previously. The frequency of the vibrations, which determines the color of the light, remains everywhere the same; but the wave length, like the velocity, varies from place to place. The magnitude of the wave length at any point P is determined geometrically by the distance between two successive wave crests, one of which is passing through P; this distance, however, must be measured along the mathematical ray which passes through P. Since in a non-homogeneous medium the mathematical ray will usually be curved, the distance between the two wave crests is taken along a curve, and it will therefore be greater than the shortest distance from the point P (on one wave crest) to the next wave crest. A distinction of this kind does not arise when the mathematical rays are straight lines.

The foregoing considerations enable us to give a new definition of the limiting conditions under which wave-optics passes over into ray-optics. We mentioned previously that these conditions are realized when the heterogeneity of the medium is not too great; and that, in the particular case where the heterogeneity is due to variations in the refractive index from place to place, the limiting conditions are satisfied when the refractive index does not vary appreciably over extensions of the order of a wave length. Now, variations in the value of the refractive index μ from place to place entail curvature for the mathematical rays. Accordingly, we may anticipate that, when the refractive index varies appreciably within volumes having dimensions of the order of a wave length, the mathematical rays will be curved to an appreciable extent within such volumes. We may therefore assert that, whenever the mathematical rays manifest an appreciable curvature within regions whose dimensions are of the order of a wave length, the conditions of ray-optics cannot be satisfied. In this event we are certain that the light will not proceed along rays. Thus, wave-optics passes into ray-optics only when, within the volumes mentioned, the mathematical rays are sufficiently straight. The definition just given for the limiting conditions will be of use in wave mechanics.

Huyghens's Construction of the Equiphase Surfaces.—Huyghens devised a geometrical method for obtaining all the equiphase surfaces from a single one of them when the refractive index of the medium is known at each point. Huyghens's geometrical construction is valid for anisotropic as well as for isotropic media, but we shall examine it here only in connection with the latter media.

Consider an isotropic medium whose refractive index μ varies continuously from place to place. We assume that the refractive index is known. Suppose that a train of waves is being propagated through the medium. Let S represent the equiphase surface with which a wave front coincides at the instant considered. Huyghens supposed that each point P of the wave front acted as a centre of disturbance and emitted wavelets. Since the medium is isotropic, the velocity V with which a wavelet will expand from a point P will be the same in all directions. Consequently, after an infinitesimal period of time dt, the wavelet will be represented by a tiny sphere whose centre is P and whose radius is $V dt$. We may compute the velocity $V = \dfrac{c}{\mu}$ from our knowledge of the refractive index at P, and we may therefore construct the sphere. If we proceed in a similar way for all the points P of the wave front, we shall obtain an infinite number of spheres whose radii will vary continuously on account of the continuous variation of the refractive index.

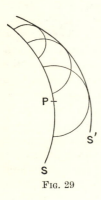

Fig. 29

According to Huyghens, the equiphase surface S', to which the wave front will move in the time dt, is defined by the envelope surface of all the tiny spheres. It would be as though the perturbations represented by the spheres manifested themselves only at their points of contact with the envelope surface. Although Huyghens was unable to account for this peculiar behavior of the wavelets, his views have since been proved correct, and so we may apply his geometrical construction to obtain the equiphase surface S'.

Having obtained the locus S' of the wave front at time dt, we may utilize this surface to determine a third equiphase surface, by following the method just outlined. Proceeding in this way, we may secure any number of successive equiphase surfaces situated as close together as we choose. The rays are then determined as usual by the trajectories which are orthogonal to the family of equiphase surfaces.

Huyghens's method of constructing the successive equiphase surfaces is intuitive rather than rigorous. But less than two centuries later, Fresnel showed that, as a result of interferences between the crests and troughs of the wavelets, the equiphase surfaces and the rays would be correctly given by Huyghens's construction (at least in those situations where ray-optics was valid). Even Fresnel's analysis failed, however, to remove all objections. Thus in Huyghens's construction, two distinct surfaces S' on either side of the initial surface S are tangent to all the spheres. Each of these two surfaces is an equiphase surface, and there is no apparent reason why a wave crest, initially at S, should move towards one of these surfaces rather than towards the other. Hence we might suppose that wave crests should start from S in both directions simultaneously. But this assumption is contradicted by observation, for it would imply that, in a beam of light, the waves travel in both directions, backwards and forwards. We are thus confronted with the difficulty of reconciling Huyghens's construction with the established fact that the waves advance only in one direction. Kirchhoff solved the problem when he integrated the fundamental mathematical equation for waves. He showed that interference effects cancel the waves in one direction and not in the other.

Huyghens's construction, however, can only be utilized when the conditions of ray-optics are satisfied. The fact is that when these conditions are not realized, the interference effects on which Huyghens's construction depends for its validity no longer occur, so that rays and equiphase surfaces become vague.

The Principles of Fermat and of Maupertuis—Fermat's Principle of Least Time, which we discussed in the last chapter, is valid only when we assume that light is a wave manifestation and that the limiting conditions of ray-optics are satisfied. The principle states that in a transparent medium the path followed by a ray of light, between a point A and a point B, will be such that the time of transit from A to B is an extremum. The reference to a ray of light in Fermat's principle shows that the principle can be applied only under the limiting conditions of ray-optics.

Now we saw in the last chapter that Fermat's principle is equivalently expressed by the Principle of the Optical Path. Let us recall this latter principle. If ds represents an element of length, and μ the refractive index of a medium from place to place, the magnitude μds is called the *optical path* over the distance ds. If now we integrate the variable quantity μds along a continuous line from A to B, we obtain

(7)
$$\int_A^B \mu ds,$$

which defines the optical path from A to B along the line in question. Fermat's principle of least time may then be expressed equivalently by the statement:

Along the ray of light extending between A and B, the optical

path $\displaystyle\int_A^B \mu ds$ will be an extremum.

When Fermat's principle is expressed in terms of the optical path, it presents a striking analogy with Maupertuis' principle in dynamics. This mechanical principle was discussed in the last chapter. We recall that Maupertuis' principle states:

A particle m, thrown from a point A with a total energy E in a conservative field of force and reaching a point B, will follow between A and B that path for which the Maupertuis action

$$(8) \qquad \int_A^B mvds$$

is an extremum (here v is the velocity of the particle).

Thus, the only difference between Maupertuis' principle and Fermat's is that the concept of action (8) takes the place of the optical path (7), and hence the momentum mv of the particle takes the place of the refractive index μ of the medium.

The resemblance between the two principles entails an interesting consequence which was first noticed by Hamilton. To understand it, let us suppose that a point-mass m is thrown time and again, with the same total energy E and from the same point O, in a conservative field of force. The direction in which the particle is thrown differs, however, at each throw. We thus obtain an infinite number of different paths traced by the particle, all of which start from the same point O and are described with the same energy E. These paths may be referred to as "paths of energy E." Let us select any one of these paths and let P be some point situated on it. It will always be possible to start the particle with the same energy E from some other point O' and in such a direction that the new path of energy E issuing from O' also passes through P. In short, by varying our choice of the initial point and by giving appropriate initial directions to the particle, we shall obtain an infinite number of different paths of energy E, issuing from different initial points O, O', O''; and all these paths will pass through the point P. . The directions along which the various paths reach the point P will of course be different, and as a result the momentum vector $\overrightarrow{mv}$ of the particle as it passes the point P will have one direction or another.

Now a simple mathematical argument shows that, though the direction of the particle's momentum as it reaches P depends on the path followed, the bare magnitude mv of this momentum will always be the same. In other words, regardless of the position of the initial point O, the particle, on following any one of the paths of energy E which pass through the point P, will reach the point P with the same value for its momentum.* The point P thus appears to be associated with a definite value of the momentum mv; and since P is any arbitrary point in the field of force, we conclude that a definite value of the momentum is associated with each point..

Next let us consider a refractive medium, the refractive index μ of which (for a radiation of given frequency) has the same numerical value at each point P as has the momentum mv in our previous illustration.† Thus, $\mu = mv$ at each point. We assume that the conditions of ray-optics are satisfied by the medium, so that a point-source of light situated in the medium will emit light under the form of rays. Fermat's principle tells us that a ray of light, passing through two given points O and P, will lie along the path for which the optical path (7) is an extremum. By utilizing this result, we may determine the precise path of the ray. Now, owing to the equality, $\mu = mv$, the path between O and P, for which the optical path (7) is an extremum, is also the path along which the Maupertuis action (8) is an extremum. Since O and P are any two points, we conclude that the paths of the rays of light in the refractive medium and the paths (of energy E) of the particle m in the field of force will be the same.

As an example, let us suppose that, in the gravitational field of the earth, stones of mass m are thrown in all directions from any arbitrary

---

* In classical mechanics the momentum mv of a particle, moving with total energy E in a conservative field of force, is equal at each point P of the particle's trajectory to

$$\sqrt{2m(E-\phi)},$$

where ϕ is the potential energy of the particle at the point P. Since in a conservative field of force the potential energy ϕ has a determinate value at each point (if we agree on its value at any one point), the magnitude of the particle's momentum at a point P depends solely on the position of the point P and on the total energy E of the motion. If, then, we consider only those particles which are moving with the same total energy E, the magnitude of the momentum at a point is seen to depend solely on the position of this point.

† It would be sufficient for our purpose to suppose that μ is proportional to the momentum.

point O but always with the same total energy E. According to our previous conclusions, we may imagine a refractive medium which surrounds the earth and in which the rays of monochromatic light issuing from this point O follow exactly the same parabolic paths as the stones. Quite generally, any path followed by a stone is a path which defines a possible ray of light, and vice-versa. Needless to say, the identity between the paths of the rays and those of the stones holds only for spatial position and does not extend to the respective velocities with which these paths are described.

The close resemblance between the principles of Fermat and of Maupertuis, and hence between waves and particles, was commented upon by Hamilton in the early part of the last century; it inspired him to construct his theory of wave mechanics. At that time, however, Hamilton's theory was regarded as a mere mathematical curiosity, resulting from a casual resemblance between the principles of Fermat and of Maupertuis. We shall have more to say on Hamilton's wave mechanics in the next chapter.

Wave Number—A magnitude often considered in optics is the *wave number*. Suppose a train of plane waves is propagated along the direction marked OA in the figure. The equiphase planes are perpendicular to this direction, and those of the equiphase planes that are associated with the same value of the phase, *e.g.*, the wave crests, are separated one from the next by the distance λ, which represents the wave length. The number of wave crests which at any instant are situated in an interval of one centimeter, measured along the ray OA, is called the "wave number." Denoting its magnitude by σ, we obviously have

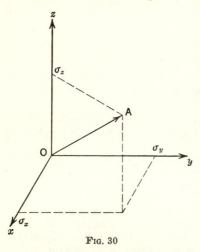

Fig. 30

$$\sigma = \frac{1}{\lambda}.$$

Suppose now we place an arrow of length σ along the direction OA. This arrow $\vec{\sigma}$ is called the *wave-number vector*. We project it on the

three coordinate axes Ox, Oy, Oz and obtain as projections three segments σ_x, σ_y, and σ_z, the lengths of which are found to represent, respectively, the numbers of wave crests intersecting the three axes over lengths
of one centimeter. If the numerical values of σ_x, σ_y, and σ_z are given,
we may deduce the value and the direction of the original vector $\vec{\sigma}$. The
wave length λ and the direction of propagation of the plane waves are
then obtained immediately. The introduction of wave numbers gives
no additional information, and we might have adhered to wave lengths.
In some cases, however, the wave number yields greater symmetry in
the formulae.

The Various Velocities—The velocity V with which we have been
concerned thus far, *i.e.*, the *wave velocity* or *phase velocity*, is the velocity
of the wave fronts of a train of sinusoidal waves. We have seen that it
is connected with the refractive index of the medium. Another velocity
which must sometimes be considered is the *propagation velocity*. It is
the velocity which appears in the mathematical equation of the waves,
namely, d'Alembert's wave equation. The propagation velocity represents the velocity with which a single pulse of light will be transmitted,
and in general it differs in value from the wave velocity. The two velocities coincide, however, whenever the conditions of ray-optics are satisfied.

A third kind of velocity is also of the utmost importance; it is the
group velocity, also called the *amplitude velocity*. Its significance will
be understood from the following illustration:—When a stone is dropped
into a pond of water, concentric waves spread out from the centre of
disturbance. The expanding waves are comprised between two concentric expanding circles. Beyond the outer circle and inside the inner
circle no motion occurs. Between the two limiting circles several intermediary wave crests may be seen, so that the disturbance as a whole is
spread at any instant over a limited region of the water's surface. The
name *group of waves* is given to the aggregate of wave crests which lie
between the two limiting circles. We must now examine more carefully
the changes that occur. The two limiting circles which bound the disturbed region do not expand with quite the same velocity. But we can
strike a mean value, and the mean velocity obtained may be said to represent the radial velocity of the disturbance from the point where the stone
was dropped. If we take a snapshot picture, we find that wave crests of
unequal height lie along circles situated between the inner and the outer
limiting circles; and we also find that the highest wave crest lies along
a circle half way between the limiting circles. When we proceed from
this highest wave crest towards either one of the limiting circles, the

successive wave crests are seen to decrease gradually in height. The individual wave crests expand faster than the limiting circles; as a result they appear to run outwards from the inner to the outer circle, increasing gradually in height and then dying down again when the outer circle is reached. We thus have two velocities to consider: the velocity of the disturbance as a whole, measured by the velocity of expansion of the limiting circles; and the velocity of the wave crests which rush from end to end of the disturbance. As we shall see, the latter velocity corresponds very approximately to what we have called the wave velocity. But the former velocity is entirely new; it is the *amplitude velocity*, or the *group velocity*, to which we referred at the top of the paragraph. The name "group velocity" serves to indicate that the new velocity is that of the group of waves as a whole.

The complicated phenomenon we have been discussing may, for theoretical purposes, be regarded as a resultant effect due to the superposition of a large number of wave trains. Thus let us imagine an aggregate of circular wave trains expanding from the point O where the stone was dropped, the wave trains being assumed to have approximately the same amplitude, or height, but to differ slightly in their wave lengths and also in their wave velocities. Furthermore, the longer waves are supposed to be travelling with speeds slightly greater than the shorter waves.

Even without calculation we may obtain a rough idea of the resultant disturbance which should be expected. Because of the differences in the wave lengths of the various wave trains, wave crests will be superposed on wave crests in some regions of the water's surface, whereas they will be superposed on wave troughs in other regions. Some regions will thus exhibit considerable disturbances while others will be practically still. The symmetry of the situation shows that the regions of disturbance and of stillness will be bounded by circles having the point O as centre. First, let us limit our attention to the appearance of the disturbance at a given instant. There will be a circle along which the wave crests of the various wave trains happen to coincide; and along this circle we shall observe a resultant wave crest of considerable height. If we proceed from this circle either inwards or outwards, we shall pass by regions where the phases of the superposed wave trains differ by increasing amounts (owing to the differences in the wave lengths of the individual wave trains). The wave crests resulting from the superposition will thus decrease progressively in height, and eventually we shall reach a region where stillness prevails. Thus, the general appearance of the phenomenon at a given instant will be that of a disturbed region bounded by two concentric circles; beyond the limiting circles no motion will appear; and between them a circular wave crest of maximum height will be seen, preceded and followed by wave crests of progressively decreasing elevation.

Next we consider the changes that will occur as time passes. Since the various wave trains differ only little in their wave velocities, the wave crest of maximum intensity will advance with the average wave velocity of the trains. But as it advances, it penetrates into regions where the phases of the wave trains differ by increasing amounts. Hence our wave crest will progressively decrease in height, and the wave crest which immediately follows it will now become the wave crest of greatest height. This second wave crest in turn will decrease in height as it advances, and the wave crest behind it will become the crest of greatest height, and so on.

Mathematical analysis alone can disclose the finer details of the resultant disturbance. Suffice it to say that the disturbance will exhibit all the peculiarities of the motions observed on the surface of a pond when a stone is dropped into the water. Interference phenomena of a similar kind occur in acoustics and in optics.

Wave Packets—Suppose that several trains of plane waves of light are superposed, the waves having slightly different frequencies ranging from red to yellow. The various trains are directed very nearly, but not quite, in the same direction. We also assume that the red waves are travelling slightly faster than the yellow ones. This situation is realized when the waves are transmitted through glass or water and, quite generally, through any medium exhibiting normal dispersion.* We shall suppose that the dispersive medium in which the waves are propagated is homogeneous, so that insofar as the medium itself is concerned the conditions of ray-optics are fulfilled. For reasons that will appear presently, the individual wave trains, assumed rigorously monochromatic, must extend throughout space. In spite of this extension of the individual trains, the wave motion resulting from their superposition may yield a single small spot of light advancing in the general direction of the

---

* A medium is said to exhibit normal dispersion when waves of lower frequency advance through it with greater speed than waves of higher frequency. Red light, for instance, progresses with greater velocity through glass than does violet light; hence glass exhibits normal dispersion. Owing to the relation between the velocity of the waves and the refractive index, violet light will be subjected to a greater refraction than red light. If the medium exhibits anomalous dispersion, the reverse situation occurs: violet light proceeds with greater speed than red light and is refracted less. Under ordinary conditions, practically all media exhibit normal dispersion; only in the immediate neighborhood of the absorbed frequencies do they manifest anomalous dispersion.

wave trains. Outside this spot of light, interference phenomena cancel the waves, and only where the mobile spot is situated does reinforcement occur. If our eyes could discern the individual waves of light, they would reveal mobile wave crests within the spot of light.

In short, the region of the disturbance contains wave crests and thus exhibits a limited group of waves, as arises when a stone is dropped into the water. But whereas our group of water waves was not limited laterally, our present group of radiation waves is limited in all directions, forming a tiny packet. The name *wave packet* is given to this particular specimen of a group of waves. Similar wave packets may be formed with sound waves and with water waves.

To revert to the changes occurring within the wave packet, we should find that, with our present assumption of normal dispersion, the wave crests inside the packet would be moving in the direction of advance of the packet with a velocity greater than that of the packet. In this respect the situation would be the same as in the group of water waves discussed above. The velocity of the wave crests is the mean wave velocity of the individual trains; and the velocity of the packet (here a spot of light) is the amplitude velocity, or the group velocity, to which we referred in the case of water waves.

The mathematical theory of wave packets was developed by Lord Rayleigh. His results may be summarized as follows: If the waves of lower frequency move more rapidly than the waves of higher frequency, the group velocity (*i.e.*, the velocity of the spot of light) will be less than the average wave velocity of the various trains, and hence will be less than the velocity of the wave crests moving within the spot of light. This is the situation which occurs in media exhibiting normal dispersion. But if the waves of higher frequency are the faster (anomalous dispersion), the group velocity will exceed the wave velocity. Finally, if we operate in a non-dispersive medium, *e.g.*, a vacuum, all the wave trains will have the same velocity c, and then the group- and the wave-velocities will be the same. This situation is very nearly realized in air, for though in all rigor the air is dispersive, in practice its dispersive properties are often negligible.

An important set of relations connects the dimensions and the velocity of a wave packet with the wave numbers and frequencies of the infinitely extended trains of monochromatic waves, which by their superposition yield the packet. Thus suppose that the frequencies of the constituent wave trains vary over a range extending from ν_0 to $\nu_0 + \Delta\nu$. Next consider the components of the various wave-number vectors. A component, such as σ_x, represents the number of wave crests that intersect the Ox axis between two points one centimeter apart. The value of σ_x will vary from one constituent wave train to another: first, because the various trains do not have the same wave length, and secondly, because their directions of advance are assumed to differ slightly. There will therefore be a spread $\Delta\sigma_x$ between the components σ_x of the various wave trains; similarly, there will be spreads $\Delta\sigma_y$ and $\Delta\sigma_z$ for the wave-number components σ_y and σ_z. We designate by Δx, Δy, Δz the dimensions of the wave packet; and by Δt the time taken by the packet to pass a fixed point situated on its line of advance. The relations connecting the wave trains and the packet are then expressed by the four important formulae:

$$(9) \qquad \begin{cases} \Delta x \ . \ \Delta\sigma_x \geqslant 1 \\ \Delta y \ . \ \Delta\sigma_y \geqslant 1 \\ \Delta z \ . \ \Delta\sigma_z \geqslant 1 \\ \Delta t \ . \ \Delta\nu \geqslant 1. \end{cases}$$

The following illustration conveys the meaning of these inequalities. Suppose we consider a number of infinitely extended wave trains having more or less the same frequencies and advancing more or less in the same direction. We assume that the medium is dispersive, so that the wave trains do not advance with exactly the same velocities. The various directions of the wave trains and the spread in their frequencies are determined by the ranges $\Delta\sigma_x$, $\Delta\sigma_y$, $\Delta\sigma_z$, $\Delta\nu$. We assume that these ranges are given. We now inquire: What kind of wave packet can be formed by the superposition of these wave trains? The inequalities (9) show that the smallest wave packet that can be generated will have dimensions, Δx, Δy, Δz, defined by

$$(10) \qquad \Delta x = \frac{1}{\Delta\sigma_x}, \quad \Delta y = \frac{1}{\Delta\sigma_y}, \quad \Delta z = \frac{1}{\Delta\sigma_z};$$

and that the time of transit Δt of this wave packet past a fixed point on the line of its advance will be

$$(11) \qquad \Delta t = \frac{1}{\Delta\nu}.$$

Suppose the packet is advancing along the x-axis. The trains of plane waves, which by their superposition generate the packet, are now also advancing very approximately in the direction of this axis, and hence the wave-number vectors $\overrightarrow{\sigma}$ of the different wave trains point in this direction. The magnitude of the wave-number vector $\overrightarrow{\sigma}$ associated with any one of the wave trains, and the magnitude σ_x of its projection on the line of advance Ox are then very nearly the same. Similarly, the spread $\Delta\sigma_x$ in the components σ_x for the various wave trains is equal to the spread $\Delta\sigma$ in the magnitudes of the wave-number vectors themselves. For our present purpose we may therefore replace the relations (10) and (11) by

$$(12) \qquad \Delta x = \frac{1}{\Delta\sigma} \text{ and } \Delta t = \frac{1}{\Delta\nu}.$$

Since the wave packet advances over a length Δx in the time Δt, the velocity with which the packet is moving, *i.e.*, the group velocity, is $\dfrac{\Delta x}{\Delta t}$.

From (12) we see that this velocity is also $\dfrac{\Delta\nu}{\Delta\sigma}$. Let us compare this group velocity with the wave velocity of the wave crests within the packet. The wave velocity, here considered, is a compromise between the wave velocities of the individual wave trains. However, as the component wave trains differ only little in velocity, we may identify the velocity of the wave crests within the packet with the wave velocity of any one of the superposed trains of waves. If we call ν, λ, and σ the frequency, wave length, and wave number of one of the wave trains, its wave velocity V is given by

$V = \nu\lambda = \dfrac{\nu}{\sigma}$; and this velocity is also, according to what has been said, the wave velocity of the wave crests within the packet. Collecting these results, we obtain for the wave velocity V and for the group velocity U (or the velocity of the packet) the expressions

$$(13) \qquad V = \frac{\nu}{\sigma}, \qquad U = \frac{\Delta\nu}{\Delta\sigma}.$$

The two velocities are thus expressed in terms of the characteristics of the component wave trains.

The wave packet considered in the foregoing discussion was derived theoretically when we assumed that the inequalities (9) were replaced by equalities. When we make this assumption, the dimensions Δx, Δy, Δz of the packet correspond to *the smallest* possible wave packet that can be generated by the superposition of the given wave trains. But to realize this smallest wave packet, or indeed to obtain any wave packet at all, we cannot superpose the wave trains arbitrarily, for only under very special conditions is the packet of minimum dimensions realized. In general, when a packet is formed, it will be of larger dimensions. According to the mathematical theory, even if we assume that the conditions are such that the smallest wave packet is formed, it will spread as it advances. The enlarged packet is, however, produced by the superposition of the same trains of waves, and so we realize that the same trains may generate packets of various sizes. Mathematical analysis shows that if the packet is not to spread as it moves, the number of wave crests in the packet at any instant must be large. In other words, the average wave length of the constituent wave trains must be small in comparison with the length of the packet. But this is merely an ideal limiting case; in practice the packet will inevitably spread. The spreading of the packet is similar to the spreading of heat from a locally heated region. We may also mention that, if the dispersive medium in which the packet is formed has a refractive index varying considerably over distances of the order of the average wave lengths of the component waves, the spreading will be more rapid, and the packet may be disrupted entirely. We conclude that the conservation of the packet is best realized when the medium satisfies the requirements of ray-optics.

The significance of the fundamental inequalities (9) may be summarized as follows: The smaller the packet, the greater must be the differences in the directions and in the wave lengths of the constituent wave trains (and vice versa); also, the smaller the time of transit of the packet past a fixed point on its path, the greater must be the differences in the frequencies of the constituent wave trains (and vice versa).* Further conclusions may be drawn. One of these is that we cannot obtain a train of plane monochromatic waves occupying only a limited volume of space. We may understand the justification for this statement by noting that if the waves in the packet were plane and monochromatic, the ranges $\Delta\sigma_x$, $\Delta\sigma_y$, and $\Delta\sigma_z$ would vanish, and the dimensions Δx, Δy, Δz of the volume occupied by the waves would be infinite, in accordance with the inequalities (9); the waves would therefore fill all space. We may also verify that a strictly monochromatic packet limited in all directions cannot be formed. In such a packet the monochromacy postulated would require $\Delta\nu = 0$, and hence according to the last relation (9) the packet would require an infinite period of time to pass a fixed point. Obviously the packet could not be of finite dimensions in the direction of its advance. Finally, we might verify that a wave packet cannot remain at rest.

Thus far we have investigated the formation of a wave packet by the superposition of different infinite trains of plane waves; we have seen that the waves may cancel everywhere except in a small volume which constitutes the packet. But in practice wave packets are not constructed by the superposition of different wave trains. The simplest way of

---

* Under this form, the theory of wave packets bears a striking resemblance to Heisenberg's uncertainty relations. In Chapter XXX the Heisenberg relations will be discussed, and we shall see that the resemblance mentioned is not accidental.

creating a packet of light is to direct a beam of light, say yellow monochromatic light, onto a screen in which a small hole is made. A shutter
can close and open the hole; and if we first assume that the shutter is
closed, and we then open it for a tiny fraction of a second, and finally
close it again, part of the light will stream through the hole. In this
way we shall obtain a packet of yellow light travelling away from the
hole. In this example the packet of yellow light is generated by the
screen and shutter, and we have not constructed it by superposing infinitely extended wave trains. Nevertheless, in theory, exactly the same
packet might have been formed had we superposed infinitely extended
wave trains differing slightly from the yellow color and also differing
slightly in direction. The numerical relations connecting the wave
trains that would have been required to produce our yellow packet are
deducible from the general relations (9). We may thus identify the
physical packet, generated by means of the screen and shutter, with the
mathematical packet generated by the superposition of infinitely extended
wave trains; and this, in spite of the fact that the physical packet has
not actually been produced by superpositions. The present method of
viewing a disturbance as a superposition of simpler ones affords a physical
illustration of Fourier's mathematical discoveries (exemplified by Fourier
series and Fourier integrals).

The identification of the mathematical packet with the physical packet
enables us to anticipate the physical behavior of the latter from the mathematical theory of the former. Thus we know that the mathematical packet
cannot be rigorously monochromatic, for it is formed by the superposition
of trains of waves differing slightly in frequency. If, then, we accept
the identification just referred to, we must assume that the physical
packet, generated by the incidence of monochromatic yellow light on
the screen and shutter, cannot itself be strictly yellow. In particular,
we must conclude that if we were to analyze its light, we should detect
green and red radiations in addition to the original yellow one. Furthermore, according to the last relation (9), the wave trains, which yield by
their superposition the mathematical wave packet, have a frequency
range $\Delta \nu$ that is the greater the shorter the time of transit Δt of the
packet past a fixed point. Now, physical wave packets which would pass
a fixed point in shorter and shorter intervals of time will be generated
if the shutter of the screen is left open for correspondingly shortened
intervals of time. We must therefore expect that the shorter the interval
of time during which the shutter remains open, the greater will be the
range of the different frequencies present in the wave packet produced.
In other words, the color of the incident monochromatic light passing

through a screen and shutter should be affected in an increasing degree by the rapidity with which the shutter is opened and closed.

A second conclusion is that the physical packet, by analogy with the mathematical one, will not advance with the wave velocity of yellow light in the medium but will move with the group velocity. If the dispersive power of the medium is high and if the dispersion is normal, the group velocity of the packet will be less than the wave velocity. Finally, we know that the mathematical packet will tend to spread. The same conclusion will therefore hold for the physical packet. This latter information, however, conveys nothing new, for we already know that light, on passing through a small aperture, will tend to scatter laterally.

These conclusions, as we shall now see, have an important bearing on the interpretation of certain optical experiments. As an example let us consider the wave velocity of light in a dispersive medium. It is connected with the refractive index of the medium by the formula (2). Suppose, then, that a wave packet of light is generated in the medium by the opening and closing of a shutter or by some equivalent means. If we measure the velocity of the packet and erroneously identify it with the wave velocity, we shall of course be faced with discrepancies; especially so, if the medium is highly dispersive, for in this case the group velocity of the packet will differ perceptibly from the wave velocity. The mistake of confusing the two velocities is one that Michelson once made when he sought to measure the wave velocity of light in the highly dispersive medium, carbon disulphide. He measured the velocity of a packet and assumed that it represented the wave velocity. As a result of this confusion he obtained inconsistent results. But when the theory of wave packets, due mainly to Lord Rayleigh, was developed, Michelson was seen to have measured the wrong velocity.

If we except the method of measuring the wave velocity of light based on the phenomenon of astronomical aberration, we find that all the earlier methods were tainted with the error noted in connection with Michelson's measurement. Thus in the methods of Römer, Fizeau, and Foucault, wave packets are formed, and the velocity measured is the velocity of the packets, not the velocity of the waves as was formerly supposed. However, since the dispersive power of the atmosphere is small, no perceptible discrepancies were detected when the velocity derived from these methods of operation was compared with the wave velocity obtained from the phenomenon of astronomical aberration. But in all rigor, for the velocity of the wave packet to coincide with the wave velocity, the medium must not be dispersive. Inasmuch as all media are more or less dispersive, a method of measurement, such as Fizeau's, will yield the wave velocity of light only if we operate *in vacuo*. Both velocities then have the same velocity c.

Young's Experiment—The experiment we are about to describe was performed in the early part of the last century. It is a typical interference experiment and was one of the first to furnish concrete evidence of the undulatory nature of light. We mentioned it incidentally at the beginning of this chapter, omitting a detailed explanation so as not to break the continuity of the presentation. But in view of the importance of interference phenomena in wave mechanics, we shall consider Young's experiment more fully.

First let us examine the phenomenon of diffraction. Suppose a parallel beam of monochromatic light is directed against a screen in which a small circular hole has been made. The behavior of the light, after passing through the hole, will depend on its wave length and on the dimensions of the hole. If the diameter of the latter is large in comparison with the wave length of the light, a parallel beam will emerge from the hole and the light will proceed along rays. If we place a sheet of paper at some distance behind the screen, we shall observe a bright circular disk having the same diameter as the circular hole. It is true that irregularities will be seen around the edge of the disk, but these will be barely perceptible.

Suppose now the diameter of the hole is decreased. A new phenomenon occurs. The light no longer emerges in a single straight beam; it also proceeds along deflected directions. On the sheet of paper placed behind the perforated screen, we now observe a bright central disk surrounded by concentric bright rings of increasing radii. The rings are called *diffraction rings,* and the entire pattern of rings is named a *diffraction pattern.* The pattern produced in any individual situation depends on the shape of the hole (circular, square, rectangular), on its dimensions, and on the wave length of the light. Finally, let the dimensions of the hole be reduced still further. The diffraction pattern tends to become obliterated, and, in its place, a more or less uniform illumination is obtained on the sheet of paper. In this case the light after passing through the hole spreads in all directions.

We are now in a position to understand Young's interference experiment. A screen with two fine holes, a and b, is placed before a point-source S of monochromatic light. In the figure, the screen is supposed to be placed perpendicularly to the page, and the holes a and b, as also the point-source S, are assumed to be in the plane of the page. Furthermore, S is equidistant from a and b. The light emitted from S streams through both holes, and if these holes are sufficiently small, the waves emerging from a and b will spread in all directions to the right of the screen. If the holes are sufficiently close together, the two beams will overlap. Thus, thanks to Young's device, two different beams of light are derived from the initial beam and become superposed, with the result that their waves fill the same region of space and thus interfere.

To examine the interference phenomenon, we proceed as before by placing a sheet of paper behind the perforated screen (the paper is indicated by the line L in the figure), and we observe the luminous pattern formed on the paper. The only difference between the phenomenon we now propose to examine and the diffraction one discussed earlier is that, in the present case, we have two holes in place of one. But this

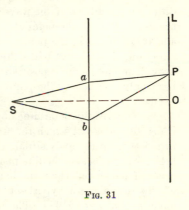

Fig. 31

change modifies the situation entirely; and we shall find that alternating bright and dark bands situated in a vertical direction are formed on the sheet of paper. The pattern thus formed is the interference pattern observed by Young.

Let us understand exactly how this interference phenomenon is brought about. We consider any point P on the sheet of paper. It receives light from both holes. The wave motion at P results from the superposition of the two sets of waves which reach P from S after having passed through the holes a and b, respectively. Since a superposition of two harmonic vibrations of the same frequency yields a harmonic vibration having this frequency, we are certain that the vibratory disturbance at P will be harmonic. The only question to be settled is the amplitude of this vibratory motion. Let us call "waves a" those waves which have passed through the hole a, and "waves b" those which have passed through the second hole. Suppose, then, that the distances SaP and SbP differ by a half wave length of the light, or, more generally, by any odd multiple of the half wave length. In this event the waves a and the waves b, on reaching the point P, will be opposite in phase. Thus, a wave crest pertaining to the waves a will always reach the point P at the precise instant a wave trough of the waves b reaches this point. We cannot expect the intensities of the two sets of waves to be exactly the same at the point P. Nevertheless we may suppose the difference in the intensities to be so small that, as a result of the opposition in the phases, the waves a and b will cancel at P. There will then be no vibratory disturbance at P, and hence no illumination. The point P therefore lies in one of the dark bands we have mentioned.

On the other hand, if the distances SaP and SbP do not differ or else differ by a multiple of the wave length λ, the two sets of waves will have the same phase at the point P. Wave crests pertaining respectively to a and to b will reach the point P simultaneously. Reinforcement now occurs, and the point P will exhibit a greater illumination than it would have done had no superposition of waves been effected, e.g., if there had been only one hole. The point P is now situated in one of the bright bands.

We may readily determine how the bands will lie on the sheet of paper. Thus, the points at which the waves a and b exhibit the same difference in phase are necessarily situated on the surface of a hyperboloid of revolution having as axis the line passing through the two holes a and b. The sheet of paper, on which the bands are formed, intersects a hyperboloid of this sort along a hyperbola, and hence the contiguous bands will lie along arcs of hyperbolae. The band which passes through the mid-point O will be straight, however, and will have the vertical direction. The other bands, though in reality curved, will appear straight over short extensions, and so the general appearance of parallel vertical bands is accounted for theoretically.

We might suppose that the same interference effect would be obtained if we replaced the source S and the perforated screen by two monochromatic point-sources of light of the same color, situated at the points a and b occupied by the former holes. As before, waves of light would be emitted from the points a and b, and their superposition on the sheet of paper should give rise to the same interference bands. But in practice, if this experiment were undertaken, no interference pattern would be formed; instead, a uniform illumination would appear. This surprising situation prompts us to examine the problem more carefully.

The Coherence of the Waves—Hitherto we have assumed that the waves emitted by a point-source of monochromatic light are regular sinusoidal waves proceeding without sudden changes. But this is not so. In actual practice, regular waves are emitted only over extremely short periods of time, say $\frac{1}{1000}$-th of a second. At the end of each one of these short periods a sudden change in the phase occurs; for instance, there may be a sudden jump from a wave crest to a wave trough or to any intermediary phase of the vibration. The successive intervals of time during which the waves are emitted in a regular way are of variable length; though exceedingly short, they are, nevertheless, usually long enough to involve many millions of vibrations.

Suppose, then, we represent the wave train at an instant by a wavy curve. If it were not for the abrupt changes in the phase, our curve would be a sinusoid without any breaks in it. But by reason of the abrupt changes, this continuous curve must be replaced by a succession of regular sinusoidal segments placed end to end. We thus obtain a discontinuous curve the general appearance of which is exhibited in the figure. The successive segments are unequal in length and are distributed at random. Each segment contains millions of wavelets (instead of the scant few we have been able to trace in the figure), and hence each

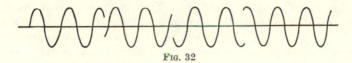

Fig. 32

segment extends over millions of wave lengths of the light. If we suppose that, on an average, one thousand of these segments is emitted every second from the luminous source, the length of a segment will be about 186 miles. The discontinuities we are here considering are not altogether unexpected, for Bohr's theory indicates that the emission of radiation is a discontinuous and not a continuous process.

The presence of sudden changes in the phase complicates our understanding of monochromatic wave trains. Let us consider a parallel beam of monochromatic waves. The waves are said to be *coherent* if the phase at any instant is the same over any cross section of the beam. More generally, let us investigate the coherence of different beams. Suppose that two parallel beams A and B of the same monochromatic light fall normally on a screen, giving rise to two luminous spots a and b. We assume that both beams are formed of coherent waves in the sense defined previously. We consider any point P_a in the first spot of light and any point P_b in the second. The phases at these two points at any instant t may, but need not, be the same. Furthermore, as time passes, the phase at P_a will undergo thousands of abrupt changes every second, and similarly for the phase at P_b. If, despite these jumps, the *difference* in the phase at P_a and at P_b remains fixed throughout time, the two beams are said to be coherent. In this event the abrupt changes of the phases at P_a and at P_b necessarily occur simultaneously and are equal in magnitude.

Owing to the random nature of the changes in the phases, we cannot expect the waves emitted from two distinct monochromatic point-sources of the same color to be coherent. Nor will the waves be coherent when they issue from different regions of the same extended luminous source, *e.g.*, the flame of a candle. Nevertheless in this latter case, the coherence of the waves will tend to be realized when the two tiny regions which emit the light are very close to each other. In some experiments we must consider two wave trains issuing from the same point-source but at different times. The two wave trains will not be coherent, for the abrupt changes in the phase do not repeat in the same way over successive intervals of time. As will be gathered from the foregoing considerations, the coherence of two beams is far from being the rule.

Whenever we wish to obtain interference effects from the superposition of two monochromatic beams, we must operate with beams that are coherent. The reason is obvious. Thus, suppose we have two point-sources a and b emitting the same monochromatic light. If at a point P and at the instant t the waves issuing from the two sources are opposite in phase, they will tend to cancel each other at this instant; and the cancellation will be complete if the amplitudes and hence the intensities of the two sets of waves are the same. But for this cancellation to betray itself to the eye by an absence of luminosity, it must endure over a certain period of time. Now, permanent cancellation will be secured only if the phases of the two sets of waves remain opposite throughout time —and this requires that the two sets of waves be coherent. Similar arguments hold if we consider a point P where the two sets of waves are in phase at a given instant. Permanent luminosity at this point will also require that the two sets of waves be coherent. In short, when the waves are coherent, we may expect that permanently bright and dark regions will be formed and hence an interference pattern produced. On the other hand, when the two sets of waves are incoherent, cancellations and reinforcements at the same point P will follow one another at random, thousands of times every second. Obscurity and luminosity will alternate with extreme rapidity at each point, and the eye will detect a mere average illumination. No interference pattern will be observed.

The perfect coherence of the superposed wave trains constitutes the ideal condition for a clearly defined interference pattern to be formed. But even when the phases of the two sets of waves are not subjected to their sudden changes simultaneously, interference phenomena may still be obtained. Thus let us suppose that as the waves of the two superposed beams reach a point P, the same abrupt changes in phase occur in both beams, but that the two sets of changes take place with a small lag in time. The difference in the phases at P cannot now remain fixed, and so permanent extinction or permanent luminosity cannot be expected. Nevertheless, if the lag in time is exceedingly small, the intervals of time during which the same difference in the phase is maintained are considerably greater than those for which random differences occur. Hence the same interference pattern will be formed, say ninety-nine times out of a hundred; and the eye will still detect a definite pattern, though it will be less clearly defined than when the waves were rigorously coherent.

These preliminary considerations enable us to obtain a better understanding of Young's experiment. In this experiment the coherence of the two sets of superposed waves is assured by the device of the screen and two holes. The waves leave the point-source S in phase, and since the distances Sa and Sb are equal, the waves reach the holes a and b in phase (Figure 31). They are then emitted from the holes with the same phase at each instant; hence if the distances aP and bP differ but little, the phases of the two sets of waves at the point P will undergo the same jumps very nearly in unison. At such points P the coherence of the waves cannot be truly rigorous, but as was explained in the previous paragraph, the coherence may be sufficiently rigorous to ensure the formation of an interference pattern.

It is now a simple matter to understand why Young's experiment would be a failure if the two holes a and b were replaced by two monochromatic point-sources of light. The reason is that the two beams issuing from a and b would be completely incoherent, for there is no connection between the emissions of radiation from two distinct point-sources. Even when Young's apparatus is retained, we may destroy the coherence of the two beams in a number of ways and thereby destroy the interference pattern. For example, let us displace the point-source S gradually from its initial position, in such a way that the two lengths Sa and Sb become increasingly unequal. The corresponding abrupt changes in the phases at a and b will no longer occur simultaneously, but will now be separated by an increasing lag in time. Consequently the two beams issuing from a and b will become increasingly incoherent. The interference pattern will gradually lose its distinctness, and when the difference in the lengths Sa and Sb attains some critical value, the pattern will be obliterated completely. Incidentally, a determination of the above critical value will enable us to compute the average frequency of the abrupt changes in the phase of the waves that issue from the point-source.

Electromagnetic Waves —The theory of optics we have so far discussed merely requires that light be a wave phenomenon in which the waves are transverse. The precise nature of the entity that is vibrating is not taken into consideration. Fresnel, however, impressed by the similarity between waves of light and the transverse waves that may arise in elastic media, assumed that waves of light were waves propagated through a hypothetical elastic medium, called the elastic ether. Maxwell, in his electromagnetic theory, maintained that waves of light were none other than waves in the electromagnetic field. The views of Fresnel and of Maxwell could have been reconciled had it been possible to identify the electromagnetic vibrations of the field with the mechanical vibrations of the ether. But, for a number of reasons, all attempts along these lines were gradually abandoned, and in Lorentz's theory, the ether was stripped of practically all its mechanical properties; the theory of relativity subsequently completed the ether's dematerialization. The electric and magnetic vectors of the electromagnetic field, being no longer ascribed to strains or displacements of the ether, were therefore regarded as fundamental.

The interpretation of plane polarized monochromatic waves of light in the electromagnetic theory is easily understood. For simplicity, we shall assume plane waves *in vacuo*, and we shall suppose that the waves are regular, *i.e.*, abrupt jumps in the phase do not occur. The direction in which the waves are advancing is the direction of the straight rays. All planes perpendicular to the direction of advance are equiphase surfaces, for over the surface of any one of these planes the phase of the vibration remains uniform.

We restrict our attention to a fixed point O in the region of the waves. According to Maxwell the vibrations of the waves of light at the point O are represented by the vibrations of the electric

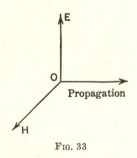

Fig. 33

and magnetic vectors $\vec{E}$ and $\vec{H}$; the directions of vibration of these vectors are perpendicular to the lines of advance, *i.e.*, to the rays (transverse waves), and in addition they are perpendicular to each other. The vibrations are harmonic, so that sinusoidal waves will result. The amplitudes of the vibrations are the same (when measured in appropriate units). As for the relative orientations of the two vibrating vectors with respect to the direction of the propagation, they are exhibited by the three mutually perpendicular vectors in Figure 33. The propagation is assumed to be taking place to the right in a horizontal direction; the electric vector $\vec{E}$ is vibrating above and below the point O, along the vertical, and the magnetic vector $\vec{H}$ then vibrates towards us and away from us. Both vectors assume their maximum elongations simultaneously, and both vanish at the same instant. We may represent the state of the vibration at any given instant for any point along the line of the propagation. The result is depicted in Figure 34. The motion of the wave is represented when we assume that the drawing is displaced rigidly to the right with the velocity c of light.

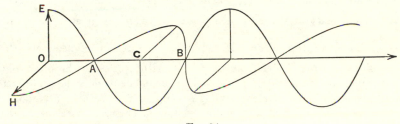

Fig. 34

The density of the electromagnetic energy at any point at a given instant is $\dfrac{E^2 + H^2}{8\pi}$, where E and H represent the magnitudes of the electric and of the magnetic vectors at the point and at the instant in question. Inasmuch as E and H vibrate simultaneously and have the same magnitude, the density of the electromagnetic energy is also measured by $\dfrac{E^2}{4\pi}$. This density will have the same value at a given instant at all points of the same equiphase surface, and therefore at all points of a plane perpendicular to the direction of advance. On the other hand, the density of the energy, at a given point, varies with time, and, at a given instant, it varies with position. At the instant when the wave train occupies the position represented by the drawing in Figure 34, the density of the energy is greatest over the equiphase planes passing through points such as O or C (these points correspond to wave crests or to wave troughs); and the density of the energy is zero over equiphase planes passing through A or B. From this account we see that the energy density is not distributed uniformly along the direction of the propagation. Now we have said that the motion of the waves through space is represented graphically if we assume that the two sinusoidal curves in the figure are displaced with uniform velocity c in the direction of the propagation, i.e., to the right in the figure. Hence, the electromagnetic energy is carried along by the waves with this velocity c. If we imagine the region of the wave motion to be sectioned into tiny cubes, some of these cubes, according to what we have just said, will contain more, and others less, energy. For a given cube to contain always the same energy, it must move along with the waves; it then describes a straight line which is perpendicular to the equiphase planes. But a line of this kind is what we have called a ray. We conclude that, in the electromagnetic theory, the rays represent the lines along which the electromagnetic energy streams with velocity c (in vacuo).

In addition to energy, the waves also carry electromagnetic momentum, the density of which at each point (in vacuo) is given by $\dfrac{EH}{4\pi c}$, or $\dfrac{E^2}{4\pi c}$; the quantities E and H, as before, represent respectively the magnitudes of the electric, and of the magnetic, vector at the point and at the instant considered. The density of the momentum is the same over any given equiphase surface and is a maximum or zero in accordance with the same rules that were explained for the energy. As a result, the distribution of momentum throughout the train of waves will be carried along by the waves with velocity c, just as occurs for the energy distribution. We may also note that the momentum density at any point and instant is $\dfrac{1}{c}$ times the energy density at the same point and instant. In contradistinction to the energy, which is a scalar, the momentum is a vector. It is directed perpendicularly to the equiphase planes and points in the direction of the propagation. To the momentum of the waves is due the pressure they exert when they fall upon matter; and when these electromagnetic waves have the frequencies of light, it is this momentum which is responsible for the pressure of light. To the energy of the waves of light is due the intensity of the light and the effects exerted on the retina.

This summary description of electromagnetic waves will suffice for our present purpose. The important point to retain is that the waves carry energy and momentum, and that in vacuo the density of the momentum is $\dfrac{1}{c}$ times the density of the energy.

CHAPTER XX

TRANSFORMATIONS, HAMILTON'S WAVE MECHANICS, CONTINUOUS GROUPS

Suppose that we have two sets of variables and that whenever the values of the variables of the first set are specified, the values of those of the second set are determined, and conversely. A correspondence is thus established between the two sets of variables. The mathematical expression of this correspondence is called a transformation. It may happen that, to every choice of values for the variables of the first set, one and only one allotment of values follows for the variables of the second set; and it may happen that the reverse is also true. In this event the correspondence between the two sets is said to be "one-to-one." Let us assume that we are dealing with a transformation which establishes a one-to-one correspondence. If we assign numerical values to the variables of the first set A, the transformation defines corresponding values for the variables of the second set A'. The variables of the set A are then said to have been transformed into those of the set A'.

Contact Transformations—Thus far we have disregarded the significance of the variables. In many cases the variables can be interpreted geometrically. As an example, we shall first consider so-called *contact transformations* in 3-dimensional space. In these transformations five variables are transformed simultaneously. Representing them by the letters x, y, z, p, q, we may suppose that x, y, z are the Cartesian coordinates of a point P, and that p and q are two parameters defining a direction D in space. When the contact transformation is applied, these five variables are transformed into five new ones x', y', z', p', q', which may be assumed to represent the Cartesian coordinates of a new point P' and the parameters of a new direction D'. The correspondence is one-to-one. We shall refer to a point P and a direction D as defining an *element*. Thus x, y, z, p, q are the five coordinates of an element; and a contact transformation is therefore said to establish a correspondence between elements x, y, z, p, q and elements x', y', z', p', q'.

Let us, then, consider a surface S. An element of this surface will be defined by any point x, y, z of the surface and by the normal p, q to the surface at this point. If, now, we apply a contact transformation to all the elements of the surface S, we shall obtain an aggregate of transformed elements (points and directions) which will usually define a second surface S' and its normals.   Thus, the first surface and its normals are transformed into a second surface and its normals.

Suppose, then, we have two surfaces S_1 and S_2 which are tangent at a point P and which exhibit a certain order of contact.*   The two surfaces obviously have an element in common; it is determined by the point P at which the surfaces touch and by the common normal to the two surfaces at P.   We apply a contact transformation to the two surfaces S_1 and S_2 and to their normal directions.   In other words we transform the elements of both surfaces.   The element which is common to the two surfaces, whether we view it as belonging to the surface S_1 or to the surface S_2, is necessarily transformed into some same element.   Consequently, the two transformed aggregates of elements will have an element in common.   In geometrical language, the two transformed surfaces S'_1 and S'_2 will have a point P' in common, and a common normal at this point.   As a result the two transformed surfaces will be tangent at the point P'.

In fine a contact transformation transforms tangent surfaces into tangent surfaces.   Furthermore, the order of contact of the two transformed surfaces S'_1 and S'_2 can be shown to be the same as that of the original surfaces S_1 and S_2.   We have here a characteristic property of contact transformations: they do not modify the order of contact.   It was on account of this important property that Sophus Lie coined the name "contact transformation."

As a further example of the properties of contact transformations, let us suppose that the elements we wish to transform are represented by a fixed point P and by all the directions D issuing from this point.   Each element (P, D) of this doubly infinite family of elements is transformed into some different element (P', D') defining a point P' and a direction D'; and the aggregate of new elements likewise forms a doubly infinite family. From the fact that we are transforming a fixed point P and all directions D, we might suppose that the transformed elements would be represented by a fixed point P' and all directions D'.   But this situation, though

---

* In analytical geometry, we are led to distinguish different degrees of tangency. Thus two surfaces may be tangent at a point P but may diverge rapidly from each other around this point.   In other cases, the points of the two surfaces in the immediate vicinity of the point of tangency may be practically coincident.   In popular language the two surfaces may be said to hug each other more or less closely around the point of tangency.   This vague notion is expressed rigorously in mathematics; it is determined by the *order of contact* of the two surfaces at the point P.   The higher the order of contact, the closer the two surfaces in the neighborhood of the point P.

possible (*i.e.*, when the contact transformation degenerates into a point transformation), is not the usual one. In general we shall find that the transformed points P' cover some surface S' or lie along some line L' (curved or straight); in such cases the transformed directions D' will be determined by the normals to the surface or to the line.

Let us observe that if we perform a contact transformation and then follow it up with the inverse transformation, we shall merely undo what we have done. Consequently, if the original transformation transforms a point and its directions into a surface or a line and its normals, the inverse transformation will transform back the surface or line and its normals into the original point and its directions. We see from this example that a contact transformation may occasionally transform a surface or line, and its normals, into a point and its directions.

For the sake of conciseness, explicit reference to the directions D and D' is sometimes dropped, and we speak of a point being transformed into a surface or into a line, and of a surface or a line being transformed into a point. Such modes of expression though convenient are not rigorous. For, as we have seen, it is not the point P by itself which is transformed into a surface or into a line and vice versa; instead it is the point P together with all the directions D which is transformed into a surface and into all the normals to the surface, or into a line and all the normals to the line, or vice versa. However, if the true state of affairs is understood, we may without danger of confusion retain our inaccurate phraseology, and we may then summarize the properties of contact transformations as follows:

A point is usually transformed into a surface or into a line. Exceptionally, it may be transformed into a point (the last situation is the rule when the contact transformation degenerates into a point transformation).

A line is usually transformed into a surface or into a point. It may also be transformed into a line.

A surface is transformed into a surface, a line, or a point.

Let us examine the transformations to which a point and a surface passing through this point will be subjected. We shall assume that the contact transformation is of a kind which transforms points into surfaces. A point P is then transformed into a surface Σ', and a surface S passing through the point P is transformed into some other surface S'. We propose to show that the two surfaces Σ' and S' will be tangent at a point. To prove this, we note that the point P is associated with all directions, and that the surface S is represented by all its points and by the normals at these points* The point P and the surface S thus have an element in common, namely, the element defined by the point P and by the normal to the surface at this point. As a result, the transformed surfaces Σ' and S' likewise must have an element in common, and hence must have a point in common and a common normal at this point. In other words the two surfaces Σ' and S' must be tangent. Thus, a surface and a point on it are transformed into two tangent surfaces.

If, then, we consider all points P on the surface S and all the surfaces Σ' into which these points are transformed, we see that each one of the surfaces Σ' is tangent to the surface S' into which the original surface S is transformed. We conclude that when a surface S is transformed into a surface S', the surface S' is the envelope of the surfaces Σ' into which the points P of S are transformed.

As a further illustration let us consider two intersecting lines. The elements of a line are defined by the points of the line and the normals to the line at these points. Obviously, therefore, two intersecting lines have an element in common, namely, the point of intersection and the common normal to the two lines at this point. If, then, our contact transformation is of a kind which transforms lines into surfaces, the two intersecting lines will necessarily be transformed into two tangent surfaces.

Finally, let us transform a line and a point on it. We assume that the transformation changes the point and also the line into surfaces. The point and the line have a simply infinite family of elements in common. These elements are defined by the point and by all the normals to the line at this point. Consequently, the two surfaces resulting from the transformation of the point and of the line will likewise have a simply infinite family of elements in common; and so the two surfaces will be tangent along a line.

One of the most celebrated contact transformations was devised by Sophus Lie. It transforms straight lines into spheres.† Two intersecting straight lines will thus be transformed into two tangent spheres. To illustrate Lie's transformation, let us consider the surface into which a

---

* At each point of the surface there are two opposite directions for the normal. To avoid ambiguity, only one of these normals is retained. For instance, if the surface is painted red on one side and blue on the other, we may agree to retain only those normals which are on the red side.

† There are usually two different non-intersecting straight lines which are transformed into the same sphere. If the transformation is reversed, the sphere is transformed back into the two straight lines.

hyperboloid will be transformed. A hyperboloid is the surface generated by a mobile straight line constrained to pass through three fixed straight lines. If we transform the three fixed lines and the mobile one by Lie's transformation, we obtain three fixed spheres and a variable sphere which remains tangent to the three fixed ones. Now the mobile straight line always lies on the surface of the hyperboloid, and hence has a simple infinity of elements in common with the hyperboloid. It follows that the variable sphere into which the mobile straight line is transformed will be tangent along a curve to the surface into which the hyperboloid is transformed. This latter surface will therefore be the envelope of the variable sphere. In short, the surface into which the hyperboloid is transformed is the envelope surface of the family of spheres which are tangent to three fixed spheres. This envelope surface is a well-known surface called Dupin's cyclide.

Point Transformations—We revert to the effect of a contact transformation on a fixed point $P(x, y, z)$ associated with all directions $D(p, q)$. We have seen that, when the point P is associated first with one direction D and then with another, we obtain different transformed points $P'(x', y', z')$ and directions $D'(p', q')$. Thus, a change in our choice of the direction D modifies the position of the transformed point P'. For this reason, the point P and the aggregate of directions D may be transformed into a surface and its normals or into a line and its normals. This condition arises because the coordinates x', y', z' of the transformed point depend on all five of the original variables x, y, z, p, q.

Next we consider a more restricted kind of transformation, called a *point transformation*. In a point transformation the direction D associated with a point P has no influence on the position of the transformed point P'. The only effect of changing the direction D will be to change the transformed direction D'. In this case, a point P will always be transformed into a point and never into a line or into a surface.* Similarly, a line will always be transformed into a line, and a surface into a surface. The confusion of categories (points, lines, surfaces), noted in the usual contact transformations, no longer occurs.

Analytically, this situation implies that the coordinates x', y', z' of the transformed point P' depend solely on the coordinates x, y, z of the original point P, and are independent of the parameters p and q of the

---

* We are restricting our attention to non-singular transformations, *i.e.*, to those that establish a one-to-one correspondence between points.

direction D.* A correspondence (assumed here to be one-to-one) thus holds between the points $P(x, y, z)$ and $P'(x', y', z')$ of space. Since the stress is placed on points, it is usual to disregard the correspondence between the transformed direction D' and the original element (P, D). But we may, if we choose, express this correspondence explicitly, and in this case the point transformation is called an "extended point transformation." We shall, however, omit consideration of extended point transformations.

We might be tempted to inquire whether point transformations have anything in common with contact transformations. The answer is in the affirmative, for point transformations, like contact transformations, do not alter the order of contact. Thus a point transformation will transform two tangent surfaces into two other tangent surfaces, and the order of the contact will be the same for both pairs of surfaces. Since the preservation of the order of contact is a distinctive characteristic of contact transformations, we are justified in viewing point transformations as forming a sub-class of the class of contact transformations.

An important sub-class of the class of all point transformations is the class of "conformal transformations." In the sequel, we shall consider only the conformal transformations of the plane. They have the property of leaving angles unchanged. As a result, any two plane curves intersecting at a given angle will be transformed into plane curves intersecting at the same angle. Another important sub-class of the class of all point transformations is the class of "projective transformations." These transformations transform planes into planes, and straight lines into straight lines. Furthermore, if we consider any four points on a straight line, the anharmonic ratios of the four transformed points and of the four original ones are the same. Anharmonic ratio is thus invariant under projective transformations. A sub-class of the class of projective transformations is the class of linear transformations. These transformations are intimately connected with the theory of matrices, and we shall examine them in a later chapter where the theory of matrices will be explained.

Applications of Transformations—The applications of transformations in pure mathematics and in theoretical physics are many and varied. In the Matrix Method, in Dirac's transformation theory, and in the theory of relativity, transformations are ubiquitous. For the present, we shall

---

* The parameters p', q' of the transformed direction D' will, however, depend on the five variables x, y, z, p, q and hence on the point P as well as on the direction D.

be concerned more particularly with illustrations afforded in the field of pure mathematics; and in view of the greater simplicity of point transformations we shall consider these transformations first.

A problem of theoretical and also of practical interest is to represent on one kind of surface a drawing which is traced on a surface of different shape. For instance, we may wish to reproduce on a plane sheet of paper a drawing made on a sphere. This particular problem arises in the construction of maps of the earth's surface, *i.e.*, geographical maps. Now an accurate reproduction of the configurations traced on a spherical surface cannot be made on a plane. If the reproduction is attempted, distortions of one kind or another are inevitable. But we may preserve some features of the configurations, for example the angles at which lines intersect. It is precisely this course which is followed by the designers of geographical maps—as may be verified, parallels and meridians which intersect at right angles on the earth's surface also intersect at right angles on maps. Quite generally, when drawings traced on one surface are reproduced on another in such a way that angles are unchanged, a *conformal representation* of the first surface on the second is said to be performed. According to this definition, maps are conformal representations of a spherical surface on a plane.

Let us suppose that by some means or other a map has been obtained. What we propose to show is that from the first map an infinite number of others can be derived. We recall that any conformal transformation of the plane transforms points and lines but does not affect angles. Consequently, if any conformal transformation is applied to the first map, the drawings of the first map will be modified but all angles will be preserved. Hence our transformed picture will still yield a map (in the sense previously defined). Thus, thanks to conformal transformations, we may derive an infinity of different maps from an initial one. The maps most widely utilized are called Mercator's projection and the stereographic projection.

Conformal transformations operating on the points x, y of a plane are intimately related to analytic functions of a complex variable and to harmonic functions. If z represents the complex variable $x + iy$, and if we consider some function $f(z)$, or $f(x + iy)$, and then separate the real from the imaginary part, we get

$$f(z) = f(x + iy) = A(x, y) + iB(x, y),$$

where $A(x, y)$ and $B(x, y)$ are two functions of the real variables x and y. If the function $f(z)$ is analytic throughout a given area of the complex plane, the functions $A(x, y)$ and $B(x, y)$ will satisfy certain relations at all points within this area. A particular consequence of these relations is that $A(x, y)$ and $B(x, y)$ will both be harmonic functions, *i.e.*, solutions of Laplace's equation in the plane.

Suppose the required relations are satisfied at all points of a given area of the plane, so that $f(z)$ is an analytic function within the area. It can then be shown that, for all points x, y of the area, the point transformation

$$(1) \qquad \begin{cases} x' = A(x, y) \\ y' = B(x, y), \end{cases}$$

which establishes a correspondence between points x, y and x', y' in the plane, is a conformal transformation. Conversely, if the transformation (1) is conformal, the function $A(x, y) + iB(x, y)$ is an analytic function of the complex variable $z = x + iy$. An immediate consequence of these considerations is that, when $f(z)$ is an analytic function of the complex variable z in a certain region of the complex plane, the correspondence between the point z and the point defined by $f(z)$ in the complex plane is conformal. It can also be shown that a harmonic function is transformed into a harmonic function when a conformal transformation is applied. We may therefore say that "the property of being harmonic" is invariant under a conformal transformation.

The latter property of invariance has been utilized by mathematicians to transform problems involving analytic functions of a complex variable (and hence harmonic functions) into others which are easier to solve. As an example let us consider Dirichlet's existence theorem in the plane. Dirichlet's theorem in 3-dimensional space was mentioned in Chapter XVI; its significance in the two-dimensional case (*i.e.*, in the plane) is much the same. The theorem states that, at all points within the area defined by an arbitrarily given closed curve, there always exists a harmonic function which assumes given values on the curve.[*] If the theorem can be proved and hence the existence of a harmonic function demonstrated, this function is easily shown to be unique, *i.e.*, it is the only function which satisfies all the requirements imposed. In the particular case where the closed curve is a circle, Dirichlet's theorem is easily proved, and furthermore, the harmonic function can be constructed. But when the closed curve is of arbitrary shape, the theorem is exceedingly diffi-

---

[*] The harmonic function must be one-valued and continuous, together with its first and second derivatives. The values which the function is to assume on the curve are supposed to vary continuously along the curve. The problem here described is the so-called interior problem; there is also an exterior Dirichlet problem.

cult to establish. Riemann erroneously believed that he had proved it. His mistake was subsequently pointed out by Weierstrass.

We now propose to show that Dirichlet's theorem and the problem of proving the existence of a conformal transformation which transforms an arbitrary closed curve C into a circle (or vice versa) are equivalent. To prove this point let us suppose that Dirichlet's theorem has been established, so that we know that there exists a function which is harmonic at all points inside any particular closed curve C and which assumes given values along the contour of the curve. Next, we suppose that we have obtained a conformal transformation which transforms the closed curve C into a circle and the points inside the closed curve into points inside the circle, the correspondence between points being one-to-one. Since a conformal transformation transforms a harmonic function into a harmonic function, we know that the harmonic function associated with the arbitrary closed curve C will be transformed into a harmonic function existing inside the circle and assuming corresponding values along the contour of the circle. The conformal transformation thus establishes a correspondence between the two harmonic functions, so that, if one function exists, the other necessarily exists.

Now we have said that the existence of the harmonic function in the case of the circle is easily proved (Dirichlet's theorem for the circle). Hence if we can establish the existence of a conformal transformation which transforms the arbitrary closed curve C into a circle (and vice versa), we shall automatically have proved the existence of a harmonic function for the curve C. Thus the proof of Dirichlet's theorem for an arbitrary closed curve C is seen to be equivalent to the problem of determining the existence of a certain conformal transformation.

The considerations here developed are due to Riemann. We have mentioned them for two reasons. Firstly, because they reveal the connection between Dirichlet's theorem and conformal transformations. Secondly, because they afford an illustration of the manner in which a mathematical problem may be transformed into one having a totally different appearance.

Next, let us examine some of the applications of projective transformations. Cayley and Klein proved that the projective transformations of the plane were closely connected with the non-Euclidean plane geometries of the classical period. The connection may be understood from the following considerations. Projective transformations usually transform a curve into some other curve. But by imposing appropriate restrictions we may single out a sub-class composed of projective transformations which have the peculiar property of transforming the points of a given circle into points situated on the same circle. The circle as a whole is then subjected to no modification or displacement by the transformations; the circle is said to be transformed into itself, or to be invariant under the transformations of the sub-class considered. Cayley proved that the projective transformations which transform an imaginary circle into itself (Cayley's absolute) are closely associated with Riemann's plane geometry. Klein proved subsequently that if a real circle is taken in place of an imaginary one, the projective transformations which transform it into itself correspond to Lobatchewski's geometry. Finally, if the fixed figure consists of the line at infinity and the two circular points at infinity, Euclidean geometry is obtained.

Projective transformations play a fundamental rôle in the theory of functions of a complex variable. Poincaré's automorphic functions, for instance, are analytic functions of the complex variable z which are transformed into themselves when z is submitted to a certain discrete set of projective transformations.

We now pass to the applications of contact transformations. The first reference to contact transformations appears in the works of Legendre and of Ampère. At a later date Hamilton and Jacobi utilized these transformations in dynamics. But the general theory of contact transformations was elaborated by Sophus Lie in his celebrated treatise on groups of continuous transformations. The concept of group was explained in Chapter XIV; its connection with transformations will be discussed at the end of the present chapter.

Contact transformations are of great assistance in many geometrical problems. For instance Lie's contact transformation, which transforms straight lines into spheres (and conversely), enables us to transform a problem dealing with spheres into one involving straight lines. A considerable simplification in our demonstrations is secured thereby.

In analytical mechanics, contact transformations are of the utmost importance, and we shall now review some of their applications in this science. In Chapter XVII we saw that the central problem of dynamics is the integration of Hamilton's equations. When these equations are integrated, the evolution of the mechanical system of interest is known. Unfortunately, a direct integration of Hamilton's equations is usually impossible, and even when it is possible in theory, practical difficulties may be considerable. Many methods were devised by the mathematicians of the last century with a view of simplifying the problem of integration. Several of these methods were mentioned in Chapter XVII. Here we shall recall only one of them.

In Hamilton's equations the variables which determine the evolution of a mechanical system are the generalized coordinates q and the generalized momenta p. The Hamiltonian function $H(q, p)$, which enters into the expression of Hamilton's equations, is usually a function of both sets of variables. If we transform the variables q and p into new variables $\bar{q}$ and $\bar{p}$ by means of an arbitrary transformation, the general form of Hamilton's equations will undergo a change. But Jacobi proved that certain privileged transformations have the property of leaving the form of the equations unchanged. Inasmuch as Hamilton's equations are often called the "canonical equations of dynamics," the name "canonical transformations" is given to the transformations just mentioned. These canonical transformations are a special kind of contact transformation.*

Suppose, then, we apply an arbitrary canonical transformation to the Hamilton equations of a given problem. The equations will retain their form, but the Hamiltonian function $H(q, p)$ will be transformed into a different function $\bar{H}(\bar{q}, \bar{p})$ of the new variables $\bar{q}$ and $\bar{p}$. If we can integrate the new Hamilton equations, the solutions of the original ones will be obtained immediately, and our problem will be solved. In general it will be no easier to integrate the new equations, and in this event we shall have derived no advantage from the transformation. But, as Jacobi showed, if we can obtain a canonical transformation which transforms the Hamiltonian function $H(q, p)$ into one of the form $\bar{H}(\bar{p})$, which contains only the variables $\bar{p}$ (or one of them alone), the transformed Hamilton equations can be integrated immediately, and hence the dynamical problem can be solved. Thus, Jacobi's method consists in replacing the direct integration of Hamilton's equations by the quest for an appropriate canonical transformation. As we shall now see, the discovery of the required canonical transformation must be approached in an indirect way. This leads to a short digression.

Prior to Jacobi's investigations, Hamilton had observed that a certain partial differential equation was closely related to the Hamilton equations. A solution of this equation represents the action function of the dynamical system. Hamilton proved that if the Hamilton equations could be integrated, a solution of the partial differential equation could be obtained. Jacobi reversed the proof by showing that if we could obtain any solu-

---

* A contact transformation in 3-dimensional space is of form

$$x' = f(x, y, z, p_x, p_y), \quad y' = \varphi(x, y, z, p_x, p_y), \quad z' = \psi(x, y, z, p_z \, p_y).$$

But it may happen that neither x' nor y' depends on z, and that z' alone depends on this variable. In particular, if z' is of form $z + f(x, y, p_x, p_y)$, we have a canonical transformation. All reference to the variables z and z' may then be dropped.

tion * of this partial differential equation, Hamilton's equations could be integrated without difficulty. Because of the complementary contributions of Hamilton and of Jacobi to the study of this equation, it is called the Hamilton-Jacobi partial differential equation. The reason why a solution of the partial differential equation permits an immediate integration of Hamilton's equations is that such a solution furnishes the canonical transformation which transforms the Hamiltonian $H(q, p)$ into the form $\bar{H}(\bar{p})$—and this is precisely the kind of canonical transformation we were seeking in the previous paragraph.

The method established by Jacobi for integrating Hamilton's equations illustrates the transformation of one mathematical problem into another: instead of attempting to integrate Hamilton's equations directly, we seek a solution of a totally different kind of equation. A similar illustration was mentioned previously when we discussed the connection between conformal transformations and Dirichlet's problem. Mathematics abounds in such transformations of problems. Often the alternative problem suggested is no simpler than the original one, but this does not detract from the theoretical interest of attacking the same problem from different directions. In any case the transformation of problems opens new vistas and establishes connections between departments of mathematics which at first sight appeared to be isolated. As was mentioned in Chapter XVII, Jacobi's method, in addition to its theoretical interest, also affords practical advantages, because on certain occasions we find that it is easier to obtain a solution of the partial differential equation than to integrate Hamilton's equations directly. But of course, Jacobi's method does not always yield practical results, for, if it did, all the problems of dynamics could be solved without our having to appeal to methods of approximation. And we know that such is not the case—as witnessed in the Problem of Three Bodies.

Infinitesimal Transformations—Canonical transformations lead to interesting interpretations of Hamilton's equations and of the evolution of a dynamical system. But first we must explain what is meant by an infinitesimal transformation. To simplify the exposition we consider point transformations. Let us suppose that all the points P on a straight line are displaced by the same distance in the same direction. The points P are thereby displaced to points P', and a one-to-one correspondence is set up between the points P and P'. We may also say that the points P have been transformed into points P'. If the common displacements PP' are of finite length, the transformation is said to be finite; and if the dis-

---

* The kind of solution here considered is a so-called ''complete integral.''

tances PP' are infinitesimal, the transformation is called infinitesimal. Since a finite displacement can be generated by the indefinite repetition of an infinitesimal one, we conclude that a finite transformation may be obtained by repeating the corresponding infinitesimal transformation an infinite number of times. The conclusion derived from this simple illustration may be extended quite generally to all point transformations and also to contact transformations, as we now propose to show.

Consider first a finite contact transformation, in 3-dimensional space. It transforms a surface S into a surface S', which may differ considerably from S in shape and in position. Now this finite transformation can be generated by the indefinite repetition of an appropriate infinitesimal contact transformation. The effect of the infinitesimal transformation on the surface S will be to transform it into a surface S_1, situated at an infinitesimal distance from S and closely resembling it in shape. We now apply the infinitesimal transformation to the new surface S_1, and we obtain a third surface S_2 differing but little from S_1. By applying the transformation to S_2, we obtain S_3 which differs but little from S_2, and so on. Thus the indefinite repetition of the infinitesimal transformation furnishes a succession of surfaces $S_1, S_2, S_3 \ldots$ Eventually we shall reach the surface S', which we previously obtained when we applied the finite contact transformation to S directly. Thus the succession of infinitesimal transformations is equivalent to the single finite one.

According to the explanations given in the earlier part of this chapter, the surface into which a surface S is transformed by a finite contact transformation is the envelope of the surfaces into which the points P of S are transformed. These conclusions are valid whether we be dealing with a finite or with an infinitesimal contact transformation. Suppose the transformation is infinitesimal, and suppose that the surfaces into which the points P of space are transformed are tiny spheres having the respective points P as centres, and radii that may vary from point to point. In this case, each surface S_n of our sequence of surfaces will be the envelope of the tiny spheres having their centres on the preceding surface S_{n-1}. If, then, we start from the given surface S and apply the infinitesimal transformation repeatedly, we shall obtain the succession of surfaces $S_1, S_2, \ldots S'$ by successive geometrical constructions of envelope surfaces.

Now, this geometrical method of obtaining a succession of surfaces from an initial surface S is precisely the one devised by Huyghens in his construction of the equiphase surfaces for optical waves propagated through an isotropic medium of variable refractive index. We recall that in Huyghens's construction (see page 286), each equiphase surface is the envelope of tiny spheres whose centres are situated on the preceding surface and whose radii are determined by the refractive index of the medium from place to place. Consider, then, an infinitesimal contact transformation which transforms the various points of space into tiny spheres having these points as centres and having the radii of Huyghens's spheres. If we apply this transformation to an initial surface S and then repeat its application to the transformed surface, and so on, we shall obtain the successive equiphase surfaces for the waves of light in the given medium.

Next, we pass to the rays of light. In ray-optics, the rays of light are well defined; they are the orthogonal trajectories of the equiphase surfaces. To obtain a ray from the contact transformation, we apply the infinitesimal transformation to an element of an equiphase surface (*i.e.*, to a point P on this surface and to the normal to the surface at this point). The infinitesimal transformation transforms this element into another determined by a point P' on a contiguous equiphase surface and by the normal to the new surface at P'. We then apply the transformation to the new element, obtaining a point P'' and the normal to the equiphase surface that passes through P''. By proceeding in this way and then joining the points $P, P', P'', \ldots$, we obtain an orthogonal trajectory of the family of equiphase surfaces, and hence a ray.

In short, we see that the propagation of waves, the equiphase surfaces, the rays of light, and Huyghens's construction are represented mathematically by contact transformations. In particular, the contact transformations which are most suitable for the representation of wave propagations are those which we have called canonical.

Infinitesimal Canonical Transformations and Hamilton's Equations—We now consider the mathematical form of an infinitesimal canonical transformation. The infinitesimal transformation changes the variables q and p into $q + dq$ and $p + dp$. The transformation is then defined by a relation between $q + dq$ and the values of q and p, and also by a relation between $p + dp$ and these variables. Calculation shows that the general form of an infinitesimal canonical transformation is

$$(2) \qquad \begin{cases} q + dq = q + \varepsilon \dfrac{\partial F(q, p)}{\partial p} \\[2ex] p + dp = p - \varepsilon \dfrac{\partial F(q, p)}{\partial q}. \end{cases}$$

Here ε is an infinitesimal constant and $F(q, p)$ an arbitrary function of q and p. If we assume that the infinitesimal transformation takes place in the infinitesimal period of time dt, we may replace ε by dt in (2). We thus obtain the infinitesimal canonical transformation in the form

(3)
$$\begin{cases} dq = \dfrac{\partial F(q, p)}{\partial p} dt \\ dp = -\dfrac{\partial F(q, p)}{\partial q} dt. \end{cases}$$

These equations determine the changes dq and dp that the variables q and p undergo as a result of the infinitesimal canonical transformation, which acts during the interval of time dt. On the other hand, Hamilton's dynamical equations (for a conservative system moving in a conservative field of force) may be written

(4)
$$\begin{cases} dq = \dfrac{\partial H(q, p)}{\partial p} dt \\ dp = -\dfrac{\partial H(q, p)}{\partial q} dt, \end{cases}$$

where $H(q, p)$ is the Hamiltonian function of the system considered. The Hamilton equations (4) determine the changes dq and dp, in the coordinates q and in the momenta p, which take place during an infinitesimal period of time dt.

A comparison of (4) and (3) shows that Hamilton's equations are infinitesimal canonical transformations characterized by the Hamiltonian function $H(q, p)$. We conclude that the state of a dynamical system at successive instants of time, 0, dt, $2dt$, $3dt$ . . . is represented by the repeated application of an infinitesimal canonical transformation to the dynamical variables q and p. The evolution of a dynamical system has thus as mathematical expression the gradual unfolding of a canonical transformation.

To integrate Hamilton's equations, we must derive the values of the variables q and p, at any instant t, from their values q_0 and p_0 at the initial instant. Consequently, the integration of Hamilton's equations is equivalent to the problem of deriving a finite canonical transformation from an infinitesimal one. We conclude that the dynamical variables q and p at any instant t are connected with the initial values q_0 and p_0

by a finite canonical transformation.* The precise transformation varies with the instant of time t considered, but all these transformations form a class in which t acts as a mere parameter.

Hamilton's Wave Mechanics—We have just seen that, owing to the Hamiltonian form of the equations of dynamics, the evolution of a mechanical system may be represented by the gradual unfolding of an infinitesimal canonical transformation. On the other hand, we have shown that the unfolding of such a transformation affords the mathematical representation of advancing wave fronts. We may infer, therefore, that a wave accompanies the evolution of a mechanical system. This interesting discovery was made by Hamilton; it constitutes the basis of his wave mechanics.

Hamilton, however, was not led to this discovery by the considerations just outlined, for his equations of dynamics were unknown to him at the time he was establishing the connection between wave propagations and contact transformations. Accordingly, Hamilton was guided by other considerations: in particular, by the similarity between Fermat's principle in wave optics and Maupertuis' principle in dynamics. We explained this similarity in Chapter XIX. We mentioned that, in a conservative field of force, particles thrown with the same total energy describe trajectories which coincide with the paths defined by rays of light in a medium of suitable refractive index. It was this clue that led Hamilton to perceive the hidden analogy between mechanics and wave optics, and thence to obtain the Hamiltonian form for the equations of dynamics. The entity which plays the part of a wave in dynamics, is the so-called Hamiltonian Action; it passes as a wave from one surface of equal Maupertuis action to another. The trajectories of the particles are orthogonal to the surfaces of equal action, just as in optics the rays of light are orthogonal to the equiphase surfaces. These points will be clarified on pages 617 to 619.

---

* Canonical transformations may as a special case degenerate into point transformations. But this situation does not arise for the canonical transformation which connects the dynamical variables q and p of a dynamical system at time t with their initial values q_0 and p_0 at time zero. The reason is easily understood. Thus the coordinates q of a dynamical system at time t necessarily depend not only on the initial configuration of the system but also on the initial momenta p_0. Hence the q's at time t depend on the initial p_0's as well as on the initial q_0's. A dependence of this sort is realized in the general canonical transformations, but not in point transformations. In the latter the q's depend solely on the q_0's.

Groups of Transformations—In Chapter XIV, we mentioned that if A, B, C, . . . represent a set of operations, and if any two operations of this set, performed in succession, are equivalent to one of the operations of the set, then our set of operations is said to have the group property. For this set of operations to form a group, the inverse of each operation must be present in the set, and also the identical operation (which leaves everything unchanged).

In Chapter XIV we discussed Galois' groups of substitutions and noted their connection with the theory of algebraic equations. Galois' groups comprise only a finite number of different substitutions. But groups may also involve an infinite number of substitutions, as for example the group which transforms the modular function into itself. A group which contains an infinite number of substitutions may be discontinuous or continuous. This distinction is clarified when we adopt a geometric representation. Thus if the substitutions of our group have for effect to transform a point x to other points x', the group will be discontinuous if none of these substitutions transforms x to a point x' only infinitesimally distant from x. In the contrary event the group is said to be continuous. A continuous group necessarily contains infinitesimal transformations. Examples of discontinuous groups are afforded by the groups which transform the modular function or Poincaré's automorphic functions into themselves. Continuous groups of transformations were first investigated by Sophus Lie.

A simple illustration of a continuous group is afforded by the class of all translations in space. Thus let us consider the operation which consists in translating all the points of space by a specified length in a specified direction. As a result of this operation, the points P of space are transferred to points P'. We next apply a second translation which displaces the points P' in some other direction over some other distance. Points P'' are thus obtained. Since two translations performed in succession are equivalent to a single translation, we might also have transferred the points P directly to the points P'' by means of a single translation. If, then, we consider the class of all translations in space, we see that any two operations of our class, when performed in succession, are equivalent to some particular operation belonging to the same class. Thus, the class of all translations betrays the group property. Furthermore, the inverse of each translation, being itself a translation, is necessarily included in the class. Finally, if the identical operation (which corresponds to all points remaining fixed) is assumed to be one of the operations of the class, all the requirements of a group are satisfied. Since all conceivable translations are comprised in the class, the group will also contain infinitesimal translations. Consequently, the group is a continuous one.

A sub-group of the group of translations in space is obtained if we consider only those translations which are parallel to a given plane. We may go further and retain only those translations that are parallel to a fixed line in the foregoing plane. We thus secure a sub-group of the former sub-group. The group of translations in space and the two sub-groups mentioned are continuous groups. But if we select a discrete set of translations from among those of a continuous group, a discontinuous sub-group can be obtained. For instance, the set of translations which displace points by integral numbers of inches along the x-axis forms a group. And since this group contains no infinitesimal operation, it is a discontinuous group.

Translations are special instances of point transformations; but quite generally the group characteristics of the set of all translations is encountered in other transformations. Thus the set of all contact transformations forms a group. The set of all point transformations forms a group. Also, a point transformation being a particular type of contact transformation, the group of all point transformations is a sub-group of the group of all contact transformations. Similarly, the group of all point transformations in p-dimensional space is a sub-group of the group of point transformations in a space having a higher number n of dimensions. Conformal transformations form a sub-group of the group of point transformations. So also do projective transformations. Subgroups of the latter are illustrated, among others, by the group of rotations and by the group of translations. The different classes of projective transformations which we have said to be associated with the geometries of Riemann, of Lobatchewski, and of Euclid form three groups, each one of which is a sub-group of the general projective group. In the special theory of relativity, the Lorentz transformations are point transformations in 4-dimensional space-time; they form a group.

Since the transformations which form a group exhibit some common property, we may expect groups to be associated with invariant relations. Such is indeed the case. Thus the various substitutions belonging to the Galois group of a given algebraic equation have a common property: they leave unchanged the numerical values of certain rational functions of the roots (see page 162). The numerical values of such functions are therefore invariant under the substitutions of the Galois group. A similar situation holds for contact transformations, viz., the order of contact of two surfaces is invariant under any of the substitutions of the group of contact transformations. An equivalent way of expressing this result is to say that united elements are transformed into united elements. Thus "the property of being united," which the elements of a certain aggregate may betray, is an invariant property under contact transformations. The group of all conformal transformations of the plane is likewise associated with a property of invariance, namely, with the invariance of the angles at which plane curves intersect. The invariance connected with the group of all projective transformations in space is exhibited by the invariance of the anharmonic ratio of any four points on a straight line.

Clearly, a sub-group of a given group will retain all the properties of invariance that hold for the group itself. But, in addition, it may exhibit further invariant properties. For instance, the group of translations and of rotations of a rigid body in space is a sub-group of the general projective group in space. As such, its substitutions will necessarily ensure the invariance of the anharmonic ratio of any four collinear points. But we also find that the plane at infinity and the imaginary conic at infinity are invariant configurations under the substitutions of this sub-group. Furthermore, the Euclidean distance between two points remains unchanged when the two points are transformed to two other points.

Next let us consider the group of rotations about a fixed point O in space. This group is a sub-group of the previous one; and, according to what we have said, we may expect it to be associated with all the former invariants. But other invariants now appear. Thus the point O, about which the rotation takes place, will obviously remain fixed when a rotation is performed. Similarly, all spheres having the point O as centre will be transformed into themselves, and hence will define invariant surfaces. In particular, the sphere of zero radius (which is also the isotropic cone) will be an invariant surface; it intersects the plane at infinity along the conic at infinity.

The three sub-groups of the projective group which are associated with the geometries of Riemann, of Lobatchewski, and of Euclid are likewise characterized by invariant configurations. Finally, the group formed by the Lorentz transformations in the special theory of relativity can be shown to be the group of rotations about a fixed point O in 4-dimensional space-time. When account is taken of the peculiar geometry of space-time (flat hyperbolic), the invariant configurations are found to be the point O, hyperboloids of revolution, the isotropic cone (here real), the plane at infinity, and the intersection of the cone with this plane.* In

---

* In the physical theory, the generatrices of the cone are the world lines of particles moving with the velocity c of light, and so the invariance of the cone under the Lorentz transformations implies the invariance of the velocity c.

addition, the hyperbolic distance between two points (Einsteinian interval) remains unchanged.

The general significance of the foregoing explanations is that groups of transformations may be characterized by their invariant properties.

The notion of invariance we have been discussing refers to individual configurations or to mathematical expressions. But this notion also occurs in connection with families of configurations. An example of a family of configurations is given, for instance, by the family of all spheres which have a common centre. Now, it may happen that the transformations of a group do not leave the individual members of the family invariant, but that they nevertheless always transform one member of the family into another, so that the family as a whole remains invariant. Of course the family as a whole also remains invariant if each one of its members is transformed into itself; but, according to what has just been explained, this circumstance, while sufficient to ensure the invariance of the family, is not necessary. A simple illustration is the following:

Consider the family of all vertical straight lines traced in a plane; the lines are perpendicular to the x-axis. If to this family we apply the transformations of the group of all translations in the plane, each vertical line is displaced parallel to itself, and hence is transformed into some other vertical line belonging to the same family. Thus, the family as a whole is invariant, though each one of its individual members is not transformed into itself.* On the other hand, we may consider that sub-group of the former group which contains only the translations parallel to the y-axis. If we apply its transformations to our family of verticals, each vertical is transformed into itself, so that not only the family but also each member of the family is invariant.

Lie's Theory of Differential Equations—One of Sophus Lie's fundamental contributions resides in the relationships he established between differential equations and groups of transformations. A sketch of the problem involved will now be given in connection with ordinary differential equations of the first order.

A differential equation of the first order is a relation which connects an unknown function of x, $e.g.$, $y(x)$, with the first derivative $\dfrac{dy}{dx}$ of this unknown function and with the variable x. Thus,

$$F\left(x, y, \frac{dy}{dx}\right) = 0$$

---

* The line at infinity is, however, always transformed into itself. It is an invariant line, and its points are invariant points.

$\left(\text{in which } F \text{ is some function of } x,\ y,\ \text{and } \dfrac{dy}{dx}\right)$ represents a first-order differential equation. To integrate this equation means to obtain the general expression of the unknown function $y(x)$. It can be shown that the general solution will contain an arbitrary constant C, so that the general solution

$$y = y(x, C)$$

will constitute a simply infinite family of functions. If we represent these functions in the Cartesian manner on a plane, we obtain a simply infinite family of plane curves.

The integration of a first-order differential equation may prove to be impossible. This statement, however, must be construed as meaning that the solution function cannot be obtained by **quadratures***; it does not mean that methods of approximation, involving expansions in series, would be futile. The situation is similar to the one noted in the theory of algebraic equations in Chapter XIV. There we saw that the solution of the general algebraic equation of the fifth degree was impossible; but we pointed out that this impossibility was restricted to algebraic solutions, i.e., to solutions in which only algebraic operations were allowed to enter. Indeed, we mentioned that Hermite, by rejecting the limitations imposed by algebraic solutions, was able to solve the general equation of the fifth degree in terms of the modular function. The integration of a differential equation by quadratures thus plays a rôle similar to the solving of an algebraic equation by radicals.

To return to our differential equations. The mathematicians of the seventeenth and eighteenth centuries, in their attempts to integrate differential equations, applied the method of quadratures. But even in the case of the simplest differential equations, namely, those of the first order, their attempts were usually unsuccessful. In special instances, however, by utilizing various ingenious devices, they were able to solve restricted types of first-order equations, e.g., the equations of Clairaut, Bernoulli, and Lagrange. Unfortunately the various methods of solution lacked unity: each equation had to be integrated in a different way, so that no general rule could be stated. Then in the last century an important advance was made by Sophus Lie. Lie revealed the hidden unity behind the piecemeal devices of the earlier mathematicians and showed why only some kinds of equations could be solved by quadratures. We shall now examine Lie's investigations.

---

*A solution furnished by a quadrature or by quadratures is a solution expressed by an integral or by more than one integral.

Let us consider a first-order differential equation. We shall suppose that the equation is transformed into itself (*i.e.*, does not change in form) when it is subjected to the transformations of an appropriate continuous group. The equation is then said to "admit the group." The significance of this situation may be brought out in another way. Thus the solution functions, $y = y(x, C)$, of the first-order differential equation define a simply infinite family of curves, called "integral curves." The transformations which transform these curves into one another are also the transformations which transform the equation into itself, and conversely. Hence the group admitted by the differential equation is also the group which leaves invariant the family of its integral curves.

Sophus Lie established the following important theorem:

All first-order differential equations admit some group; but only when the group of the differential equation can be found is it possible to integrate the equation by quadratures. The problem of finding the group or of integrating the equations by quadratures is thus fundamentally one and the same.

Thanks to Lie's theorem, the reason why the earlier mathematicians had been successful in integrating certain first-order differential equations, but had failed with others, became clear; and in this way a unified scheme replaced disconnected discoveries.

We have dwelt at some length on first-order differential equations because of the greater simplicity of the exposition. But Lie's methods extend to the more complicated cases of differential equations of any order. Quite generally, each additional group of transformations which a differential equation may admit simplifies the problem of integrating the equation. Lie's general theory applies also to systems of differential equations and to partial differential equations. The integration of the latter is often connected with groups of contact transformations.

CHAPTER XXI

THERMODYNAMICS

THERMODYNAMICS deals with heat regarded as a form of energy and with the relationships of heat energy to all other forms, whether mechanical, chemical, or electrical. Thermodynamics is a phenomenological science: the phenomena and magnitudes which are its concern are of a kind that can be observed on the macroscopic level of common experience; direct relationships are established among the observed magnitudes, and no appeal is made to hypothetical invisible occurrences. To this rejection of speculative occurrences, thermodynamics owes its strength—but also its weakness.

Heat and Temperature—Two concepts which play an essential part in thermodynamical discussions are those of heat and temperature. In ordinary conversation heat and temperature are often viewed as practically synonymous; and we speak indifferently of the heat of the fire or of its high temperature. But in thermodynamics, as in all the sciences, the popular meanings attributed to words are usually refined, and unless these refinements are understood, confusions are apt to rise. In thermodynamics heat is regarded as a form of energy; it is an extensive magnitude, which is measured by means of a calorimeter and is reckoned in calories. Temperature is an intensive magnitude, which is measured by the thermometer and is reckoned in degrees. The following illustrations help to clarify further the difference between the two concepts.

Every substance (unless it be at the absolute zero) contains some heat energy. But the quantity of heat contained in a given substance does not depend solely on the temperature of the substance; it also depends on the quantity of substance involved, on its physical state, and on its nature. Two grams of water at a given temperature contain twice as much heat as one gram at the same temperature. Thus, for the same substance in the same phase, or state, at the same temperature, the heat content is proportional to the mass of the substance considered. The foregoing example illustrates the extensive character of heat, as opposed to the intensive character of temperature. As a further illustration, consider one gram of water and one gram of ice at the same temperature. Though the temperature is the same, the water contains

more heat than the ice. The justification for this statement results from
the observation that heat must be furnished to the ice at 0° Centigrade
if we wish to transform the ice into water at the same temperature. For
similar reasons one gram of water vapor contains more heat than one
gram of water at the same temperature. Next let us consider two different substances, *e.g.*, a gram of oil and a gram of water at the same temperature. When we attempt to increase the temperatures of both
substances by the same number of degrees, we find that less heat must
be furnished to the oil than to the water. Thus, the same increase in the
temperature is not accompanied by the same increase in the heat content.

The most striking difference between heat and temperature is illustrated when changes in the physical state of a substance are involved.
For instance, we might suppose that when heat is furnished to a substance, its temperature should necessarily increase. Yet this belief
would be erroneous. If we place a mixture of ice and water over the
fire, the temperature of the mixture does not change. The sole change
that takes place consists in a melting of the ice, and only when all the
ice has disappeared does the temperature begin to rise. In some cases
the application of heat may even lower the temperature. If we pour a
highly volatile liquid into an extremely hot cup, part of the liquid
evaporates immediately, but the temperature of the remaining liquid,
far from being increased, is lowered to such a degree that freezing may
result. These illustrations show that heat and temperature are far from
synonymous.

The Principle of Conservation of Energy—There are three
fundamental principles in thermodynamics. These are the Principle of
Conservation of Energy (or first principle), the Principle of Entropy
(or second principle), and Nernst's Heat Theorem (or third principle).
First we shall be concerned with the principle of conservation.

We saw in Chapter XIV that the total mechanical energy of an
isolated and conservative system is conserved. But we also noted that
a conservative system is a mere abstraction, and that in practice frictional effects and inelastic impacts dissipate the total mechanical energy.
This dissipation of mechanical energy occurs daily before our eyes. Thus
let a wheel be set spinning about an axis. We may assume that the
wheel and bearings form an isolated system, and yet we know that the
wheel eventually comes to a stop. Its total energy (here kinetic energy
of rotation) vanishes completely.

Leibnitz and Lavoisier ventured the suspicion that the kinetic energy
which disappears is in reality transmitted to the atoms of matter. But

this suspicion was merely a guess and led to no further consequences. In the earlier part of the nineteenth century Rumford was impressed by the tremendous quantity of heat that appeared when a hole was bored in a metal. He surmised that the mechanical work, or energy, consumed in the boring process was converted into heat by friction. Such views were, however, incompatible with the ideas of the time. In those days, heat was regarded as a fluid substance (caloric) which might pass from here to there but which could neither be created nor destroyed. It was therefore impossible to suppose that mechanical energy could be transformed into heat. Even Carnot, the originator of the second law of thermodynamics, believed that heat was a substance. But at a later date Joule conducted experiments on the dissipation of mechanical energy by friction. He found that the amount of heat produced bore always the same quantitative relationship to the mechanical energy dissipated. As a result of this discovery, the suspicion entertained by Leibnitz, Lavoisier, and Rumford was proved correct. Heat thus came to be recognized as one of the forms of energy, and it lost its status as a substance that can neither be created nor destroyed. A vanishing of mechanical energy does not always lead solely to the production of heat. Thus the mechanical energy of a waterfall may be transformed into an electric current, which, in turn, may be utilized to generate light or to bring about chemical decompositions. If, then, all those existents into which mechanical energy may be transformed are to be regarded as forms of energy, we must agree that energy can assume many different forms. Electric energy, chemical energy, light energy, and many others were thus added to the list.

The principle of conservation of energy states that, in an isolated system, the sum total of all the various forms of energy remains constant and hence is conserved. Each individual kind of energy present in the system does not, however, remain unchanged in magnitude, for transformations from one form to another may quite well take place. It is merely the sum total that is conserved. From this standpoint energy has the properties of a substance.

The distinction we made in mechanics between isolated systems which are, and those which are not, conservative no longer arises in thermodynamics. The distinction is valid only when mechanical energy alone is considered; but from the standpoint of thermodynamics, where all forms of energy are countenanced, every isolated system is necessarily conservative. The principle of conservation furnishes no information on the direction in which transformations of energy will occur. In particular, it does not tell us whether heat energy can be transformed into mechanical energy as easily as the latter may be transformed into the former. Problems of this sort, dealing with the direction in which transformations occur, will be examined when the second principle of thermodynamics is investigated.

Gases afford convenient illustrations of the principle of conservation. Suppose a gas is contained in a rigid cylinder the walls of which are impervious to heat. The volume occupied by the gas may be changed by our acting on a piston. If the force of compression exerted on the piston is less than the pressure of the gas, the gas expands, performing work against the force. But work is mechanical energy, and according to the principle of conservation this mechanical energy cannot have arisen out of nothing. It may be shown that the mechanical energy results from the transformation of an equal amount of the internal energy of the gas, and that the drop in the internal energy betrays itself by a fall in the temperature of the gas. Similarly, a compression of the gas increases its internal energy and thereby its temperature. In these illustrations only the changes in the internal energy come into consideration; the total value of the internal energy present in the gas is irrelevant. In point of fact the methods of thermodynamics furnish no means of computing the absolute, or total, value of the energy contained in an isolated system. The special theory of relativity, however, by its identification of mass with energy, sheds some light on the matter.

Reversible Transformations—The considerations we are about to develop are intimately connected with two major categories of transformations: those that are performed "reversibly" and those that take place in an "irreversible" manner. The prototype of reversible transformations is to be found in the behavior of the conservative systems of dynamics. If, in any such system, the velocities of all the component parts be reversed at any instant, the system will go through exactly the same succession of configurations, but in the reverse order. For instance, the solar system is very approximately a conservative one. If at any instant the velocities of all the planets, of all the satellites, and of the sun could be reversed, the solar system would pass in inverse order through exactly the same sequence of configurations. Accordingly, we say that the succession of changes undergone by the solar system is performed reversibly. Let us observe, however, that the solar system, or any other mechanical system, is not truly conservative since frictional influences cannot be avoided in practice. Hence no mechanical process can ever be truly reversible: rigorous reversibility is thus an ideal limit, and irreversibility is the rule.

The distinction between reversible and irreversible processes is of particular importance in thermodynamics. A thermodynamical process is said to be taking place reversibly when it is susceptible of occurring in either direction. In the contrary event the process is said to be taking place irreversibly. The following illustrations will clarify the distinction.

Suppose we place a small piece of ice in a large volume of hot water. The ice melts and soon disappears, and the water remains hot during the process of melting; a change, or transformation, has taken place. Let us consider the reverse succession of states, such as would be illustrated in a motion picture film turned backwards. We should now start with hot water, and, while the water remained warm, ice would gradually form. But this reverse succession of states cannot be obtained in practice, and for this reason we say that our original transformation of ice into water was performed irreversibly.

We may, however, operate in another way. Thus suppose a piece of ice is placed not in warm water, but in water at 0° Centigrade (the pressure to which the mixture of ice and water is subjected is assumed to be the pressure of the atmosphere). No change occurs. Let us now increase very slightly the temperature of the surrounding air, while keeping the pressure fixed. We shall find that the ice gradually melts. Conversely, if we decrease the temperature very slightly below 0° Centigrade, the reverse transformation will take place, *i.e.*, water will freeze. In all rigor neither of these transformations is performed reversibly. For instance, when the temperature is above 0°, ice becomes water and water does not transform into ice. But it must be observed that a change in the temperature from a value only infinitesimally greater than 0° to one only infinitesimally below, will generate a reversal of the transformation. Hence, at the limit, we are justified in saying that, when the temperature is in the immediate vicinity of 0°, the transformation between ice and water may take place equally well in one direction or the other. We shall, therefore, agree to say that under such conditions the transformation is performed *reversibly*. If we contrast the previous illustration with the experiment in which a piece of ice was thrown into hot water, an important difference appears. In the case of the ice and the hot water a slight change in the temperature was not sufficient to produce ice at the expense of water; instead, a considerable drop in temperature was required.

The illustration we have examined for a reversible change in thermodynamics involves the temperature. But we might also consider cases in which the temperature is fixed and the pressure slightly varied. Thus in our example of the mixture of ice and water at 0° Centigrade and under atmospheric pressure, let us keep the temperature fixed and make the pressure vary. The slightest increase in the pressure will cause the ice to melt; the slightest decrease will generate the reverse transformation.

The two illustrations just given show that a slight variation in the causes suffices to reverse the direction of a reversible transformation. We conclude, therefore, that the causes which produce a reversible transformation must be extremely weak (*e.g.*, very small differences in temperature or pressure). But then it follows that a transformation taking place reversibly must be proceeding extremely slowly. Furthermore, since a very small variation in the causes may generate a change in one direction or another, a substance undergoing a reversible transformation is at all instants in a state which differs only infinitesimally from one of indifferent equilibrium. This situation has its analogue in mechanics. For instance, if a ball is placed on a perfectly horizontal and frictionless table, it will remain motionless and will be in a position of equilibrium. The state of equilibrium is of the kind called indifferent, because the slightest force exerted on the ball to the right or to the left will cause it to move in one direction or the other.

Work—In mechanics, when a constant force is pulling a body, say along a straight line, the work done by the force during the displacement is defined by the product of the force and the displacement. We must distinguish, however, between the work done by the force and the work done by a system. The following example will make this point clear.

Let us suppose that a gas is compressed in a cylinder by means of a piston assumed frictionless and massless. The gas exerts a force against the piston, and if we release the piston, it will recede. The force multiplied by the displacement of the piston defines a certain amount of work, so that the expansion of the gas would appear to generate work on the surroundings. But in point of fact, the gas system need not be delivering work to the surrounding medium. For the expansion of the gas to generate work, the recession of the piston must be resisted by an opposing force. If the cylinder is placed in a vacuum, no opposition of any sort is developed and no work will be performed by the expanding gas. If, on the other hand, some opposing force, whether it be due to a spring or to the atmospheric pressure, is acting on the piston, the gas will generate work against the outside force during its expansion. According to the principle of conservation of energy, the sum total of the work performed by the gas and of the energy contained in the gas must remain the same at any instant. Consequently, if the gas is performing work during its expansion, its internal energy is being transformed into work as the expansion proceeds.

In mechanics, when we are dealing with conservative systems, the only forms of energy to be considered are the kinetic and the potential varieties. These forms, commonly called "mechanical energy," may be transformed completely into work. Conversely, by expending work on a conservative mechanical system, we may increase its store of mechanical energy. In either case the transformation from mechanical energy into work, and vice versa, can be performed completely, so that work and mechanical energy are interchangeable. The name "available energy" is often given to any form of energy which can be transformed totally into work. According to this definition it is obvious that mechanical energy is always of the available kind. But as we shall see presently, not all forms of energy are available energy.

Carnot's Principle—We have said that heat is a form of energy. The next step is to determine its relationship to mechanical energy. We know that mechanical energy can be transformed totally into heat. For example, a flywheel in rotation eventually comes to a stop; its mechanical energy is converted into heat by friction. But, for heat to be available energy, the converse must also be true, i.e., it must be possible to transform heat totally into work, or mechanical energy. We shall now see that heat does not satisfy this requirement.

The first clue in this matter was afforded by Carnot in his theoretical study of the steam engine. Carnot made the important observation that work cannot be derived from heat unless this heat be allowed to fall from a higher to a lower temperature. This statement constitutes "Carnot's Principle."

In the steam engine heat falls from the hotter boiler to the colder condenser; and from this heat-fall is obtained the work, or mechanical energy, delivered by the engine. When we recognize that heat is a form of energy and when we take the principle of conservation of energy into account, the source of the mechanical energy furnished by the engine is clear: this mechanical energy must result from the transformation of the falling heat. Carnot, however, initiated his investigations before the nature of heat was properly understood, and so he shared the then prevalent belief that heat was an indestructible substance. For this reason he did not suspect that the work generated by the engine might merely represent the heat in another form. According to Carnot the fall of heat in the engine was in many respects analogous to a waterfall which sets a waterwheel into rotation and thereby furnishes work. In this hydrodynamical simile, the work produced is generated not by the water as such, but by its fall. Carnot therefore concluded that the work furnished by the engine was generated not by the heat as such, but by its fall.

If Carnot's mistaken interpretation be accepted, the quantity of heat reaching the colder condenser should be equal to the quantity falling from the hotter boiler. Direct measurements soon proved that this expectation was incorrect, for they showed that a considerable amount of heat disappeared entirely during the fall. To return to our simile of the waterfall, it would be as though only a part of the water that started from the mountain ridge reached the valley; the remainder would vanish. The discovery that some of the heat disappeared in the steam engine afforded further evidence that heat was not an indestructible substance but was merely one of the forms which energy could assume. The revision of Carnot's original interpretation was the work of Clausius and Kelvin, but Carnot's fundamental observation on the connection between a fall of heat and the generation of work was correct, so that the general principle uncovered by him still bears his name. Carnot also made other discoveries of great importance which we shall now review.

Carnot's other discoveries will best be understood if we examine a particular illustration. Let us suppose we have two hot bodies at the same temperature, e.g., two hot bricks. If we place the bricks in contact, no flow of heat arises, and hence the heat energy contained in the bricks cannot be transformed into mechanical energy; it is thus utterly unavailable. Very different would be the situation if, in place of the two bricks containing heat energy, we were dealing with a mechanical system having mechanical energy. In this case all the energy would be available and could be transformed into work if we harnessed the system in a suitable way. Next let us assume that the bricks are at different temperatures. If they are placed in contact, a flow of heat falls by conduction from the hotter to the colder body; and even if the bricks are separated and are placed in vacuo, the flow occurs through radiation. In either case the bricks eventually reach the same temperature, which we may assume to be midway between the two initial temperatures; from then on, no further flow occurs. In spite of this heat-fall no mechanical energy is generated. We conclude that a fall of heat is not sufficient by itself to generate work. But the point brought out by Carnot is that work *might* have been obtained had we harnessed the fall in an appropriate way.

The steam engine constitutes one of the devices that may be utilized to harness a fall of heat. In the steam engine the two bricks of our previous illustration are replaced by the boiler and the condenser. Heat is constantly supplied to the boiler by the furnace, and this heat then falls to the colder condenser. Since heat is constantly supplied to the boiler, the temperature of the boiler does not decrease; similarly the colder condenser remains at a fixed lower temperature owing to radiation. For this reason, provided the furnace be kept alive, the fall of heat continues indefinitely and does not come to a stop as it would in the case of two bricks left to themselves. The mechanism of the steam engine ensures the transformation of a part of the falling heat into mechanical energy; in this sense the fall is harnessed.

We must now determine what percentage of the falling heat is transformed into work. By means of a theoretical argument it can be shown that even when the heat-fall is harnessed in the most perfect manner conceivable, only a fraction of the falling heat energy can be transformed into work. The fraction that can be thus transformed is called the *available heat energy*; the remainder, which falls without generating work, is the *unavailable heat energy*. This discovery furnishes the answer to a question we raised previously. We inquired whether heat energy was available energy. We now see that heat energy is not entirely of the available kind.

The precise quantitative relation between the available and the unavailable heat energy in any particular case was established by Clausius and Kelvin. We shall restrict ourselves to a bare statement of their results.

Let us call T_1 and T_2 the absolute temperatures of the hotter and of the colder body at the instant of interest, and let us consider the amount of heat that falls during one second. What Kelvin proved was that under ideally perfect conditions of operation, only a fraction, $1 - \dfrac{T_2}{T_1}$, of the falling heat can ever be transformed into work, and is thus available energy. The remaining percentage of the falling heat, *i.e.*, $\dfrac{T_2}{T_1}$, is unavailable; it falls without undergoing any transformation.

We may apply these results to the case of the two bricks. During the first second the percentage, $1 - \dfrac{T_2}{T_1}$, of the falling heat may be converted into work (provided the fall be harnessed). During the following second, more heat falls and more work may be extracted; but the temperatures of the two bricks will now have come closer together, with the result that the fall will be less important and less heat will flow. Aside from this difference, the same percentage of the falling heat will be available. Eventually both bricks will attain the same temperature, and the phenomenon will be at an end. If then we add the amounts of available energy present in each successive fall, we shall have the total available energy involved in the fall.

Let us make clear that the available energy refers to the heat energy that can *under the most ideal conditions* be transformed into mechanical energy, or work. In practice, a perfect harnessing of the fall is impossible, and only a part of the available energy is converted into work; the remainder of the available energy, which we have failed to utilize, falls with the unavailable energy proper, and when it has fallen, it becomes unavailable. As an extreme case let us suppose that no harnessing device is applied. In this event the heat falls by conduction or radiation; no work is accomplished; and the available energy, which we have now failed to avail ourselves of, becomes unavailable and is irretrievably lost. We are thus led to differentiate between energy in the most general sense, and available energy. The former has many of the attributes of a substance, for according to the law of conservation if it disappears here it reappears there (albeit in a different form). But available energy has no such conservative properties, since unless it be saved and stored, it becomes unavailable.

A further point is that the entire concept of available energy has a definite meaning only with respect to a given situation. For instance, if we are dealing with a single hot body, there is no available energy. If now we place the hot body in contact with a colder one, available energy comes into existence; and the colder the cold body, the greater the amount of available energy present. It is thus apparent that the distinction we have made between available and unavailable energy does not connote any intrinsic difference between two kinds of energy; it merely refers to a difference dependent on the conditions contemplated.

Another feature of great importance established by Carnot is the nature of the requirements that must be satisfied if the entire amount of available energy is to be converted into work. Carnot showed, by means of a theoretical argument on the steam engine, that the maximum of efficiency is realized when the motion of the engine proceeds with infinite slowness and when only infinitesimal differences of temperature and of pressure are present between bodies in contact. In these circumstances the fall of heat also proceeds with extreme slowness. Carnot's conditions of maximum efficiency are none other than those which are associated with all changes operating reversibly. Hence we may say, quite generally, that the most efficient way of utilizing a fall of heat to generate mechanical energy is to operate under reversible conditions.

But even under ideal reversible conditions and when no engineering losses of any kind are taken into account, only that percentage of the falling heat which we have called the available energy is converted into work. How small this percentage is for ordinary differences of temperature may be gathered from the following illustration: If the hotter body has the temperature of boiling water ($T_1 = 373°$ in absolute measure) and the colder body is kept in ice ($T_2 = 273°$), the ratio of the available energy to the total amount of heat that falls every second is only $1 - \dfrac{273}{373}$, *i.e.*, approximately $\dfrac{1}{4}$. The remainder of the falling heat, *i.e.*, ¾, is unavailable in any case. All the conclusions to which we have been led hold true regardless of the particular harnessing device we may utilize (*e.g.*, a steam engine, a hot air engine, or any other kind of machine). Our conclusions are thus seen to betray a general law of Nature and not a mere peculiarity of the steam engine. Therein lies their importance.

Thus far we have examined a spontaneous fall of heat from a hotter to a colder body, and we have seen that this fall involves a certain amount of available energy. The direction of the flow of heat may, however, be reversed, and we may make heat pass from a colder body to a hotter one. But to do this, work must be expended. As an example in point, suppose we have two bodies at the same temperature. By performing work on the system we may withdraw heat from one of the bodies and deliver it to the other. The body losing heat will become colder and colder while the second body increases in temperature.* The result of our efforts will be to create a difference in temperature where a uniform temperature originally prevailed. As Carnot showed, the reverse passage of heat from a colder to a hotter body may be obtained in a steam engine by expending work on the engine and constraining it to operate backwards. All the changes that take place when the engine is in normal operation are then reversed; in particular, heat passes from the colder condenser to the hotter boiler. By means of a theoretical discussion Carnot proved that, if we wish to transfer a given amount of heat from

---

* It is this reverse passage of heat which is illustrated in refrigerating machines.

the colder to the hotter body while expending a minimum amount of work, we must operate with infinite slowness, *i.e.*, reversibly.

A comparison of the two processes is instructive. Suppose we start with two bodies at temperatures T_1 and T_2, the first body being the hotter. By harnessing the heat-fall and by operating reversibly, we may salvage all the available energy, converting it into mechanical energy. When the fall is completed, the two bodies are at the common temperature T_0. We now attempt to reverse the process by expending work. Our aim is to restore the intial conditions, with the bodies at temperatures T_1 and T_2 respectively. Carnot's discussion shows that this may be done and that, provided we operate reversibly, the work we shall have to expend is precisely equal to the maximum amount of work we might have obtained from the original fall. Suppose, then, that the two processes are performed in succession. We first allow the heat to fall and we salvage all the available energy, storing it for example as potential energy in a compressed spring. According to what has just been said, the mechanical energy thus stored is precisely sufficient to reverse the fall and to restore the initial situation. Let us utilize this stored energy to reverse the fall. We shall then be back in the initial situation. We may then allow the fall to occur afresh. In theory we might continue in this way indefinitely. During each fall the available heat energy would be converted into the same amount of mechanical energy, and during the reverse process the mechanical energy would be transformed back into available heat energy. Every step and every transformation occurring in one process would take place in the opposite direction in the reverse process.

In practice, however, ideal reversible operations are impossible; and, just as in the direct process not all the available energy can be converted into mechanical energy, so also during the inverse process not all the work we expend can be consumed in raising heat from the colder to the hotter body. In practice, if the back-and-forth process is attempted, there will be a decrease in the available energy at each successive cycle; eventually, the two bodies will be at a common temperature and there will be no mechanical energy stored in the spring.

Spontaneous Transformations—The name "spontaneous", or "natural", transformations is given to those processes which occur in Nature without our having to expend work. The fall of heat from a hotter to a colder body is an example of a spontaneous transformation; the reverse flow, though possible, can be obtained only by the expenditure of work and is therefore not a spontaneous process.

Many other illustrations of spontaneous processes may be given. Here are a few taken at random:

(a) A gas, which at an initial instant occupies only a part of the volume at its disposal, immediately expands so as to fill the entire volume. This phenomenon is often described by the statement: A gas expands spontaneously into a vacuum. The reverse change, namely, a spontaneous contraction of the gas, is not observed.

(b) If two different gases are placed in two communicating vessels, we find that the two gases gradually intermingle and yield a homogeneous mixture. The reverse passage from a uniform mixture to a separation of the two gases does not take place spontaneously.

(c) A piece of sugar placed in a glass of water dissolves, yielding a uniformly sweet solution. The reverse passage from the solution to the lump of sugar does not occur spontaneously.

(d) Fluorine and hydrogen combine spontaneously under ordinary conditions of pressure and of temperature, forming hydrofluoric acid. But under the same conditions, the acid does not decompose into fluorine and hydrogen of its own accord.

All such illustrations show that the direction in which a spontaneous change takes place is well defined. In other words, if we take a motion picture film of a spontaneous transformation, we find that, on reversing the film, we obtain a picture which is not observed in Nature.

The considerations developed in connection with the fall of heat apply to spontaneous changes generally. Thus in any spontaneous change there is a certain amount of available energy at our disposal, but this available energy can be utilized only if we are able to harness the change. The totality of the available energy can be salvaged and transformed into work, or mechanical energy, provided our harnessing device causes the change to occur reversibly (*i.e.*, infinitely slowly). If the change is poorly harnessed, so that it occurs only in a semi-reversible way, a part of the available energy may be salvaged, but the remainder is irretrievably lost, being degraded into an unavailable form. The extreme case of degradation arises when no harnessing device is applied; all the available energy is then degraded.

The method of harnessing the various spontaneous changes depends on the change studied. In the spontaneous change exhibited by the fall of heat, we have seen that the steam engine serves as a harnessing mechanism. In the case of a gas expanding into a vacuum, the harnessing is secured when we resist the expansion: we cause the expanding gas to press against a piston, thereby compelling it to perform work. By this means the available energy can be utilized. A similar method of harnessing can be applied to the dissolution of a piece of sugar in water. We place the sugar and water in a cylinder in which a porous piston can be moved. The pores of the piston are assumed sufficiently large to allow the water molecules to pass freely, but they are small enough to arrest the larger sugar molecules. The sugar molecules then behave like the molecules of a gas, colliding against the piston and causing it to recede. If we oppose the recession of the piston, constraining it to occur with extreme slowness, we shall be under the conditions of reversible operation, and we may recover in the form of work the available energy involved in the dissolution of the sugar. Finally, a chemical reaction, such as that of the combination of hydrogen and fluorine, may be harnessed in theory by causing the reaction to proceed in a galvanic cell. The available energy connected with the reaction is delivered under the form of an electric current, and this latter type of energy is equivalent to mechanical energy, or work.

When discussing the spontaneous fall of heat, we mentioned that the fall could be reversed by the expenditure of work. A similar reversal can be secured in any of the spontaneous changes. To take one illustration, let us consider the dissolution of sugar in water. By pushing against the porous piston, we may compress the sugar molecules into one part of the cylinder, for the water molecules pass freely through the piston. All the sugar molecules will thus be concentrated into a small space, and pure water will remain on the other side of the piston. In short, the work we have expended by pushing forward the piston against the pressure of the sugar solution has enabled us to reverse the original natural change and restore the available energy. A similar situation is presented in a chemical reaction. We may reverse the spontaneous combination of fluorine and hydrogen by directing an electric current through hydrofluoric acid; but to generate this current, we must expend work.

In all these reversals, the work we have to expend, under the most favorable conditions of reversible operation, is equal to the work that might have been derived in the direct process. In practice, however, owing to inevitable irreversibility, it will be impossible to transform all the work into available energy; part of it will always be degraded.

The Second Principle of Thermodynamics—Suppose we have an isolated system in which changes are occurring. The changes to be considered are of two kinds. First, there are the spontaneous changes, such as are illustrated by the expansion of a gas into a vacuum or by the fall of heat from a hotter to a colder body. Secondly, we have the inverse changes, which can be brought about by the conversion of mechanical energy into other forms of available energy. We must also consider how these changes take place—irreversibly or reversibly.

Let us first examine the spontaneous changes. We have seen that they are associated with a certain amount of available energy, which in theory may be transformed totally into mechanical energy and hence into a form of energy which is itself available. The transformation is brought about by harnessing the change and by operating reversibly (extremely slowly). However, since perfectly reversible operations are impossible, only a part of the available energy can be thus transformed in practice, so that some of the available energy will become unavailable. Suppose, then, that spontaneous changes are occurring in an isolated system. As a result of the foregoing considerations, we conclude that while the changes are proceeding, the total amount of available energy within the system will always be decreasing (or at best, in the limiting case of ideally reversible changes, will remain stationary). In any case the available energy can never increase. We next pass to the reverse changes. Let us suppose that our isolated system is at uniform temperature and that it contains mechanical energy stored, for example, in a compressed spring. The mechanical energy, if it is released, may cause heat to pass from one part of the system to another, with the result that local differences in temperature are generated in the system. Available heat energy is thus brought into existence. But during the process, the mechanical energy stored in the spring will disappear, and unless the change be performed reversibly (a practical impossibility), the mechanical energy that disappears will more than offset the available energy that is generated. Thus, the sum total of the available energy in all its forms will decrease as a result of the change, or, at best, will remain stationary (reversible operation).

From this analysis it appears that the total amount of available energy in an isolated system can never increase. Either it will decrease as the changes proceed, or else, as an extreme limiting case (when the changes occur reversibly) it will remain stationary. We have here one of the modes of expression of the second principle of thermodynamics; and since perfectly reversible changes may be dismissed as impossible in practice, the second principle can be stated more concisely as follows:

Any change that takes place in an isolated system is accompanied by a decrease in the total available energy of the system.

The second principle of thermodynamics enables us to determine under what conditions further changes in an isolated system are impossible. We must first observe that some changes are excluded by the restrictions imposed on the system. For instance, if the system has a fixed volume, we cannot contemplate changes which would alter this volume. We also know, according to the principle of conservation of energy, that the total energy content of the system must remain constant. We cannot therefore countenance changes which would alter the total energy. Let us now apply the second principle. It tells us that a contemplated change is possible only if it results in a decrease of the available energy. As time passes, changes may continue until all the available energy is degraded into unavailable forms. When this situation is realized, further spontaneous changes become impossible, and the system is said to be in *thermodynamical equilibrium.* The equilibrium is stable if all conceivable changes from the given configuration are accompanied by an increase in the available energy. The equilibrium is indifferent if none of the conceivable changes yields a decrease in the available energy, but one at least of these changes entails no modification in the value of the available energy.

If we apply the second principle of thermodynamics to the entire universe viewed as forming an isolated system, we are led to the conclusion that eventually all change must come to an end. Differences of temperature and of pressure will be evened out; all motion will cease as a result of frictional influences; and the universe will die of inanition. These gloomy conclusions are necessary consequences of the second principle of thermodynamics, and their correctness is thus dependent on that of the second principle. In this connection we must remember that the second principle is a mere generalization from commonplace experience: heat flows from a hotter to a colder body, and so on. But from our present standpoint there is no rational necessity for this unidirectional passage of heat, and a man who was unfamiliar with heat conduction might perfectly well countenance a spontaneous reversal of the flow. Inasmuch as extrapolations from the commonplace level of macroscopic experience to the more remote ones are always dangerous, the only claim we can safely make is that the second principle is valid so long as we remain on the commonplace level. Indeed the wisdom of adhering to this cautious attitude is amply demonstrated in the kinetic theory of gases, where the remoter level of microscopic masses is explored. In this remoter level the second principle is seen to have a mere statistical validity. Conceivably, in other levels still more remote, there may be other influences at work, reversing the action of the second principle. A decision on this matter is any man's guess. But in view of the possible existence of such influences, the safer course is to refrain from attaching too much importance to those wider implications of the second principle which refer to the universe as a whole.

It is instructive to contrast the second principle of thermodynamics with the first principle, *i.e.*, with the principle of conservation of energy. The first principle makes no distinction between the different forms of energy; it merely states that the total amount of energy in an isolated system remains constant. The second principle then adds that the available energy will be transformed into unavailable energy. Thus the second principle forces us to recognize that the various forms of energy differ in status.

Energy and Quality—The second principle of thermodynamics introduces the notion of quality in connection with energy; we are thus led to establish a hierarchy of the various forms of energy. We shall agree that if two different kinds of energy *A* and *B* can be transformed completely (in theory) one into the other, then both these forms of energy are of the same grade. On the other hand, if the form *A* can be transformed completely into the form *B* whereas the reverse transformation can never be complete, then the form *A* will be regarded as of higher grade than the form *B*. For instance, mechanical energy can be completely transformed (in theory) into the energy of the electric current, and vice versa. These two forms of energy are thus of the same grade. They are both forms of available energy and, as such, are of the highest grade possible.

We now consider heat energy. Mechanical energy may be transformed totally into heat, *e.g.*, by friction; but the reverse transformation can never be complete. It is true that, by allowing the heat to fall to a lower temperature, we may convert a part of it into mechanical energy, but never more than a part will be transformed.* We conclude that heat is a lower, or a more degraded, form of energy. A distinction may also be made between heat at different temperatures. As an illustration we consider two bodies *A* and *B* which are at different temperatures but which contain the same amount of heat. Let *A* be at a higher temperature than *B*. If we allow the heat to fall from the body *A* to a body *C* colder than either *A* or *B*, and if we repeat the same experiment with the bodies *B* and *C*, we know, according to Carnot's findings, that a greater amount of heat can be transformed into work in the first experiment than in the second. We may therefore say that heat at a high temperature is of

---

* The entire amount of heat could be transformed into work if the heat were to fall to the temperature of the absolute zero. We exclude this eventuality as impracticable.

higher grade than heat at a lower temperature. Heat is one of the lowest forms of energy; and so also are the latent heats of fusion and evaporation. Chemical energy, *i.e.*, the energy that is liberated in a chemical reaction, occupies an intermediary position; part of it is high-grade and the residue low-grade. Light is also a low-grade form of energy.

Whenever energy passes from a high-grade into a low-grade form, some available energy is degraded and becomes unavailable. For instance, when mechanical energy is converted into heat by friction, available mechanical energy is degraded, since heat energy is never entirely available. Similarly when heat falls by conduction from a higher to a lower temperature, a certain amount of available energy is degraded; this available energy is the energy we might have transformed into work had the fall been harnessed, as in the steam engine.

From these illustrations we gather that the degradation of energy, and the transformation of available energy into unavailable forms, express the same condition. Consequently, the second principle of thermodynamics may be stated:

> The changes occurring in an isolated system result in a degradation of energy.

The second principle is therefore sometimes called the "Principle of Degradation of Energy."

In practice we cannot avoid some measure of degradation, even when high-grade forms of energy are transformed *inter se*. Friction, viscosity, the Joule effect (whereby an electric current heats the wire through which it is passing), and magnetic hysteresis are among the agents of degradation. Insofar as we may wish to generate mechanical energy, it will be to our interest to prevent degradation whenever possible. Thus by means of the steam engine we harness a fall of heat and save from degradation at least a part of the available energy. But of course, while the steam engine is operating, coal or oil is being burned to heat the boiler; and the degradation attendant on these processes of combustion is far greater than the degradation that is obviated through the operation of the engine. The galvanic cell is far more efficient. Thanks to it, the available energy in a chemical reaction (which, if the reaction occurred spontaneously, would be degraded into heat) is saved and transformed into high-grade electric energy.

In any transformation of energy where it is not our primary purpose to obtain heat, we must avoid its generation. For instance, if we wish to transform electrical energy into light-energy, it will be to our advantage to effect the transformation directly. The earlier electric lamps were unable to secure this result, for the light was produced by incandescence and hence by the transformation of the electric energy into heat. Modern lamps have overcome this defect with increasing efficiency. But even today we cannot effectively rival the cold light of the fire-fly.

Fires are agents of degradation because they degrade into heat the available energy which is released when oxygen and carbon combine. Green plants, by reversing this reaction (under the influence of sunlight), counteract degradation and restore available energy. At first sight we might suppose that the chlorophyllian action of plants violates the second principle of thermodynamics. But our supposition would be incorrect; our error would arise from a failure to apply the second principle correctly. The principle only applies to isolated systems, and the available energy that is degraded according to the principle refers to the sum total of the available energies in the entire system. The principle does not apply to local changes within the system. For instance in the steam engine, high-grade mechanical work is obtained at the expense of low-grade heat, so that regeneration, and not degradation, is taking place. But this is only part of the process, for concurrently with the first transformation, heat is falling from a higher to a lower temperature and is thereby suffering degradation. The second principle would be at fault only if the regeneration occurring in the first process exceeded the degradation taking place in the second; and such is not the case.

The same analysis must be followed in connection with the chlorophyllian action of green plants. We must consider not only the plant and the surrounding air, but also the source that emits the light. The plant, the air, and the source of light form a single isolated system, and it is to the processes occurring in this total system that the second principle of thermodynamics must be applied. When this course is followed, the sum total of the available energies is not found to increase, though the decrease is considerably smaller than it would be were no chlorophyl present in the leaves of the plant.

Entropy—We have seen that any change occurring in an isolated system necessarily entails a decrease in the available energy of the system. An immediate consequence is that the system will be in a state of thermodynamical equilibrium whenever the available energy is a minimum. Now the foregoing conclusions may be expressed in an equivalent way in terms of a new thermodynamical magnitude, which was introduced by Clausius and to which the name "entropy" was given.

Consider a given mass of gas in some particular state. The gas then occupies a definite volume v, exerts a definite pressure p, and is at a definite absolute temperature T. Any two of these three magnitudes suffice to determine the third, so that we may select, for example, the volume v and the temperature T to define the state of our given mass of gas. We propose to make the gas pass from some initial state to some final one. We may secure this passage by compressing the gas or allowing it to expand, and also by delivering or subtracting heat. There is always an infinite number of different ways in which we may pass from the initial to the final state. To take a specific illustration let us suppose that in the final state the volume is greater but the temperature is the same; and let us assume that the passage from the initial to the final state is obtained by allowing the gas to expand at constant temperature.

Now this general statement of the sequence of changes which we propose to make the gas follow, from the initial to the final state, is not sufficiently precise to determine the transition unambiguously, for we shall see that the same sequence of changes may be followed in different ways, *viz.*, reversibly or irreversibly.

For example, we may compel the expansion to take place against an equilibrating pressure and hence with extreme slowness (reversibly); or again, we may allow the expansion to proceed unresisted, against no pressure (as in a vacuum). In the former case, the gas performs work during the expansion, and so we must supply heat to it in order to maintain a constant temperature. In the second case, where the expansion is not resisted, no work is done by the gas and its temperature does not fall.* The final state we wished to obtain is thus reached without our having to furnish heat. These two modes of passage from the same initial to the same final state comprise exactly the same sequence of intermediary states, but they differ in an important respect: the first passage is performed reversibly, the second irreversibly. Any number of other modes of transition could be considered. We might start by compressing the gas or by cooling it, or we might alternate compressions and expansions. In any event, regardless of the particular transition from the initial to the final state, the passage could always be made to occur reversibly or irreversibly, at pleasure.

Suppose, then, we fix our choice on some particular transition, or "path," extending between two prescribed terminal states. For convenience we assume that the continuous path is split up in thought into a discrete succession of states, each intermediary state differing only in-

---

* This statement applies to a perfect gas. In the case of a real gas a lowering of the temperature would result.

finitesimally from its predecessor. In other words, the volume and the temperature of the gas differ but slightly from one intermediary state to the next. The discrete succession of states may be secured by modifying the volume in a minute degree at each step and supplying (or withdrawing) some appropriate small quantity dQ of heat. The amount of heat transferred at each step is assumed so small that it leaves the temperature of the gas approximately unchanged during the transference. According to this procedure we shall begin by supplying (or withdrawing) a quantity dQ of heat when we start from the initial state. The temperature of the gas is then T. At the same time we may modify the volume slightly. We thus obtain the next intermediary state, for which the temperature is, say, T' (where T' differs but little from T). We now furnish (or withdraw) some other small quantity dQ' of heat and modify the volume slightly, and so on, till the final state is reached.

Clausius considered the ratio of the quantity of heat, supplied (or withdrawn) at each step, to the temperature of the gas in the momentary state of interest. He added all these ratios, obtaining the sum

$$\frac{dQ}{T} + \frac{dQ'}{T'} + \ldots\ldots, \text{ or } \Sigma \frac{dQ}{T} \text{ for short.}$$

In this expression the dQ's are assumed positive if heat is received by the gas and negative if the gas is giving out heat. Clausius proved that the above sum, extended to all the intermediary states of the chain from the initial to the final state, has a value which depends not only on the terminal states, but also on our choice of the sequence of intermediary states. Furthermore, he proved that the value of the sum depends on the way in which the small transformations from one intermediary state to the next are performed, i.e., on whether the changes occur reversibly or irreversibly. Inasmuch as there is an infinite number of different ways of passing from the initial to the final state, and since different values are obtained for Clausius's sum according to the path followed and to the method of transition, it would appear that the value of this sum cannot measure any intrinsic difference in the extreme states. But Clausius proved that if the successive transitions are made to occur reversibly, the value of the sum will always be the same, regardless of the intermediary succession of states that may be contemplated. Thus, when we restrict our attention to reversible changes, the value of the sum depends solely on the terminal states. For this reason it measures some kind of an intrinsic difference between these two states, a difference which Clausius called a difference in *entropy*. Entropy is a highly abstract concept, the significance of which will become clearer when particular examples are discussed.

Clausius's definition gives only the difference in the entropy of the gas in the two terminal states; it tells us nothing of the absolute value of the entropy. The absolute value of the entropy can be determined only if we are able to agree on the state of the gas which corresponds to zero entropy. For then, the difference between the values of the entropy of the gas, in any given state A and in the zero state, automatically becomes the absolute value of the entropy in the state A. The third principle of thermodynamics, to be discussed presently, attempts to define the state of zero entropy and thereby to determine an absolute value of the entropy. However, the determination of an absolute value is not a pressing matter, because in the majority of cases all we need be concerned with are differences in entropy. Absolute values may therefore be left in abeyance for the present.

Entropy is an extensive magnitude. Thus the entropy of two grams of gas in a given state (*e.g.*, defined by the temperature and pressure) is twice that of one gram of the gas in the same state. Hence a given mass of gas at specified temperature and pressure may be said to contain a certain amount of entropy. But entropy, as we shall see, is not a substance, for it is not conserved. Quite generally, if a body at temperature T receives a quantity dQ of heat, we say that the body has received an amount $\dfrac{dQ}{T}$ of entropy. If heat is surrendered by the body, entropy is lost. It is not necessary to specify that the gain or the loss of heat occurs reversibly in this case. For instance, if the transfer of heat by conduction were to occur reversibly, the source surrendering the heat would have to be at a temperature only infinitesimally greater than the body receiving the heat. But obviously, since dQ and T refer to the body which receives the heat, the value of $\dfrac{dQ}{T}$ would be exactly the same even if the source were considerably hotter.

We have illustrated the concept of entropy in reference to a gas. But the conclusions to which we have been led are general; they apply to any substance or system of bodies. The total change in the entropy of an isolated system, when the system passes from one state to another, is defined by the sum total of the changes in entropy of the various constituents of the system.

The Principle of Entropy—The second principle of thermodynamics is compressed in the statement: Any change occurring reversibly in an isolated system leaves the total available energy constant; any irreversible change causes the total available energy to decrease. As will be explained presently, this same principle may be expressed in terms of the entropy; it is then called the "principle of entropy." The principle of entropy states:

> Any reversible change occurring in an isolated system leaves unchanged the value of the total entropy of the system, and any irreversible change causes an increase in the total entropy.

Since all changes are more or less irreversible, we need retain only the latter part of this statement and say that all changes necessarily entail an increase of the total entropy of the isolated system. According to this new presentation of the second principle, an increase in the entropy accompanies a decrease in the available energy or, what comes to the same, an increase in the unavailable energy. We may also say that the increase in the entropy is associated with a degradation of energy.

The conditions of equilibrium for an isolated system are easily interpreted by means of the entropy. Thus an isolated system can undergo no change (compatible with the restrictions imposed on the system) when the available energy of the system is a minimum. We may transcribe this conclusion by saying that the system can undergo no further change when its total entropy has attained a maximum value. If, in the state of interest, all small changes permitted by the restrictions imposed on the system result in a decrease of the entropy, the equilibrium is stable. If among the possible changes none can result in an increase of the entropy, but one of the changes, at least, leaves the value of the entropy unmodified, the equilibrium is indifferent. The final state of heat-death and rest which the second principle predicts for the universe as a whole, viewed as an isolated system, is equivalently expressed by the statement that the total entropy of the universe will increase to a maximum value and then proceed no further.

Thus far we have made no attempt to justify the equivalence of the principle of entropy and of the second law of thermodynamics in its original form. The following illustrations will, however, show that the two statements of the principle are indeed equivalent. As a first example let us consider a heavy fly wheel revolving with friction around an axis. The mechanical, or available, energy of the wheel is degraded progressively into heat, which manifests itself in the bearings; the wheel comes to a stop. This degradation of available energy illustrates the second principle. Let us see whether the requirements of the principle of entropy are satisfied in this example. The flywheel loses no heat while it is slowing down, for the heat that appears through friction is not drawn from the heat contained in the wheel but results from the transformation of the wheel's kinetic energy. The wheel itself therefore loses no entropy. On the other hand, the bearings become warm and thus receive heat, and so their entropy is increased. The total entropy of the system thus continues to increase until the wheel comes to a stop. The principle of entropy is verified.

As a second illustration, consider the fall of heat from a hotter to a colder body by conduction. We know that this fall is in agreement with the second principle. Let us then verify that it also agrees with the principle of entropy. The verification will be obtained if we can show that the fall of heat causes an increase in the total entropy of the two bodies considered. We call Q the amount of heat that falls during, say, the first second, and T_1 and T_2 the temperatures of the hotter and colder bodies respectively. The hotter body loses entropy $\dfrac{Q}{T_1}$ while the colder body gains entropy $\dfrac{Q}{T_2}$. The total change in entropy for the two bodies is thus

$$\frac{Q}{T_2} - \frac{Q}{T_1} \; ;$$

and since $T_1 > T_2$, the expression just written has a positive value. The total change in the entropy, due to the fall of heat, therefore represents an increase; hence the principle of entropy is verified. We also see that the reverse passage of heat, from the colder to the hotter body, would entail a decrease in the entropy. A spontaneous passage of this sort is therefore forbidden by the principle of entropy. For similar reasons the principle of entropy prohibits a spontaneous generation of local differences of temperature in a medium initially at uniform temperature. In short, in all these illustrations the application of the principle of entropy leads to the same conclusions as does the second principle of thermodynamics expressed in terms of the available energy.

Incidentally we may verify that a change occurring reversibly entails no change in the total entropy. For instance, in the fall of heat by conduction from a warmer body at temperature T_1 to a colder body at temperature T_2, let us suppose that the higher temperature exceeds the lower one by an infinitesimal amount. Under these conditions the fall of heat tends to take place reversibly. When a quantity Q of heat falls, the total change in the entropy of the system formed by the two bodies is given as before by $\dfrac{Q}{T_2} - \dfrac{Q}{T_1}$. But now, since T_1 and T_2 are practically the same, the foregoing expression vanishes; hence the value of the total entropy does not change.

As a further illustration of the principle of entropy, let us consider the passage of heat that can be forced from a colder to a hotter body by the expenditure of work. When heat is made to pass from a colder to a hotter body, the total entropy of the two decreases, but at the same time the work we expend in order to bring about the passage of heat is itself transformed into heat which appears on the hotter body. The hotter body receives entropy from this source, and the entropy thus received is sufficient to offset the loss first mentioned. The net result is that when all the changes are taken into consideration, the total entropy is found to have increased or, at best, to have remained stationary (reversible operation). The principle of entropy is verified.

We may show that Carnot's principle is a consequence of the principle of entropy. Suppose we have an isolated hot body. Since there is no colder body to which the heat may fall, no work can be generated. Suppose, however, that contrary to Carnot's principle, work could be derived from the heat energy. The hot body would then have to lose heat and therefore entropy, and our total system would be suffering a decrease in entropy; and this is contrary to the principle of entropy. Thus, Carnot's principle and the principle of entropy are in agreement. We now place a colder body near the hotter one and obtain a flow of heat. We shall see that the law of entropy leads to Kelvin's quantitative results. We call T_1 and T_2 the temperatures of the hotter and colder bodies respectively. Let Q_1 represent the amount of heat that falls from the hotter body every second, and let Q_2 measure the heat received by the colder body. The change in the sum total of the entropy for the two bodies is

$$\frac{Q_2}{T_2} - \frac{Q_1}{T_1}.$$

The principle of entropy requires that this change be positive or vanishing (reversible operation). We must therefore have

$$\frac{Q_2}{T_2} - \frac{Q_1}{T_1} \geqslant O, \text{ whence } Q_1 - Q_2 \leqslant Q_1\left(1 - \frac{T_2}{T_1}\right).$$

Now the heat $Q_1 - Q_2$ which disappears during the fall is that part of the heat which is transformed into work. Hence the foregoing inequality shows that, under the most favorable conditions, only the fraction $1 - \dfrac{T_2}{T_1}$ of the heat which leaves the hot body can be transformed into work. We also see that the most favorable conditions are realized when the operation is reversible, and that the slightest trace of irreversibility decreases the amount of work that can be obtained. Complete irreversibility occurs when the heat falls by conduction. In this last case none of the heat vanishes during the fall (*i.e.*, $Q_1 - Q_2 = 0$), and no work is obtained. In short, Carnot's results are seen to be a' consequence of the principle of entropy. We might also have shown that the expansion of a gas into a vacuum furnishes an illustration of the principle of entropy.

Let us examine the bearing of the principle of entropy on the idealized conservative dynamical systems studied in analytical mechanics. Consider a pendulum swinging without friction in the earth's gravitational field—the pendulum and the earth jointly form a conservative, isolated dynamical system. We know that the pendulum will continue to swing indefinitely. We also know that the motion is occurring reversibly, so that the total entropy of the system will remain constant during the motion. Though this conclusion does not violate the principle of entropy, it does show that the principle cannot be utilized to predict the direction in which the system will evolve. Furthermore, since the entropy of the mechanical system remains constant, it cannot assume a maximum value in any particular configuration of the system, and hence the criterion of equilibrium furnished by the principle of entropy cannot be applied.*

To study the possible changes and the equilibrium conditions of isolated conservative mechanical systems, we must appeal to the principles of dynamics. Thus, if the system has kinetic energy, changes will continue to occur indefinitely. The same is true if the system is initially at rest in a configuration for which the potential energy is not a minimum: the system will then start to evolve towards configurations of lower potential energy. Finally, the system will remain permanently at rest, and hence will be in equilibrium, if it is initially at rest in a configuration of minimum potential energy. Since, when the system is at rest, its kinetic energy vanishes, we conclude that the system will be in a state of equilibrium when its total energy is a minimum. For example, a pendulum will remain at rest if we release it when its centre of gravity is at the lowest possible point; the total energy of the pendulum is then a minimum. From the standpoint of equilibrium, the total energy of a mechanical system thus plays a part similar to that of the entropy of a thermodynamical system.

The upshot of this discussion is that the principle of entropy has no bearing on the conservative mechanical systems studied in dynamics.

---

* The conclusions stated in the text apply only to simple mechanical systems. When the complication of the system is very great, its entropy is no longer constant and the principle of entropy becomes relevant once more. See Chapter XXII.

In practice, however, rigorously conservative mechanical systems being unattainable idealizations, the principle of entropy is of general application.

The principle of entropy summarizes the results of ordinary observation. But, in addition, the principle has led to many remarkable theoretical discoveries which were totally unsuspected when they were first stated. Some of these will now be mentioned briefly.

The Phase Rule—One of the most interesting applications of the principle of entropy is illustrated by the *Phase Rule,* discovered by Gibbs. It exemplifies that peculiar beauty which characterizes all of Gibbs's work.

A few definitions are necessary. A chemical substance, *e.g.,* water, can exist under ordinary conditions in three different forms: the solid form (ice), the liquid form (water), and the vapor form (water vapor). These forms are called *phases.* Other substances, such as sulphur, exhibit a larger number of phases. Thus, under ordinary conditions, sulphur can exist in the liquid or in the vapor phase or in two different crystalline forms constituting two different solid phases. Entropy considerations arise in changes of phase, as indeed in all physical changes. When a piece of ice melts, it absorbs heat energy. This heat energy is used up entirely to secure the passage from the solid to the liquid phase. No part of it serves to increase the temperature of the melting ice, which remains unchanged during the process of fusion. Let us call Q the quantity of heat absorbed by the ice during the fusion. If we represent by T the constant temperature of the ice, the entropy acquired is $\frac{Q}{T}$. Thus, even though the temperature does not change, entropy has been accumulated owing to the change in phase. Similar absorptions or rejections of entropy accompany all other changes of phase, such as evaporation and sublimation.

Another important notion is that of the *degrees of freedom* of a thermodynamical system. A given mass of gas in a given state has a definite volume, temperature, and pressure. But when any two of these three magnitudes are specified, the third is automatically determined. Since the state of the gas is defined by means of two independent magnitudes, we say, by analogy with the terminology of mechanics, that the gas system has two degrees of freedom.

With these preliminaries disposed of, we pass to the equilibrium of phases. Suppose we have a mixture of ice and water. The system is said to be in thermodynamical equilibrium if none of the ice melts into water, and if none of the water freezes into ice. Equilibrium is thus manifested by the absence of change. Obviously, a piece of ice placed in hot water does not represent a system in equilibrium.

Direct experiment has furnished considerable information on these equilibrium phenomena. As an example let us revert to the mixture of ice and water. Usually a mixture of this kind is in contact with the air and consequently with water vapor. To obviate the presence of the water vapor, we shall assume that the mixture fills a closed cylinder. If the temperature of the mixture is 0° Centigrade and the pressure one atmosphere, the ice and water are in thermodynamical equilibrium; the ice subsists and the water does not solidify. The slightest change, however, in the temperature or in the pressure will cause the equilibrium to be destroyed. For example, if we keep the temperature unchanged but increase the pressure by pushing against a piston, the ice will turn into water. The mixture is no longer in equilibrium. On the other hand, equilibrium will be restored under this increased pressure provided we lower the temperature by a suitable amount. Experiment shows more generally that at each definite pressure (within a certain range) there is a corresponding temperature of equilibrium for the ice and water. Conversely, if the temperature is specified, there is a corresponding equilibrium value for the pressure.

Next we examine the equilibrium of the two phases from the standpoint of the degrees of freedom. We have said that a gas has two degrees of freedom because both the pressure and the temperature of a given mass of gas can be prescribed at pleasure. On the other hand, if our mixture of ice and water is to be in equilibrium, the temperature and the pressure cannot both be prescribed arbitrarily: the numerical value of only one of the two variables can be chosen freely; the value of the other variable is then determined by this choice. For this reason, we say that the system formed by a mixture of ice and water in equilibrium has but one degree of freedom. Other cases of equilibrium will be considered presently.

The conclusions which we have just stated follow from direct experiment and make no demands on any theoretical knowledge. But Gibbs, in his "phase rule," obtained a formula which determines the number of degrees of freedom of any system formed of one or more substances and phases in equilibrium. Gibbs's demonstration of his formula was based on the principle of entropy. We have seen that according to this principle, if any conceivable change which is compatible with the restrictions imposed on an isolated system would result in a decrease of the system's entropy, no change can occur; the system will be in stable equilibrium. Even if, among the conceivable changes, some would leave the value of the entropy unaltered, spontaneous changes will still fail to appear. Consequently, the system is also in equilibrium in this case, but the equilibrium is now indifferent. The problem of determining under what conditions a system will be in thermodynamical equilibrium throws us back therefore on the problem of deciding under what conditions the entropy will be a maximum.

Guided by these considerations Gibbs established the following remarkable formula: He proved that, if a system formed of n different substances in k different phases is in a state of thermodynamical equilibrium, the number of degrees of freedom of the system in equilibrium is defined by

$$n - k + 2.$$

This is the celebrated phase rule.*

As an example we revert to the mixture of ice and water. Here we have a single substance in two phases, and so we must set $n = 1$ and

---

* There is an interesting analogy between the phase rule and the theory of the connectivity of surfaces. For reasons into which we shall not enter here, a surface such as a plane sheet of paper is called simply-connected. If a hole is made in the paper, a doubly-connected surface is obtained. If now we effect a cut in this perforated paper along a line extending from the hole to the exterior edge, the surface becomes simply connected again. A sectioning of this sort is called a cross-cut. Surfaces of higher connectivity may also be considered. For example, the surface of a doughnut is a triply-connected surface.

According to the theory of connectivity, if p is the number of distinct simply-connected surfaces into which a surface of connectivity N can be resolved by means of q cross-cuts, we have the relation

$$q - p + 2 = N.$$

This relation coincides with that of the phase rule, if we correlate q with n, p with k, and N with the number of degrees of freedom of the thermodynamical system. Consequently it will be possible to represent the phase rule by a paper model.

For example, the phase rule requires that one substance ($n = 1$) in two different phases ($k = 2$) should have one degree of freedom. This result corresponds to the fact that for one cross-cut ($q = 1$) to yield two simply-connected surfaces ($p = 2$), the surface that is cut must be simply connected ($N = 1$). Similarly, two substances in two phases have two degrees of freedom; and the analogue of this situation is that for two cross-cuts ($q = 2$) to yield two simply-connected surfaces ($p = 2$), the original surface must be doubly connected ($N = 2$). Our method of representation (unless supplemented) does not differentiate, however, between the case where the number of degrees of freedom is zero or is negative.

Dr. Silberstein has drawn my attention to a similar correlation between the phase rule and Euler's theorem on convex polyhedra. According to Euler's theorem we have

$$e - f + 2 = c,$$

where e, f, and c are the numbers of edges, faces, and corners of a polyhedron.

$k = 2$ in Gibbs's formula. The value of Gibbs's expression $n + 2 - k$ is then 1. We conclude that the system has one degree a freedom, a result already mentioned.

Next let us consider the equilibrium of three different phases of the same substance, *e.g.*, ice, water, and water vapor. Here we have $n = 1$ and $k = 3$, and Gibbs's formula yields zero for the degrees of freedom of the system. This means that neither the pressure nor the temperature can be assigned arbitrarily. There is but one specific pressure and one specific temperature at which all three phases can coexist in equilibrium.

If we assume one substance and four phases, Gibbs's formula gives -1 for the degrees of freedom. This result is meaningless, for the number of degrees of freedom cannot be negative. We conclude that, regardless of what the pressure and temperature may be, four different phases of the same substance (*e.g.*, the four phases mentioned for sulphur) cannot possibly coexist in equilibrium. If, then, a recipient contains a substance in four different phases, we may expect one of the phases to be transformed gradually into the other three. The three remaining phases may then be in equilibrium at a certain temperature and pressure, as explained in the previous illustration.

Finally, we consider a system formed of two different substances in four phases. Here $n = 2$ and $k = 4$, and Gibbs's rule yields zero for the number of degrees of freedom. Hence there is but one temperature and pressure at which all four phases can coexist (the temperature and the pressure of equilibrium). Suppose, then, the mixture is placed in a container. The system will pass of its own accord to the state of equilibrium, for this state corresponds to maximum entropy. The temperature of the mixture, therefore, moves automatically to the temperature of equilibrium.

As an illustration let us suppose that salt is sprinkled on ice. Some of the salt dissolves in the water adhering to the ice, and some of the salt persists in the solid phase. In addition, owing to evaporation, water vapor is formed. Altogether then we have two components (water and salt) and four phases; namely, ice, solid salt, solution of salt in water, and water vapor. The situation is the one we have just been discussing. Consequently, the mixture tends to assume automatically the temperature of equilibrium which experiment shows to be $-22°$ Centigrade. This lowering of the temperature is often utilized for refrigeration purposes; our mixture constitutes a "freezing mixture."

The great beauty of the phase rule, added to its importance in physics and in chemistry, has made it one of the most celebrated laws of theoretical physics.

Thermochemistry—Thermochemistry studies chemical reactions with particular regard to the thermal changes which accompany them. In the early days of chemistry the generation and the absorption of heat displayed in chemical reactions were well known, but no particular importance was attributed to these heat manifestations. When the law of conservation of energy was accepted, it became obvious that the heat appearing in a chemical reaction must result from the transformation of internal energy already present in the reacting substances. With a view of determining the changes in value of the internal energy in the various reactions, Berthelot conducted a large number of experiments and measured the amounts of heat evolved. But not until the investigations of Gibbs and of Helmholtz, was the important rôle played by the principle of entropy properly understood.

Chemical reactions are particular examples of spontaneous changes, and so we may apply to them the fundamental principles of thermodynamics with which we are familiar. For instance, if chemical substances are placed in an isolated enclosure, we may be certain that no reaction can occur unless the total entropy of the system would increase as a result of the reaction. If, then, we have determined the relative entropies of different substances and compounds, we shall be able to tell whether a prospective reaction is possible or impossible in principle. The theoretical study of chemical processes is facilitated when we assume that the temperature and the volume occupied by the reacting substances are kept fixed during the reaction. Let us suppose that the reacting substances and the products of the reaction are gases. We introduce the initial gases into a rigid cylinder which they fill completely. Constancy in the volume is thus assured; the gases cannot expand (or contract) and consequently can perform no work on the surroundings during the reaction. We next examine a means of securing the constancy of the temperature. This leads us to inquire into the meaning of a "thermostat." A thermostat is a body which can yield and absorb a practically unlimited amount of heat without varying in temperature. A mixture of ice and water is an example: if we supply heat to the mixture, some of the ice melts but the temperature remains at 0° Centigrade; if we withdraw heat, some of the water freezes but the temperature still remains unchanged. If, then, in this mixture we place the cylinder containing the reacting substances, the temperature of the substances will remain at 0° Centigrade during the reaction. Other kinds of thermostats would ensure the constancy of other temperatures.

Suppose we place our cylinder in a thermostat. A chemical reaction occurs in the cylinder, and the volume and temperature remain constant.

Let us call U_1 and U_2 the internal energies of the initial substances and of the final products respectively. In the majority of cases heat is generated during the reaction and is then absorbed by the thermostat. According to the principle of conservation of energy, the heat liberated must have been drawn from the reacting substances. And since these have undergone no expansion and have performed no work on the surroundings, the heat liberated is necessarily equal to the difference between the internal energies of the initial substances and those of the final products; thus this heat energy must be measured by $U_1 - U_2$.

In the earlier investigations the part played by the second law of thermodynamics was poorly understood, and it was thought that the drop $U_1 - U_2$ in the internal energies measured what is sometimes called the "driving force" of the chemical reaction. According to these ideas, for a chemical reaction to take place, the difference $U_1 - U_2$ would have to be positive; in other words, the initial store of energy would have to be greater than the final one, and consequently, heat should be liberated by the reaction. The opinions of the day were expressed by the statement:

> Of all possible chemical processes which can proceed without the aid of external agency (*i.e.*, spontaneously), that process always takes place which is accompanied by the greatest liberation of heat.

This erroneous law is called "Berthelot's law of maximum work." At first sight Berthelot's law seems plausible, for it appears to have a close analogue in mechanics, *viz.*, a mass abandoned at rest will start to move of its own accord only if its potential energy would fall to a lower value as a result of the displacement. In other words, the motion will be spontaneous only if it is susceptible of generating work and hence heat (by friction). But analogies are often deceptive, and further consideration soon proved that Berthelot's law was untenable. Some chemical reactions, far from generating heat, absorb heat and thereby generate cold (endothermic reactions). In such cases the internal energy U_2 of the final products obviously exceeds the internal energy U_1 of the initial substances. In the mechanical analogy it would be as though a weight were to rise of its own accord. Since this spontaneous motion of the weight is impossible, we must recognize that there is a profound difference between chemical reactions and the aforementioned dynamical illustration.

Helmholtz and Gibbs elucidated the true state of affairs by means of the principle of entropy. To apply the principle, we observe that the thermostat and the cylinder containing the chemical substances form together an isolated system. We are therefore certain that a reaction can proceed spontaneously only in that direction which involves an increase in the total entropy of the system. Let us then compute the entropy change that will arise. We first consider the reacting substances and disregard the thermostat. Let S_1 and S_2 represent the total entropies of the initial and of the final substances respectively. The change in the entropy is thus $S_2 - S_1$. To obtain the change in entropy of the isolated system, which also contains the thermostat, we must add to $S_2 - S_1$ the change in the entropy of the thermostat. If the thermostat, whose temperature T is fixed, absorbs a quantity Q of heat, its entropy increases by $\dfrac{Q}{T}$; and if the thermostat supplies heat to the chemical substances (endothermic reaction), it loses an amount $\dfrac{Q}{T}$ of entropy. Both situations may be described by saying that the thermostat suffers a change $\dfrac{Q}{T}$ in entropy (a negative value being assigned to Q in the case of a loss of entropy). The total change in entropy of our isolated system as a result of the reaction is then

$$(1) \qquad \frac{Q}{T} + S_2 - S_1 .$$

Now according to the principle of conservation of energy, the total energy contained in the isolated system can undergo no change during the reaction. This implies that the heat energy, which we may assume has been absorbed by the thermostat, must be equal to the drop $U_1 - U_2$ in the internal energy when we pass from the initial substances to the final products. We may therefore, in (1), replace Q by $U_1 - U_2$; and we thus obtain for the total change in the entropy of the isolated system the expression

$$(2) \qquad \frac{U_1 - U_2}{T} + S_2 - S_1 .$$

The principle of entropy states that for a spontaneous reaction to arise, the change (2) in entropy must be an increase and hence must be positive. We thus obtain from (2), after multiplying it by T, the condition

$$(3) \qquad U_1 - U_2 - T(S_1 - S_2) > 0,$$

or equivalently

$$(4) \qquad (U_1 - TS_1) > (U_2 + TS_2).$$

The relation (4) shows that, for a chemical reaction occurring at constant temperature and in a fixed volume to be possible, the magnitude defined by $U - TS$ must be greater for the initial substances than for the final products. The name *free energy* is given to this expression,[*] and we may summarize Gibbs's results by saying that a spontaneous chemical reaction can arise only if it be accompanied by a drop in the free energy.[†] For this reason we may identify the drop in the free energy with the driving force of the reaction (when the reaction occurs at constant temperature in a fixed volume). The extent of the drop also measures the amount of available energy present in the reaction: *i.e.*, the amount of mechanical energy that might be derived from the reaction if we could make it occur under ideally reversible conditions (extremely slowly). It will be noted that the absolute value of the free energy, either of the initial substances or of the final products, cannot be obtained at this point; for we have seen that neither the internal energy U nor the entropy S are known in absolute value. But all that we are concerned with here is a difference in the free energies, so that our ignorance of the absolute values is no drawback in the present circumstance.

Gibbs's theory is compatible with endothermic reactions. In these, since heat is absorbed during the reaction, the difference $U_1 - U_2$ in the internal energies must be negative. But even in this case the expression on the left-hand side of (3) (which represents the driving force) may be positive and the reaction proceed in consequence. All that is required is that S_1 (the entropy of the initial substances) should have a value sufficiently smaller than S_2 (the entropy of the final products). Thus, endothermic reactions, which were a paradox under Berthelot's law, become comprehensible under the revised ideas.

A further comparison of the two laws is instructive. In Berthelot's, the possibility of a reaction arising spontaneously is expressed by the condition

$$(5) \qquad\qquad U_1 > U_2.$$

---

[*] Sometimes the entire expression on the left of (3) is called the free energy of the reaction. It represents in effect, the available energy when the reaction occurs at constant temperature and at constant volume.

[†] Usually the reaction is assumed to occur under atmospheric pressure p. The volumes v_1 *and* v_2 of the initial and final substances may now differ, and work is involved by this change in volume. This circumstance complicates somewhat the inequality (4). We now obtain $U_1 - TS_1 + pv_1 > U_2 - TS_2 + pv_2$ as a necessary condition for the reaction to be possible.

Now, if we contrast this condition with the revised ones (3) or (4), we see that Berthelot's law becomes correct provided $S_1 = S_2$, *i.e.*, provided the entropies of the initial and final substances happen to be the same. Berthelot's law also becomes correct if $T = 0$, *i.e.*, if the reaction occurs in the neighborhood of the absolute zero.

The Galvanic Cell—A chemical reaction arising spontaneously always occurs in a highly irreversible manner and so is accompanied by a considerable degradation of the available, or free, energy. To convert the available energy into mechanical or electrical energy without degradation, we must proceed reversibly and hence extremely slowly. A device that provides this result is the galvanic cell. Its effect is, so to speak, to harness the reaction. The cell, by harnessing the reaction, derives mechanical energy from it, and this energy appears in the form of an electric current. In much the same way the steam engine, by harnessing a fall of heat, furnishes work.

Let us examine a particular illustration. We place a piece of zinc in a solution of copper sulphate. A chemical reaction occurs irreversibly; zinc sulphate and copper are obtained, and heat is liberated. Next we attempt to harness the reaction, compelling it to take place reversibly (with extreme slowness). This may be done by making the reaction proceed in a galvanic cell. Thus, we insert two electrodes, one of zinc the other of copper, in a jar containing the solution of copper sulphate. A difference in electrical potential is generated in the two electrodes, but of course unless we connect them by a wire and thus close the circuit, no electric current passes. If the circuit is not closed, the reaction occurs irreversibly, exactly as it did before; the available energy is degraded into heat and no work is obtained. But if we close the circuit, an electric current will pass. The chemical reaction is now yielding electrical energy and therefore work. This electrical energy results from the transformation of the available, or free, energy (now no longer totally degraded into heat). Also, the reaction will proceed more slowly, as always arises when we approximate more and more to conditions of reversibility.

As yet, however, reversibility is not rigorous. To obtain the ideal reversible conditions, we must set up a counter-electromotive force in the wire. The effect of the counter-electromotive force is to decrease the current and at the same time decrease the rapidity with which the available energy is being transformed into electrical energy. The chemical reaction will also slow down in consequence. At the limit, we may assume that the current is vanishingly small; the reaction then proceeds with extreme slowness, and practically none of the available energy is degraded into heat. After a sufficiently long time, all the available energy is converted into electrical energy and the reaction is completed. Since, under these reversible conditions of operation, the available energy is no longer degraded into heat, the quantity of heat appearing in the cell will fall below the amount delivered in the spontaneous reaction. The heat value of the available energy being the heat value of the expression on the left side of (3), we conclude that the heat appearing in the cell under reversible conditions of operation is

$$(6) \qquad U_1 - U_2 - [(U_1 - TS_1) - (U_2 - TS_2)];$$

and this is

$$(7) \qquad T(S_1 - S_2).$$

According to whether the magnitude (7) is positive, negative, or vanishing, the reaction occurring reversibly in the cell will deliver heat or absorb heat or give rise to no thermal manifestation. Incidentally, we see that if we determine experimentally the amount of heat which appears during the reaction, the formula (7) enables us to deduce the difference in the entropy contents of the initial substances and of the products of the reaction.*

Finally, if we apply to the cell a counter-electromotive force greater than that generated by the reaction, thereby constraining a current to pass through the cell, we shall reverse the direction of the spontaneous reaction and cause the zinc and the copper sulphate to pass over into copper and zinc sulphate. The reversal is similar to the one obtained

---

* We mentioned that Berthelot's erroneous law coincides with the correct law when the entropies of the initial substances and of the final products are the same. When this condition is realized, we see from the formula (7) that no heat will be evolved or absorbed by a chemical reaction taking place reversibly at constant volume and temperature. Thus, Berthelot's law requires that this failure of heat to manifest itself should be the rule in all chemical reactions performed under the conditions stated. Now the reaction discussed in the text (copper sulphate and zinc) is the one that occurs in the Daniell cell; and for this reason, in Berthelot's day, it was one of the best known among the reactions that could be made to take place reversibly. It was, accordingly, one of the first reactions to serve as a test of Berthelot's law. But it so happens that in this reaction the entropies of the initial substances and of the final products are very nearly the same; consequently Berthelot's law, though incorrect, appeared to be verified. Partly for this reason, the erroneous nature of Berthelot's law was not recognized earlier.

when we make a current pass through water; the water is decomposed into oxygen and hydrogen.

Chemical Equilibrium—If one of the products of a chemical reaction is a gas, it will escape, and in this event the reaction will proceed until one of the reacting substances is exhausted. But we shall now assume that the reaction occurs in a closed volume so that nothing can escape. Operating in this way, Ste. Claire Deville in 1864 found that some chemical reactions do not proceed until exhaustion; instead, a kind of equilibrium arises between the initial substances and the products of the reaction. Such reactions are of extreme importance, especially so since it is suspected that all chemical reactions are fundamentally of this type. A simple illustration is the following:

At suitable temperature and pressure, oxygen, sulphur dioxide, and sulphur trioxide are gases. Under the catalytic action * of platinum black, the first two combine to form the third. Suppose the reaction occurs in a rigid container maintained at a constant temperature T. We find that the reaction does not proceed until exhaustion: it ceases when a certain relationship is satisfied by the concentrations of the two reacting gases and of their product, the sulphur trioxide gas. A state of chemical equilibrium is then said to be attained. In particular, if we write $[O_2]$, $[SO_2]$, $[SO_3]$ for the concentrations of the three gases, equilibrium will arise when the relation

$$(8) \qquad \frac{[SO_3]^2}{[SO_2]^2 \times [O_2]} = K$$

is realized. Here K is a constant, the value of which depends on the nature of the chemical substances considered, on the temperature, and usually also on the pressure or on the volume. However, if we assume that our gases are perfect, the constant K no longer depends on the pressure or on the volume, but solely on the temperature and on the chemical nature of the substances. The constant K is called the *equilibrium constant*.

Suppose now that, while leaving the volume and temperature unchanged, we pump more oxygen or sulphur dioxide into the volume. The concentration of the oxygen or sulphur dioxide increases, and hence the value of the denominator in our equilibrium relation (8) is increased. The value of the ratio (8) thus becomes smaller than the value of the

---

* In a catalytic action, the catalyzing substance (here platinum black) plays only a subsidiary rôle and seems to be unaffected by the reaction.

equilibrium constant K, and as a result the reaction will proceed afresh, and more sulphur trioxide will be formed at the expense of the oxygen and sulphur dioxide. The value of the ratio increases as the reaction proceeds, till it eventually attains the equilibrium value. The reaction then ceases once more. Conversely, if we introduce more sulphur trioxide into the volume, the reaction will proceed in the opposite direction: some of the sulphur trioxide is decomposed, or dissociated, into oxygen and sulphur dioxide. The reaction is thus reversible, proceeding in one direction or the other until the ratio of the concentrations attains the equilibrium value.

Thus far we have assumed the volume and temperature fixed. But suppose the temperature is changed. The value of the equilibrium constant is modified thereby, and the reaction will proceed in one direction or the other until the ratio of the concentrations attains the new equilibrium value. Similarly, if at constant temperature we vary the volume, e.g., if we halve the volume by compression, equilibrium will be destroyed. We may understand the reason for this statement by reverting to our equilibrium formula. The value of the equilibrium constant will be unaffected by the compression (since it depends only on the temperature); but, by halving the volume, we double all the concentrations, and the effect of this change will be to multiply the expression on the left-hand side of the formula (8) by $\dfrac{2^2}{2^2.2}$, i.e., $\dfrac{1}{2}$. Since the value of the ratio is now one-half the equilibrium value K, the reaction must be resumed in the original direction in order to restore the conditions of equilibrium. Thus more oxygen and sulphur dioxide will combine.

The general relationship (8) satisfied by the concentrations was discovered empirically by Guldberg and Waage, in 1868. They called it the *Law of Mass Action*. But the theoretical justification for these phenomena of chemical equilibrium could not be furnished before the law of entropy was properly understood. The theory is due to van't Hoff. Its connection with the law of entropy is fairly obvious, for the equilibrium phenomenon is similar to the equilibrium of phases studied by Gibbs. We may state the situation briefly by recalling that the driving force of a chemical reaction (occurring at constant temperature and in a fixed volume) is given by the difference between the free energies of the initial substances and of the final products. When therefore the temperature and concentrations are such that this difference in the free energies vanishes, the driving force no longer exists, and the reaction cannot proceed either in one direction or in the other. In other words, any change from the state of equilibrium would result in a decrease in entropy; and this decrease cannot occur spontaneously.

Reverting to our former example, we note that the higher the value of the equilibrium constant, the more complete will be the combination of the oxygen and sulphur dioxide to form the sulphur trioxide. Such a combination is favored when the temperature is lowered, because the value of the equilibrium constant increases under these conditions. In theory, could the equilibrium constant become infinite, the reaction would proceed till exhaustion. It is thought probable that all chemical reactions are in reality of the reversible type; but in many cases the value of the equilibrium constant may be high, and in this event the reaction will proceed so near to exhaustion that we may fail to detect any remnant of the initial substances. Furthermore, we must remember that, for equilibrium to arise, the products of the reaction must be prevented from escaping. A large number of reactions occurring in the living organism are manifestly of the equilibrium type. For instance, haemoglobin combines with oxygen in the lungs to form oxyhaemoglobin; the reverse reaction takes place in the capillary veins, where the concentrations of the relevant substances are different.

The Third Principle of Thermodynamics—In our discussion of thermodynamical problems involving the notion of entropy, we have been concerned only with the *changes* in the entropy of a substance or system; absolute values have been ignored. But situations arise where absolute values must be considered. The absolute entropy, or more simply the entropy, of a substance, in a given state at specified pressure and temperature, is determined in theory by the entropy of the substance in some arbitrarily selected standard state at some prescribed pressure and temperature, say, in some solid state at the absolute zero of temperature. Now the Second Principle of Thermodynamics, though it introduces the notion of entropy, furnishes no information on absolute values; and we might quite well suppose that the entropy of a substance at the absolute zero is positive or negative, finite or infinite. The Third Principle of Thermodynamics, as extended by Planck, enables us to secure absolute values. This principle states that the entropy of all crystalline bodies vanishes at the absolute zero.

But before we discuss Planck's statement of the Third Principle we must examine Nernst's earlier presentation (Nernst's Heat Theorem).

We have seen that the driving force of a chemical reaction, occurring at constant volume and temperature, is measured by the difference in the free energies of the initial substances and of the final products. For the reaction to be possible· the change in the free energy must result in a drop. We recall that the expression of this drop in the free energy is

$$U_1 - U_2 - T(S_1 - S_2).$$

A series of measurements performed by Nernst indicated that in chemical reactions (and more generally in transformations) involving liquids and solids, the difference $S_1 - S_2$ in the entropies of the initial substances and final products tended to vanish at low temperatures. Nernst surmised therefore that $S_1 - S_2$ would vanish entirely at the absolute zero. This assumption constitutes Nernst's Heat Theorem. It furnishes one of the expressions of the Third Principle.

An immediate consequence of Nernst's assumption is that the curves (see Figure 35), which represent the drop in the free energy and the

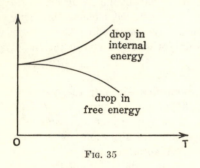

drop $U_1 - U_2$ in the internal energy, must meet and have a common horizontal tangent when the temperature of the absolute zero is attained. The fact that the tangent to the upper curve is horizontal implies that the specific heats of the initial substances and of the final products are the same when the transformation, or reaction, takes place at the absolute zero.

Fig. 35

Nernst believed that the vanishing of $S_1 - S_2$ at the absolute zero would hold for all condensed systems (*i.e.*, liquids * as well as solids), but today we have reasons to believe that this vanishing occurs only for crystalline solids. Nernst's Heat Theorem may then be stated thus:

In the immediate vicinity of the absolute zero, all transformations involving crystalline solids proceed without change of entropy, and also without change in specific heat.

Nernst's theorem has important applications, especially in connection with equilibrium problems.

Planck's extension of Nernst's results consists in the assumption that not only will the differences in the entropies and in the specific

---

* The term "liquid" applies here not only to ordinary liquids but also to supercooled ones, such as glasses.

heats of the initial and final substances vanish at the absolute zero, but that the entropies and specific heats themselves will also vanish. The final expression of the Third Principle thus assumes the following form:

> The entropy content and the specific heat of a crystalline solid vanish at the absolute zero of temperature.

Up to this point we have restricted our attention to crystalline solids. Supercooled liquids, such as glasses, do not satisfy the Third Principle insofar as the vanishing of the entropy is concerned; there is reason to believe that their entropy has a finite, non-vanishing value at the absolute zero. Next, let us consider gases. The Third Principle cannot apply to real gases, for we know that a real gas is condensed into a liquid and a solid before the temperature of the absolute zero is attained. This objection loses its force, however, in the case of perfect gases, because the classical equation of state of perfect gases * shows that there can be no condensation. At all events, if the principle does apply to perfect gases, as Nernst believed, the classical equation of state must certainly undergo a change in form at low temperatures. The accompanying change which would result for the properties of a perfect gas at low temperatures is sometimes referred to as the "degeneration of gases." †

Until comparatively recently, however, the existence of degeneration was regarded as extremely speculative. But the New Statistics of perfect gases (see page 926) shows that degeneration does occur at low temperatures, and that the entropy and specific heat of the gas do vanish at the absolute zero, in complete accord with the requirements of the Third Principle. In view of these more recent discoveries, we appear to be justified in claiming that the Third Principle presumably applies to perfect gases.

We shall see in later chapters that the Third Principle is intimately connected with the quantum theory; it symbolizes, so to speak, some of the macroscopic effects occasioned by the microscopic quantum occurrences.

Conclusions on Thermodynamics—For lack of space we have been unable to touch on more than a few of the many applications of thermodynamics, but what little has been said suffices to give an idea of the vast field covered by that science. Nevertheless in spite of its achievements,

---

* See page 379.

† This same name is often given to the quantizing of the translational motions of the gas molecules.

thermodynamics suffers from the limitations common to all phenomeno-logical theories. Because it restricts its attention to the macroscopic properties of bodies, it fails to anticipate many phenomena which find their interpretation in the interplay of underlying microscopic processes, and which have since been clarified by the more speculative theories of the hidden-occurrence type.

For instance from the standpoint of thermodynamics, there is no more reason to cast doubt on the absolute validity of the principle of entropy than on that of the conservation of energy. And yet today the kinetic theory of gases (at one time regarded as a pure speculation) has shown that the principle of entropy is only a statistical principle. In chemical reactions also, conditions arise which it is not in the province of thermodynamics to interpret. Thus the free energy of water is con-siderably less than that of a mixture of oxygen and hydrogen; conse-quently, the two gases should combine violently to produce water. This they undoubtedly do, but to start the reaction we must apply a match. There seems to be a kind of frictional resistance which must be overcome before the reaction can arise. Thermodynamics cannot attempt to in-terpret this condition. Similarly chlorine and hydrogen should combine violently. They do so, but only under the effect of some stimulus such as sunlight. In other cases, chemical reactions, which should occur according to thermodynamics, need still greater prompting. The rate at which chemical reactions proceed is also an important consideration. The methods of thermodynamics do not give any information on this score, and other methods of investigation must be adopted.

There is also the large class of photo-chemical reactions, in which light-energy is required to bring about the reaction. Here the light acts no longer as a mere stimulus, but as an integral part of the reaction. The effect of light in loosening the bonds between the atoms of chlorine and silver (a phenomenon utilized in photography) is a case in point; and so is the chlorophyllian action of green plants under the influence of sunlight. In these photo-chemical processes the free energy of the prod-ucts of the reaction may exceed that of the initial substances. In such cases we are certain, on the basis of the laws of thermodynamics, that the reactions cannot occur spontaneously. The fact that they occur none the less does not invalidate the thermodynamical laws, for the re-actions do not occur spontaneously; they draw on the energy of the light and therefore occur under the application of external energy.

Thermodynamics, however, does not go far enough in its interpreta-tion of these phenomena. Our understanding of them is beginning to be clarified only since the advent of the quantum theory; and that theory introduces hidden occurrences of a hypothetical nature, which are foreign to the spirit of thermodynamics. The fundamental law of photochemistry was stated in 1905 by Einstein and Stern. Planck's constant h enters into the expression of the law. In the interpretation of these photo-chemical reactions, a curious form of energy transmission, called a "collision of the second kind," often appears to play an important part. Finally, there are certain constants which occur in thermodynamics, the so-called "chemical constants"; and only thanks to the quantum theory has it become possible to determine their value theoretically.

These criticisms in no wise diminish the debt that science owes to thermodynamics, but they do show that no uniform rule of scientific method can be set as a model to be followed blindly in all cases.

CHAPTER XXII

THE CLASSICAL KINETIC THEORY OF GASES

"The question of the utility of atomistic conceptions is, of course, quite foreign to the fact on which Kirchhoff has insisted that our theories are to Nature what symbols are to the things they represent, or what the letters of the alphabet are to the human voice, or what notes are to music; and also to the question of deciding whether it would not be better to call our theories mere descriptions, so as to stress what they are in relation to Nature."

L. BOLTZMANN.

THE simple ratios in which substances enter into chemical combination was the controlling fact which induced chemists to adopt the atomic hypothesis for matter. A limited number of different kinds of atoms were believed to exist, and these were assumed to constitute the ultimate building blocks of the universe. Insofar as the earlier chemists were aware, the atoms were indivisible. At all events they appeared to behave as such in chemistry. Atoms of different nature could combine and form more or less stable structures called "molecules." Such molecules constituted the smallest possible particles of the compound substances. In many cases, however, atoms of the same element were likewise found to combine among themselves; and the name molecule was retained for these other kinds of stable structures so formed. Thus, a molecule came to mean any aggregate of atoms forming a stable structure, regardless of whether the atoms involved were of the same kind or not.

For future reference we shall now state the formulae accepted today for some of the molecules. The molecule of water is formed of two atoms of hydrogen and one of oxygen. Hydrogen gas, at ordinary temperatures and pressures, is composed of molecules each of which contains two hydrogen atoms. Similarly nitrogen gas is formed of molecules comprising two nitrogen atoms, and oxygen gas of molecules containing two oxygen atoms. On the other hand, the so-called rare gases, such as helium, neon, argon, are composed of single atoms. The name molecule is retained, however, in all cases, with the result that the atom of a rare gas is also called its molecule.

Molecules are often classified according to the number of atoms they contain. The molecules of the rare gases are called "monoatomic" because they contain only one atom. Molecules containing two atoms are

374

called "diatomic," and those that contain three or more atoms are classed as "polyatomic."

One of the first concerns of the earlier chemists was to establish the correct formulae of the various molecules. In this attempt considerable difficulties were encountered. As an illustration let us consider the case of the molecule of water. Experiment shows that eight grams of oxygen combine with one gram of hydrogen to produce nine grams of water. Obviously, then, a molecule of water must be composed of atoms of oxygen and hydrogen, and the total mass of the oxygen atoms present must be eight times the total mass of the hydrogen atoms. But this information in itself is insufficient to furnish the exact numbers of oxygen and hydrogen atoms in the molecule. Additional information is required before the formula of water can be determined.

In their attempts to establish the formulae of the various molecules, the earlier chemists were therefore compelled to rely on clues rather than on direct experiments. A large number of clues were considered, and thanks to these clues an elaborate system of checks and cross-checks became available. One of the most important clues was Avogadro's hypothesis.

Avogadro's Hypothesis—In the early part of the last century Avogadro advanced the hypothesis that equal volumes of different gases, at the same pressure and temperature, contained the same number of molecules.

Today, as a result of experiments we shall mention presently, Avogadro's hypothesis has been fully vindicated. But at the time it was advanced, it was little better than a guess. Indeed the entire atomic hypothesis was regarded as highly speculative, and obviously Avogadro's hypothesis could have no meaning if the atomic constitution of matter were rejected. At all events many of the earlier chemists accepted Avogadro's hypothesis, and with its help established the formulae of the simpler molecules and the so-called atomic masses of the various atoms. The manner in which the atomic masses were obtained will be understood from the following considerations.

The Atomic Masses—Experiment shows that equal volumes of hydrogen, nitrogen, and oxygen at the same pressure and temperature have masses proportional to the numbers 1, 14, 16. If we accept Avogadro's hypothesis, we know that each one of our three volumes contains the

---

*Strictly speaking the masses are proportional to 1, slightly less than 14, and slightly less than 16   We have adopted exact integral values in the text so as to simplify the presentation.

same number of molecules. Hence the molecules of nitrogen and of oxygen, respectively, must be 14 times and 16 times more massive than the hydrogen molecule. Furthermore, if we agree that the molecules of our three gases are diatomic, we conclude that the atoms of nitrogen and oxygen are 14 times and 16 times more massive than the hydrogen atom. Determinations of this sort showed that the hydrogen atom was the least massive of the atoms. The mass of the hydrogen atom was therefore taken as standard, or as unit, and the masses of all the other atoms were expressed in terms of it. The mass of the hydrogen atom was accordingly represented by the number 1, and hence the masses of the atoms of nitrogen and oxygen were represented by the numbers 14 and 16. These numbers are called the atomic masses of the elements concerned. Let us observe, however, that the atomic masses do not measure the actual masses of the atoms, but only magnitudes proportional thereto.

We now pass to the molecules. Since the hydrogen molecule was known to be diatomic, and hence twice as massive as the hydrogen atom, it was allotted the number 2. Similarly the molecules of nitrogen and of oxygen were credited with the numbers 28 and 32, respectively. The numbers 2, 28, and 32 are called the molecular masses (or weights) of hydrogen, nitrogen, and oxygen. Like the atomic masses, these molecular masses determine only relative values.

The Gram-Molecule—In the laboratory we usually deal with relatively large amounts of gases and not with only one or two molecules. Chemists therefore found it advantageous to define standard quantities of the various gases. Now the number 2 had been allocated to the molecular mass of hydrogen, and so the earlier chemists agreed to view 2 grams of hydrogen as representative of the standard quantity of this gas. Since 28 grams of nitrogen and 32 grams of oxygen occupy the same volume as 2 grams of hydrogen (at the same pressure and temperature), the foregoing quantities of nitrogen and of oxygen were taken to define the standard quantities of these gases.

It will be noted that these standard quantities of the various gases have masses which are numerically equal in grams to the molecular masses of the respective gases. For this reason, the standard quantities of our three gases were called gram-molecules of the gases considered.

Avogadro's Number N—The gram-molecules of the various gases, at the same pressure and temperature, occupy the same volume. Consequently, according to Avogadro's hypothesis these gram-molecules contain the same number of molecules. This number is called "Avogadro's number'' and is usually denoted by the letter N. As we have seen, the definition of the gram-molecule originated from the study of gases. But we may now give a more general definition of it: A gram-molecule of a substance, gaseous, liquid, or solid, is by definition a quantity containing N molecules of the substance. This definition covers, of course, the case of compound substances as well as of simple ones.

If Avogadro's hypothesis turned out to be invalid, the accepted chemical formulae and all the results mentioned in the previous pages would have to be amended. The whole of chemistry would have to be revised. The importance of testing Avogadro's hypothesis by direct experiment is thus obvious. We may note in this connection that the validity of Avogadro's hypothesis would be established if the gram-molecules of the various substances were shown to contain the same number of molecules, and it would not be necessary to determine the exact value of this number N (Avogadro's number). However, in recent times experimenters have been able not only to justify Avogadro's hypothesis, but also to establish the value of N. The value generally accepted is

$$N = 6.06 \times 10^{23}.$$

Some of the experiments and measurements which have determined the value of N will be mentioned later. Of course in none of these experiments is it possible to make a direct count of the number of molecules in a given mass of gas; consequently, when we say that we have determined the value of N by experiment, we mean that we have obtained this value by indirect methods. In practice we measure other physical magnitudes and then deduce N from these measurements. Now this procedure requires that we utilize some system of relations connecting N with the other physical magnitudes; in other words we must rely on some physical theory. Hence N is not determined by measurements alone but only by measurements supplemented by an appropriate physical theory.

At this point the following obvious objection may be made: Since the determination of N is contingent on the acceptance of some physical theory, and since all physical theories are highly speculative, the value we obtain for N must necessarily itself be speculative and of doubtful value.

The force of this objection cannot be disputed. But it is mitigated by the fact that many different theories pertaining to different departments of physics (the kinetic theory of gases, optical theory, and the quantum theory) can be utilized to derive the value of N from the measurement of appropriate physical magnitudes. And it so happens that each one of these theories yields exactly the same value for N. Unless, then, we propose to attribute this concordance to sheer chance, we are justified in claiming that, despite the speculative nature of the individual theories, the cumulative evidence points to the correctness of the value we have obtained for N, and hence also shows that this number is not a mere fiction. To this extent, therefore, Avogadro's hypothesis and all its consequences are vindicated.

The discovery of the exact value of Avogadro's number N entails other important consequences. We recall that the earlier chemists were ignorant of the precise values of the atomic and molecular masses; all they knew were relative values. But now that the value of N is established the precise values can be determined. Thus, since N defines the number of molecules present in 2 grams of hydrogen, or 28 grams of nitrogen, etc., the precise masses in grams of the molecules of hydrogen, nitrogen, and oxygen are given respectively by

$$\frac{2}{N}, \frac{28}{N}, \frac{32}{N} \text{ grams.}$$

Similarly, since the molecules here considered are diatomic, the precise masses of the atoms of hydrogen, nitrogen, and oxygen must be credited with one-half the values written above.

The Gas Laws—The earlier physicists, Boyle, Charles, and Gay-Lussac, established simple relationships connecting the volume, pressure, and temperature of a gas. Their results were compressed into empirical laws bearing their names. The inevitable imperfections of human measurements led these earlier physicists to overlook small discrepancies which would have marred the formal simplicity of the laws obtained. Subsequent research has shown, however, that undeniable discrepancies always exist, though in certain cases they may be exceedingly small. Physicists thus came to distinguish between perfect gases and real gases. The name perfect gases was given to ideal gases which would satisfy rigorously the gas laws of the earlier investigators; as for the real gases, they were the ordinary gases we deal with in practice. In the present chapter we shall confine our attention to the perfect gases, because to a high degree of approximation the real gases behave like the perfect ones—at least when they are studied at pressures and temperatures far removed from their points of liquefaction. In many other departments of physics we may observe the same tendency to concentrate on idealized situations rather than on actual ones. Thus perfectly elastic bodies are subjected to theoretical study, even though we know that such bodies exemplify mere ideal limiting cases and have no empirical existence. Likewise in thermodynamics, we utilize demonstrations which involve perfectly reversible transformations and perfectly frictionless surfaces.

The laws for perfect gases may be compressed into a single relation connecting the volume v, the pressure p, and the absolute temperature T of a gas. This relation is called "the equation of state" of perfect gases, or more simply "the gas law." We shall assume that a gram-molecule of a perfect gas is considered. The equation of state, or gas law, is then expressed by

$$pv = RT.$$

The constant R, called the "gas constant," is the same for all perfect gases. To determine the value of the gas constant R, we consider a gram-molecule of any perfect gas, and measure its volume, pressure, and temperature; the value of the gas constant is then derived from the gas law written above. Thus the value of R may be determined without any reference to Avogadro's number or indeed to Avogadro's general hypothesis.

The gas law shows that when any two among the three magnitudes p, v, and T are specified, the third magnitude is determined. Accordingly we say that a gas system has two degrees of freedom. This point was mentioned in connection with Gibbs's phase rule. In place of the gas constant R, we shall often introduce another constant designated by k. It is defined by

$$k = \frac{R}{N},$$

where N is Avogadro's number. There is no danger of confusing the two gas constants, for we shall always represent them by the different letters R and k. Whereas the gas constant R can be determined without any reference to Avogadro's number, the same is not true of the gas constant k. But of course when Avogadro's number N is known, the relation $k = \frac{R}{N}$ enables us to derive the value of k from that of R.

If we introduce the gas constant k into the gas law for one gram-molecule, we obtain

$$pv = NkT.$$

Suppose, then, we take some other mass of gas, containing n molecules instead of N as in the gram-molecule. If both masses of gas are at the same pressure and temperature, Avogadro's hypothesis requires that their volumes be proportional to the numbers of molecules they contain. Hence for our sample containing n molecules, the gas law will be

$$pv = nkT.$$

Incidentally this last formula shows that the constant k may be interpreted as the gas constant for a hypothetical gas containing only one molecule (*i.e.*, $n = 1$).

Specific Heats—The specific heat of a gas (at temperature T) is defined by the amount of heat we must furnish to one gram-molecule of the gas at temperature T in order to increase this temperature by one degree.* (Sometimes one gram of the gas is taken in place of a gram-molecule.) If the volume of the gas is kept fixed during the heating process, the specific heat is called the "specific heat at constant volume." If, on the other hand, the pressure is maintained constant (the volume therefore changes), we have "the specific heat at constant pressure." The two specific heats differ in value. In the sequel we shall consider only the former.

Experimental measurements show that at normal temperatures and pressures the specific heat of a perfect gas is a constant; it does not vary with the temperature. In other words, regardless of what the initial temperature of a given gas may be, the same amount of heat is required to increase this temperature by one degree. The specific heat may, however, differ from one gas to another. The controlling factor appears to reside in the number of atoms present in the molecule of the gas. In the case of the monoatomic, diatomic, and polyatomic gases, respectively, the specific heats are found to be proportional to the numbers 3, 5, 6. But at extremely low temperatures, as recent research has shown, the specific heats of diatomic and polyatomic gases tend to coincide with that of the monoatomic gases.

---

* It would be more rigorous to define the specific heat at temperature T by $\dfrac{dQ}{dT}$, where dQ represents the amount of heat we must surrender to the gas in order to increase its temperature by the infinitesimal amount dT. The definition given in the text yields the mean value of $\dfrac{dQ}{dT}$ over a range of one degree (*i.e.*, from T to $T + 1$). In this chapter, however, the latter definition of the specific heat will suffice for our purpose.

The Kinetic Theory of Gases—Daniel Bernoulli, in the eighteenth century, advanced the idea that gases are formed of elastic molecules rushing hither and thither at enormous speeds, colliding and rebounding according to the laws of mechanics. But like many ideas that are not pursued mathematically, no definite opinion could be passed on Bernoulli's hypothesis until it was investigated quantitatively. The honor of having founded the kinetic theory is usually credited to Maxwell and to Boltzmann, who placed it on a mathematical basis. Let us first observe that the kinetic theory of gases is not a necessary consequence of the atomic hypothesis. If we were guided solely by the atomic hypothesis, we might suppose that gases are formed of atoms or molecules which repel one another; the repulsions of the molecules would then account for the pressure the gas exerts against the walls of a container. The originality of Bernoulli's hypothesis is that this pressure is assumed to arise not from mutual repulsions, but from the impacts of the massive molecules on the walls of the container.

At the time Maxwell and Boltzmann were developing the kinetic theory, the elementary properties of gases, mentioned in the previous chapter, were known, and so also was the law of entropy. The gas laws, the specific heats characteristic of the various kinds of gases, and the formulae of the more usual gas molecules had likewise been established. On the other hand, the third principle of thermodynamics, the decrease of the specific heats at exceedingly low temperatures, and the exact value of Avogadro's number N, had not been discovered. We mention these points to stress that a large part of the information available to modern physicists was unknown to the founders of the kinetic theory. For the present, however, we shall merely examine to what extent Bernoulli's idea was able to account for the elementary properties of gases.

(a) One of the most conspicuous properties of gases is their compressibility. Bernoulli's scheme appears to account for this property easily enough. The molecules of the gas are separated by vacant spaces and there is nothing to prevent their being compressed into a smaller total volume.

(b) A second property consists in the pressure a gas exerts against the walls of the container. This pressure may, in Bernoulli's conception, be attributed to the incessant bombardment of the gas molecules against the walls. It also follows that if a partition separating a gas from a vacuum is removed, the molecules, being unresisted in their motion, will rush into the vacuum; the gas will therefore expand into the largest volume available. This expansion is in accord with the observed properties of gases.

(c) When a gas is compressed, its temperature rises. We may account for this phenomenon by noting that when we advance a piston against a gas enclosed in a cylinder, the velocity of rebound of the molecules from the piston will be increased. Some of the molecules thus acquire added kinetic energy, which, as a result of subsequent collisions, will be communicated in part to the other molecules. We have then but to establish a link between the kinetic energy of the molecules and the temperature of the gas, and the heating by compression is accounted for. Similar arguments show why a gas should cool when expanded and hence when its molecules rebound from a retreating piston.

(d) A perfect gas expanding into a vacuum does not decrease in temperature. Bernoulli's conception also accounts for this phenomenon. The molecules now surrender no kinetic energy, in contradistinction to their behaviour when rebounding from a retreating piston. Hence, when they rush into the vacuum no kinetic energy is lost: the temperature therefore remains unchanged.

(e) Local differences in temperature and pressure automatically become evened out in a gas; this is a fact of observation and is required by the law of entropy. Bernoulli's conception seems to be in harmony with these occurrences, for owing to the randomness of the molecular motions, uniformity in the motion may be expected eventually.

From these general considerations it would appear that Bernoulli's hypothesis is not impossible. We must, however, examine the situation more carefully before jumping at conclusions. Other properties of gases, such as their specific heats and their viscosity, will also have to be considered. Above all, a precise quantitative, mathematical investigation must be undertaken before the kinetic hypothesis can acquire any scientific standing.

Probabilities—Elementary notions of the calculus of probabilities are required in the kinetic theory. We shall summarize these briefly.

Suppose a die is thrown ten times. We inquire: What is the probability of the 6-spot appearing three times in succession at any stage of the game?

Our question cannot be answered until additional information has been given. This point is easily understood when we note that the answer to our probability problem cannot be the same if the die is loaded or if it is evenly balanced. For instance, we may be led to assume that the die is evenly balanced, so that each side has the same probability of appearing. On the other hand, we may be tempted to believe that the die is loaded and that the probability of the one-spot turning up is five times greater than that of any one of the other faces appearing. Precise information on this score is a necessary preliminary to any further investigation; it serves to define the so-called *a-priori* probabilities of the die's various faces appearing. Only when the *a-priori* probabilities have been agreed upon can we proceed to answer the probability problem originally set. The answer involves calculation and cannot be given offhand. The type of calculation required in the various cases forms the subject matter of a branch of pure mathematics called the "calculus of probabilities," investigated by Jacques Bernoulli, Laplace, Gauss, and others.

Our illustration shows that two different kinds of probabilities are introduced in our problem. Firstly, we have the so-called *a-priori* probabilities, which define the probabilities of the various faces of the die appearing. These probabilities must be stipulated at the start. Secondly, we have the calculated probabilities, which are deducible from the *a-priori* probabilities. In a certain sense we may compare the *a-priori* probabilities to the postulates of geometry. According to the choice of our postulates we obtain one geometry or another. Similarly in probability problems, the calculated probability for a given problem depends on the *a-priori* probabilities initially assumed. In many respects this division of probabilities into two kinds is artificial, because we might always regard the *a-priori* probabilities as the calculated probabilities deduced from some more fundamental set of *a-priori* probabilities. The distinction is, however, justified in practice, for in any case we must start from certain assumptions; otherwise we should fall into an unending regress.

The probabilities we have just considered are of the discontinuous, or discrete, type. We shall now examine continuous probabilities. Here also, both *a-priori*, and calculated, probabilities arise. To take a definite case, let us suppose that someone, unseen by us, makes a mark with a pencil on a sheet of paper. We wish to decide on the *a-priori* probability of the mark being situated here or there. We have no idea of what the *a-priori* probability truly is, because, for all we know, the person who made the mark may have had a preference for the centre or for a corner of the paper. At all events we must make a choice, and for argument's sake we shall assume that the *a-priori* probability is the same for all points of the paper.

But a difficulty arises when we wish to assign a value to this *a-priori* probability, a difficulty which did not occur in the discrete case. Its nature is easily understood. Thus the probability of the pencil mark being found somewhere on the sheet of paper is supposed to be a certainty and hence is credited with the value 1. Since the paper contains an infinite number of points, the probability that the mark will be found at one special point of the infinite aggregate is $\frac{1}{\infty}$, *i.e.*, 0. We are thus led to assign a zero value to the *a-priori* probability of finding the mark at any given point—a result which cannot be utilized. The difficulty is circumvented if we cease to consider the *a-priori* probability of the pencil mark· being situated at a mathematical point, and replace this notion by that of the *a-priori* probability of the mark being situated in a given area. We divide the sheet of paper into tiny juxtaposed squares, all of them of the same size, and we agree on the *a-priori* probabilities of the mark being found in the respective squares.

Since we have agreed to consider the case of equiprobability, we must assume that there is the same probability of the mark being found in any one of the squares. Let us, then, suppose that the area of the paper has been subdivided into 10000 equal squares. The *a-priori* probability of the mark being situated in a given square is $\frac{1}{10000}$, a finite non-vanishing amount. We are thus rid of the difficulty of the zero probability mentioned previously.

Having stipulated the *a-priori* probabilities, we pass to a problem of calculated probabilities. Suppose that two marks are made on the sheet of paper. We now inquire: What is the probable distance between the two points? Or what is the probability of this distance being less than one inch? Such questions involve the calculus of probabilities and require mathematics. In short, we see that whether the probabilities be discrete or continuous, the same distinction between *a-priori* probabilities and calculated probabilities arises.

The Foundations of the Kinetic Theory—Consider a gas in a rigid container. We assume that the gas is formed of a large number N * of molecules, moving about at random, colliding and rebounding against one another and against the walls of the container. If we assume that the molecules actually enter into collision and are not merely repelled by forces before they can come into contact, we must suppose that these molecules are perfectly elastic. The fact is that if the molecules were not perfectly elastic, their kinetic energies would be diminished after each collision (as occurs in the impact of two billiard balls that are not perfectly elastic), and after a sufficient number of collisions, the mole-

---

* In what follows, the number N will be assumed to be any large number, not necessarily Avogadro's number (unless otherwise specified).

cules would come to rest. If a gravitational field were present, the molecules would fall inert to the bottom of the vessel. From the standpoint of the kinetic theory this condition would imply that a gas, placed in an isolated enclosure and submitted to a gravitational field, would eventually drop in temperature to the absolute zero, and that a vacuum would be formed automatically. Since these anticipations are contradicted by experience, we must assume that if the molecules actually collide, they must be perfectly elastic. The hypothesis of perfectly elastic molecules is of itself extremely unsatisfying. It may be dispensed with if we suppose that, during the so-called collisions, the molecules never come into direct contact because they are subjected to mutual repulsions which become enormous at small distances.

In any case whether actual collisions occur or not, our aggregate of N molecules, moving about in a rigid container, constitutes an isolated and conservative mechanical system; its total energy (here kinetic alone) remains constant during the motion. In theory, it would be possible to determine the positions and motions of all the molecules at any instant provided the positions and velocities at any initial instant were specified.* Unfortunately the mathematical problem is far too difficult to be solved, and the future of the system cannot be determined rigorously. The only way of attacking the problem is to introduce probability considerations. The application of the calculus of probabilities to complicated mechanical systems forms a new science which Gibbs named "Statistical Mechanics." Originally conceived with a view of elucidating the difficulties of the kinetic theory, statistical mechanics has since been applied to a large number of physical problems and also to situations arising in astronomy.

We shall first examine certain preliminary points in connection with a gas system. In the kinetic theory of gases a distinction is made between the *microscopic state* and the *macroscopic state* of the gas system. The "microscopic state" of a gas at any instant is defined by the instantaneous positions and momenta (in magnitude and in direction) of the various molecules. If any change is made in the molecular distribution, a new microscopic state is obtained. Many of the different microscopic states, however, will be indistinguishable from the standpoint of macroscopic observation. For example, if the positions and motions of two or more molecules are interchanged, a new microscopic state will result, yet from the macroscopic point of view no change will be detected in the condition of the gas. Accordingly, we may say that although the microscopic state

---

* The dimensions of the molecules would also have to be known.

of the gas is no longer the same, its "macroscopic state" has not changed. We are thus led to associate a large number of different microscopic states with each macroscopic state of the gas.

Now in practice all that our observations can detect are changes in the macroscopic state of the gas. Consequently, insofar as the kinetic theory attempts to predict observable effects, it must furnish information on the macroscopic states. Since, at best, the theory can only deal with probabilities, its aim will be to establish the probabilities of the various macroscopic states. These probabilities cannot be computed, however, until the probabilities of the so-called microscopic states are known. For this reason our first concern will be to compute the probabilities of the microscopic states. If, then, we can determine the probabilities of occurrence of all those microscopic states which pertain to the same macroscopic state, we shall be in a position to deduce the probability of this macroscopic state. For example, if all the microscopic states have the same probability, the probability of a given macroscopic state will be proportional to the number of different microscopic states which it contains. (This, of course, would not be the case if the microscopic states were not all equally probable.) The probabilities of the microscopic states thus constitute, so to speak, the *a-priori* probabilities in our problem; the probabilities of the macroscopic states play the part of calculated probabilities.

These considerations show that the first step in statistical mechanics is to determine the probabilities of the various microscopic states. The fact that these probabilities play the part of *a-priori* probabilities does not necessarily imply that they are of so fundamental a nature that they cannot be calculated. It is true that in many situations the *a-priori* probabilities appear to be fundamental. For example, when a coin is tossed, the *a-priori* probabilities are represented by the probabilities of heads or of tails appearing. We cannot calculate these probabilities because to do so we should have to take into consideration the act of throwing and the idiosyncrasies of the person tossing the coin; and the difficulty would be too great. Accordingly, the values we assign to the *a-priori* probabilities in this case are dictated by other considerations, in particular by the principle of sufficient reason. We therefore decide on equiprobability and assert that the probability of heads or of tails appearing is the same. But in statistical mechanics we are able to calculate the *a-priori* probabilities because the laws which regulate the processes involved are the well-known laws of mechanics. In short, the probabilities of the microscopic states are *a-priori* only in that they enable us to calculate the probabilities of the macroscopic states, but they are not *a-priori* in the sense of being fundamental. These points will be clarified presently.

We shall now consider a problem which might appear to have no connection with statistical mechanics but which in point of fact is intimately connected with Gibbs's method of establishing the probabilities of the microscopic states. The problem we have in view concerns a fluid in steady motion. A fluid is said to be in steady motion when the magnitude and direction of its velocity, at any fixed point, remains unchanged in the course of time. We shall assume that the fluid is formed of molecules; in this case the steadiness of the flow implies that the successive molecules which pass through a fixed point always have the same velocity (in magnitude and in direction).

Suppose a fluid in steady motion is contained in a vessel which it fills completely. For the purpose of discussion we shall assume that one of the molecules can be distinguished from the others; we shall refer to this molecule as the red molecule. At the initial instant this molecule is known to be situated at some specific point of the fluid; then in the course of time it is carried along to other parts of the vessel. The problem we are setting ourselves is as follows: Under what conditions will all positions of the red molecule in the fluid be equally probable after a long period of time has elapsed?

Inasmuch as our problem involves continuous probabilities, we shall proceed as was explained in a previous paragraph. Accordingly, we shall imagine that the extension (here a volume) is subdivided into a large number of elementary extensions, *e.g.*, into tiny, juxtaposed cubes of equal size.

Now in view of the steadiness of the flow, the number of molecules in any given cube will remain unchanged in the course of time. Obviously, however, unless the fluid is incompressible, the different cubes will contain different numbers of molecules; and in this case there will be a greater probability of our finding the red molecule in a closely packed cube than in a sparsely filled one. We conclude that for equiprobability to hold, a necessary condition is that all the cubes should contain the same number of molecules, and this implies that the fluid must be incompressible.

This condition, however, though necessary, is not sufficient to guarantee equiprobability. The fact is that, if equiprobability is to hold, the red molecule must pass through every cube, and we cannot be certain that this situation will be realized. Let us examine the situation more attentively.

Three main possibilities must be considered. The trajectory described by the red molecule may be a closed path described periodically, or it may be restricted to a surface, or it may lie along a curve that never closes and fills the entire volume. In the last case alone will the trajectory pass through every cube in the volume. An example of the first situation is afforded when the vessel is a straight cylinder and the liquid rotates like a rigid body about the axis of the cylinder; each molecule then describes a horizontal circle or else remains at rest on the axis. In this case equiprobability cannot hold, for the red molecule, which describes one of the circles, cannot pass through a cube that is not intersected by the circle. A similar condition occurs when the trajectory covers a surface. Only when the trajectory passes through every cube of the volume, and hence fills the entire volume, will equiprobability be realized.

We are thus led to the following conclusions:

All positions of the red molecule will be equally probable after a long period of time, provided

1. The fluid in permanent motion is incompressible.
2. The trajectory of a molecule never closes, and fills the entire volume of the vessel.

These conditions will be encountered in the foundations of the kinetic theory of gases. Maxwell referred to the second condition as the "Hypothesis of the Continuity of Path," and Boltzmann called it the "Ergodic Hypothesis." *

The foregoing discussion of a hydrodynamical problem will enable us to obtain a clearer insight into Gibbs's method of establishing the probabilities of the microscopic states of a gas system, or equivalently of a mechanical system formed of N molecules.

The Phase Space—We are dealing with a mechanical system of N molecules. Hence, to determine the configuration of the system at any instant, we must specify the positions of these molecules. Three coordinates x, y, z are required for each molecule,† so that $3N$ coordinates

---

* A necessary condition for the ergodic hypothesis to be valid is that the integrals determining the trajectories be non-uniform functions. It will be noted that the present situation is the opposite, so to speak, of that encountered in the problem of Three Bodies. In this latter problem a general solution by quadratures was impossible because of the presence of non-uniform integrals.

† Since the molecules are assumed spherical, their orientation in space does not come into consideration.

are required in all. This requirement expresses the fact that our system has $3N$ degrees of freedom. In order to specify the microscopic state of the system at any instant, we must also specify the momenta of the molecules. The momentum of a single molecule is determined by three components, p_x, p_y, p_z, i.e., by the three projections of the momentum on the coordinate axes. Altogether, then, $3N$ components of momentum will have to be specified for the N molecules. If we designate quite generally the coordinates of position by the letter q and the components, or coordinates, of momentum by the letter p, we see that the configuration and motion of the gas at any instant will be determined by $6N$ magnitudes, half of which are q's, and the other half p's. Thus the $6N$ magnitudes

$$q_1, q_2, q_3 \ . \ . \ . \ . \ q_{3N}, p_1, p_2, p_3 \ . \ . \ . \ . \ p_{3N}$$

will determine the microscopic state of the gas at the instant considered.

Gibbs treated these $6N$ magnitudes as the coordinates of a single point situated in a space of $6N$ dimensions. He called this artificial space the "phase space"; and the point P, situated in this space and determined by the $6N$ coordinates, was called the "phase point." This phase point, by its position in the phase space at any instant, defines at this instant the positions and motions of all the individual molecules, and hence defines the instantaneous microscopic state of the gas. As the molecules of the gas move about in the container and as their momenta vary, the values of the q's and p's change, and so the phase point moves in the phase space. The phase point cannot, however, move anywhere in the phase space for two reasons. Firstly, because the coordinates of position q can vary only within the limits imposed by the container in which the gas is placed. Secondly, because, the mechanical system formed by the gas molecules being a conservative one, the energy (here kinetic alone) of this system necessarily retains its initial value during the motion, and so a restriction is thereby placed on the relative values that the momenta p of the molecules may assume. For instance, it is obvious that the total energy would not retain the same fixed value if at one instant all the molecules were moving with extremely high velocities, and then at a later instant they were all moving with reduced speeds. But except for these restrictions, the phase point P is free to move anywhere in the phase space,* or, what comes to the same, the microscopic states may be all those that are compatible with the dimensions of the vessel and with the total energy of the molecules.

---

* More precisely, the phase point will be free to move over the surface of a limited cylindrical surface of $6N - 1$ dimensions in the phase space.

Gibbs, in his study of a gas, proceeded in an indirect way. Thus let us suppose that the gas we wish to examine is enclosed in a vessel and that its total energy is E. Instead of confining his attention to this particular gas system, Gibbs considered a large number of similar systems simultaneously; all these gases were composed of the same number N of molecules as the gas of interest and they were placed in vessels having the same shape. The energies of the different systems were not quite the same, however. We shall assume that the range in the energies extends from E to $E + dE$, where dE is very small. The microscopic state of each one of these gases is defined by a phase point situated in the same phase space; and as the gases evolve, their respective phase points stream through a limited volume of the phase space. The volume cannot be visualized easily on account of the large number of dimensions of the phase space, but we may describe it roughly as the volume contained between two coaxial cylinders having very nearly the same radii. The volume has thus very little width. Furthermore it is limited above and below by planes.

In what follows, we shall suppose that at the initial instant the microscopic states of the various gas systems differ very little from one another, so that the phase points will be distributed uniformly throughout the phase space. The phase points will then start to stream between the two cylinders like the molecules of a fluid. For convenience we shall refer to this fluid as the "phase fluid."

Now we know very little about the motions of the phase points because the equations of dynamics, which control the evolutions of the gas systems, cannot be solved. However, the bare fact that the dynamical equations control the motions, enabled Gibbs and Boltzmann to prove that the phase fluid would stream like *an incompressible fluid in steady motion.*

The steadiness of the flow is established by showing that the density of distribution of the phase points about any fixed point P in the fluid does not vary with time.

The incompressibility of the fluid is expressed by two celebrated theorems, which Gibbs called "the conservation of extension in phase" and "the conservation of density in phase."

The statement of Gibbs's theorems will be facilitated if we first consider a simple illustration. We imagine an arbitrary volume in the phase space, and we restrict our attention to the phase points situated within this volume at the initial instant. For convenience we shall say that this aggregate of phase points occupies the volume. The phase points now move along their trajectories, and since their mutual distances change during the motion, the volume occupied by the points will vary in shape from one instant to another. Gibbs's theorem of the conservation of extension in phase may then be stated:

> The volume occupied by an aggregate of phase points will remain constant in size as the phase points stream through the phase space; the shape of the volume will vary, however.

The theorem of the conservation of density in phase can be derived from the former theorem. It may be stated:

> The density in the distribution of the aggregate of phase points will remain constant during the motion.

The method whereby these theorems are derived from the equations of dynamics will be mentioned presently.

Gibbs's two theorems show that there is no tendency on the part of the phase points to concentrate or to become rarefied in certain regions of the phase space. Consequently, if the distribution of the phase points is uniform initially, it will remain so throughout time.

The proof of the phase fluid's incompressibility has necessitated the simultaneous consideration of an aggregate of gas systems differing slightly in energy. But now that the incompressibility has been established, we may restrict our attention to the systems of energy E, a large number of which are included in our former aggregate. The phase points of these latter systems move along the surface of the limited cylinder E that bounds the phase space; they never penetrate into the space between the two cylinders. In their aggregate these phase points (which we shall call phase points E) constitute a fluid film which moves with steady motion over the cylinder E. Furthermore, since the width of the phase space between the limiting cylinders E and $E + dE$ is constant, and since the phase fluid in this space is incompressible, we may show that the film moving over the surface E is likewise incompressible. We conclude that the phase points E stream over the cylinder as though they were the molecules of an incompressible fluid film in steady motion.

These preliminary results enable us to investigate the probabilities of the various microscopic states of a gas. We select any one of the gas systems of energy E. The phase point which defines the microscopic state of this gas at any instant is one of the phase points E streaming over the cylinder. The different microscopic states of the gas will therefore be equally probable or have different probabilities according to whether or not all positions of the phase point on the cylinder are equally probable. Now the motion of the phase point over the cylinder is similar to that of the red molecule in our former hydrodynamical example of the fluid in steady motion. The only difference is that our phase point is moving in a hyperspace (the surface of the cylinder) whereas the red molecule was moving in ordinary space. This difference, however, is purely formal and does not affect the problem. If we refer to the results stated in connection with the red molecule, we are led to the following conclusions:

In view of the incompressibility of the fluid film, all positions of the phase point on its trajectory will be equally probable. Since, however, this trajectory need not cover the entire cylinder and may vary in shape according to the initial position of the phase point, we cannot establish definite probabilities for the points of the cylinder. On the other hand, definite probabilities will be secured if we postulate the ergodic hypothesis and assume that the trajectory covers the entire cylinder. In this case the phase point will pass in the course of time all over the cylinder,* and hence all points on the cylinder will be equally probable.

Let us express these results in terms of the microscopic states. If the ergodic hypothesis is not satisfied, a gas system will pass periodically only through some of the microscopic states. These particular states will then be equiprobable and the other states will be impossible. Since, however, the precise cycle of states through which the gas may pass will vary according to conditions, the probabilities of the states as a whole will be indeterminate. But if the ergodic hypothesis is satisfied, the gas will pass through all the microscopic states in the course of time, and all the probabilities will be determinate. All the various microscopic states will then be equally probable.

So as to render the probabilities determinate, Boltzmann accepted the ergodic hypothesis. In this way the *a-priori* probabilities of the microscopic states were obtained, and from them the macroscopic probabilities were derived.

Inasmuch as the assumption of determinate values for the probabilities of the microscopic states is essential in the construction of a statistical mechanics, we conclude that the kinetic theory can be regarded as valid only insofar as the ergodic hypothesis is justified. Unfortunately, all attempts to demonstrate the correctness of this hypothesis by rigorous mathematical means have failed.

It cannot be said, however, that the ergodic hypothesis is incorrect, but only that it has never been proved. On the other hand, in all the rela-

---

* The trajectory will not pass through every point of the cylinder, but it will pass through each one of the tiny surface elements into which the surface of the cylinder is assumed to be subdivided.

tively simple situations which have been studied and in which rigorous mathematical means of investigation have been available, the ergodic hypothesis has been proved incorrect. Kelvin gave a number of examples of such situations and, on the strength of them, rejected the kinetic theory of gases. Kelvin's attitude is especially interesting when we recall his natural inclination towards mechanical models of physical phenomena. The kinetic theory was also combated by Mach and Ostwald. But since these investigators were hostile to all theories involving invisible occurrences, their opposition to the kinetic theory is not surprising.

The impossibility of proving the ergodic hypothesis must not obscure the fact that this hypothesis constitutes only one of the two conditions that must be satisfied if all microscopic states are to be equally probable. The other condition (*i.e.*, the incompressibility of the phase fluid) is at least as important and, in contradistinction to the ergodic hypothesis, it can be proved to be a necessary consequence of the laws of dynamics. Let us see how this proof can be secured.

In Chapter XVII we mentioned the existence of mathematical expressions which Poincaré called "integral invariants." Integral invariants in dynamics are connected with the motion not of a single dynamical system, but with the aggregate of the motions of a large number of similar systems started from initial conditions that differ only slightly. A characteristic of integral invariants is that their numerical values remain constant as the motions proceed. As such, they express a permanent relationship connecting motions that differ only slightly. The first of these integral invariants in dynamics was discovered by Liouville, and the statement of its constancy is called Liouville's theorem. It is the invariance of Liouville's integral, as applied to an aggregate of phase points differing only little in their positions, that affords the mathematical expression of the conservation of extension in phase and establishes thereby the incompressibility of the phase fluid. Now the invariance of Liouville's integral is traceable to the peculiar *form* of Hamilton's equations of dynamics.* Hence we conclude that Gibbs's theorems are direct consequences of the form of these equations.

Equivalent results may be obtained by other methods, *e.g.*, by considering the rôle played by Jacobi's multiplier in the equations of flow of a fluid. We recall that a multiplier (see the method of the Last Multi-

---

* Since the mechanical equations of the theory of relativity may also be thrown into the Hamiltonian form, we conclude that the application of the theory of relativity to statistical mechanics will likewise entail the incompressibility of the phase fluid, and therefore, when account is taken of the ergodic hypothesis, will be consistent with the equiprobability of the microscopic states.

plier) is the function introduced by Jacobi in his attempts to facilitate the integration of Hamilton's equations. But Jacobi's multiplier has other applications. Thus the density from place to place of a fluid in steady motion is furnished by the solution of a partial differential equation which is none other than the equation of Jacobi's multiplier. Whenever this equation admits a constant for its solution, the density of the fluid in steady motion may remain uniform throughout the volume, and therefore the fluid is incompressible. Now the multiplier which represents the density of the phase points in the phase fluid is the multiplier connected with Hamilton's equations; and we mentioned in Chapter XVII that the peculiar form of these equations enables us to take any constant as a mutiplier. Thus it is because Hamilton's equations admit a constant as multiplier that the phase fluid is incompressible. Incidentally, when the ergodic hypothesis is taken into account, the constant multiplier, which measures the density of the phase points, also represents the probabilities of the microscopic states, or rather the densities of these probabilities. Hence the constancy of the multiplier entails the equiprobability of these states.

All these results are interesting because they show that the equiprobability of the microscopic states is a mere consequence of the peculiar form of Hamilton's equations of dynamics (and of the ergodic hypothesis), and that it has no greater *a-priori* justification than have the classical mechanical laws themselves. If the form of these laws happened to be different, equiprobability would not hold, and the classical kinetic theory of gases would give place to a different one.

The Macroscopic Probabilities—As time passes, the molecules fly hither and thither, and the gas system evolves from one microscopic state to another. We pointed out (page 385) that many of these microscopic states would be indistinguishable from a macroscopic standpoint and would thus pertain to the same macroscopic state of the gas. We also said that the probabilities of the various macroscopic states could not be calculated before those of the microscopic states were known. Since the microscopic states are now known to be equally probable, the probability of the gas being in some specific macroscopic state, at any instant, is obviously proportional to the number of microscopic states which this macroscopic state contains. To calculate the relative probabilities of the various macroscopic states, we must therefore compute the numbers of microscopic states which they embrace respectively. Boltzmann computes these numbers in the following way.

The position of a gas molecule is determined when the three coordinates x, y, z of its centre are specified. Its motion is determined in magnitude and in direction when the three components p_x, p_y, p_z of its momentum are given. We may therefore represent the position and motion of this particular molecule by means of a phase point, of coordinates x, y, z, p_x, p_y, p_z, situated in a space of six dimensions. (This phase space is not the one previously considered, for the former was a space of $6N$ dimensions.) Proceeding in a similar way for all the N molecules which are supposed to constitute the gas, we obtain N different phase points distributed in our 6-dimensional phase space. As in all cases where continuous probabilities are used, the probability of a point being situated at a given point is zero; and, to obviate this difficulty, we are led to subdivide the extension into tiny volumes and to consider the probabilities of a point or points being situated within one tiny volume or another. Let us then subdivide our 6-dimensional phase space into a large number of tiny 6-dimensional cubes, all of them equal in volume. A definite microscopic state will be defined, say, by the presence of the phase points a, b, and c in the first cell, by the phase points d and e in the second cell, and so on. The cells are assumed so small that the exact positions of the phase points within them are held to be of no importance. Suppose now we exchange some of the phase points of the different cells. The exchanges, by modifying the individualities of the phase points in the cells, give rise to new microscopic states; but since the numbers of phase points in the various cells remain unaffected, no change occurs in the macroscopic state of the gas.

The number W of different ways in which the phase points may be exchanged, and hence the number of different microscopic states pertaining to the macroscopic state of interest, is easily found to be

$$W = \frac{N!}{a_1!\, a_2!\, a_3! \,.\, .\, .\, .\, .},$$

where N is the total number of molecules in the gas and where a_1, a_2, a_3, . . . represent the numbers of phase points situated in the first, second, third, and following cells, respectively.

The most probable macroscopic state, for a gas containing N molecules and having a given energy (hence temperature), is the state for which the distribution of phase points among the various cells makes the number W a maximum. (The possible distribution of phase points is assumed restricted so as to be compatible with the total kinetic energy ascribed to the gas system.) In the course of time the most probable macroscopic state is realized more often than the less probable states. Furthermore, it can be shown that the most probable state has a probability far greater than that of any of the states which differ perceptibly from it. As a result, the most probable state will be realized practically at all instants, and the gas system will thus have a semblance of permanency when viewed from the macroscopic standpoint of ordinary observation. This most probable macroscopic state, which exhibits the appearance of permanency, is called the state of *statistical equilibrium*. The word equilibrium is here introduced, because although we do not have equilibrium in the motionless sense of the word, yet the high measure of permanency of the most probable macroscopic state endows it with one of the attributes of an ordinary state of equilibrium. The statistical stability of this privileged state was also investigated in another way by Boltzmann. Boltzmann made a direct study of the collisions of molecules assumed perfectly spherical and elastic, and showed that as a result of the collisions a certain magnitude, which he called H, would tend to decrease until it had attained a minimum value H_m. The rate of change of H was also investigated. The value of the difference $H - H_m$ was shown to measure the departure of the actual macroscopic state of the gas from the state of statistical equilibrium.

Temperature—Common experience shows that when a gas is heated, its expansive force increases. Since pressure, in the kinetic theory, is attributed to the impacts of the molecules against the walls of the container, we may anticipate that the hotter a gas, the more rapidly will its molecules be moving and the greater will be their kinetic energy of translation. Calculation shows that when account is taken of the gas laws, the sum total of the kinetic energy of translation of all the N molecules present in a gas must have as value

$$\frac{3}{2} NkT,^*$$

where k is the gas constant and T the absolute temperature. In particular, if we are dealing with a gram-molecule, N is Avogadro's number, and Nk becomes the other gas constant R. In this event, our result may be written

$$\frac{3}{2} RT = \begin{cases} \text{total kinetic energy of translation of all the molecules} \\ \text{in a gram-molecule.} \end{cases}$$

---

* This value is correct if we are dealing with a perfect, monoatomic gas. On page 407 we shall see that in the case of a diatomic or polyatomic gas, $\frac{3}{2} NkT$ represents only an average value.

The foregoing formula confirms our qualitative anticipations by show-ing that the kinetic energy, and hence the velocities of the molecules, must increase with the temperature T. The velocities of the molecules may be deduced from this formula. Thus we first note that the left-hand side in our formula contains none but known magnitudes, and that the right hand side is $N \dfrac{mv^2}{2}$ (where m is the mass of a molecule and v the average velocity). Now Nm is the total mass of the gram-molecule of gas and is therefore known. Consequently, the numerical value of the average velocity v may be obtained. The formula shows that the smaller the mass m of the molecules, the greater will be the value of their average velocity for a given temperature T.

The Concept of Heat in the Kinetic Theory—We have just seen that the temperature of a gas is proportional to the total translational energy of all the molecules. This result must be clarified, however, for temperature and heat come into consideration only when the mole-cules are moving at random, as occurs in the state of statistical equi-librium. For example, let us suppose that on a summer day the air is perfectly still with not a breath of wind stirring. The molecules of the air are moving at terrific speeds; but precisely because these motions are in random directions, no work can be extracted from them; a windmill, for instance, will remain motionless. If a slight breeze arises, the wind-mill is set turning and work may be performed. Yet the velocities of the molecules will have increased only slightly. The breeze has merely superposed a general uniform motion on the molecules as a whole, and this superposed motion is insignificant in comparison with the original molecular velocities. Plainly, the efficacy of the breeze in generating work is not due to the added velocities given to some of the molecules (while the velocities of other molecules are decreased); it is due solely to the more orderly motion that the breeze imposes. The temperature of the air is not increased by this superposed orderly motion; and so we see that the energy contributed by the breeze is not heat energy; it is mechanical energy.

Suppose, then, that by some chance all the molecules in a certain region should happen to be moving simultaneously in the same direction. When we realize that at ordinary temperatures the velocities of these molecules approximate those of rifle bullets and that even a hurricane blowing at no more than 100 miles an hour uproots trees, we may well imagine the catastrophic effects that would result from the molecular motions. The heat energy of the air would now be transformed into mechanical energy; the air would have no definite temperature, for its energy would be entirely mechanical. In short, from the standpoint of the kinetic theory, heat energy is disorganized random mechanical energy; whereas mechanical energy proper is directed, ordered. We also see that there is no sense in speaking of heat and temperature when only one molecule is concerned, for the element of randomness is now lacking, and the kinetic energy of the single isolated molecule becomes mechanical energy. Heat and temperature are thus macroscopic concepts suggested by our senses; they are of considerable use on the level of common experience, but they cease to have any precise meaning in the microscopic world.

The Law of Entropy—Thermodynamics asserts that the entropy of an isolated system always tends to increase till it reaches the maximum value compatible with the restrictions placed on the system. The kinetic theory teaches us that the gas system will pass from the less probable to the most probable macroscopic state, *i.e.*, to the state of statistical equilibrium. If, then, we accept the kinetic theory, we see that the progressive increase in the entropy of a gas, for example when the gas expands into a vacuum, accompanies the increase in the probabilities of the successive macroscopic states assumed by the gas. In particular, the state of maximum entropy must correspond to the state of maximum probability, *i.e.*, to the state of statistical equilibrium. The originators of the kinetic theory were thus faced with the necessity of showing that the identification of the state of maximum entropy with that of statistical equilibrium did not entail any obvious absurdity. For instance, in the state of maximum entropy thermodynamical considerations show that the temperature, pressure, and density of the gas are uniform throughout the volume. The same conditions should therefore be realized in the state of statistical equilibrium. Calculation proves that this requirement is indeed satisfied by the kinetic theory. To this extent at least, the kinetic theory appears to be in agreement with observation.

The kinetic theory requires, however, that we abandon our belief in the absolute validity of the principle of entropy. It shows that occasionally a less probable state may arise, and that even when the most probable state of statistical equilibrium is reached, there is always a possibility that some of the subsequent states will be among the highly improbable ones. This fact implies that the entropy may decrease of its own accord—a conclusion which contradicts the law of entropy when viewed as a law of absolute validity. The contradiction may be avoided, however, if we regard the law of entropy as a law of only approximate validity, expressing probabilities and not certainties. Such was the attitude championed by Maxwell, Boltzmann, and Gibbs; today it is universally accepted.

We must therefore assume that if we wait long enough, there is no reason why a gas should not contract spontaneously, leaving a vacuum behind it; there is no reason why the molecules of the air, which in the state of statistical equilibrium are moving at random in all directions, should not every now and then be moving in the same direction. If this unidirectional motion should occur for all the molecules situated beneath a table, calculation shows that, owing to the tremendous speeds of the molecules, the table would be hurled against the ceiling. The fact that such phenomena have never been observed does not militate against the kinetic theory, for calculation shows that they can happen only on exceedingly rare occasions. According to calculation, only once in trillions of centuries should a gas contained in a volume of one litre (under standard conditions of temperature) contract of its own accord, thereby generating a vacuum in the vase. Small wonder, then, that the phenomenon has never been observed.

Let us make clear why the phenomenon of the table suddenly rising under the molecular impacts would constitute a violation of the law of entropy. Firstly, the sudden rushing of the molecules in the same direction would tend to create a vacuum under the table, with the result that an inequality of density would arise spontaneously; and this would violate the law of entropy. But there is also another way of understanding the matter. Prior to the simultaneous rise of the molecules, their motions would be of the random type, and the energy present would be heat energy. But when the molecules move upwards simultaneously, the heat energy is transformed totally into mechanical energy, which is expended in lifting the table. This spontaneous transformation of heat energy into mechanical energy, without an accompanying fall of heat, obviously violates Carnot's principle and hence violates a particular case of the principle of entropy. Nevertheless, as we have explained, the possibility of the principle of entropy being violated under ordinary conditions is so slight that it has no practical interest. We are therefore perfectly safe in applying the principle of entropy in thermodynamics without fear of error. On the other hand, the possibility of improbable states arising, and hence of the entropy decreasing, will be much more favorable when small numbers of molecules are considered, and we shall then find that the improbable occurrences will no longer be a matter of speculation but may actually take place under our very eyes. In the early days of the kinetic theory, however, the particular phenomena to which we are here referring (Brownian movements and fluctuations) were unknown, and the theory could receive no support from this quarter.

Boltzmann, having sensed the intimate connection between the growth in the entropy of a gas and the increase in the probabilities of its successive macroscopic states, proceeded to obtain the exact expression of the entropy in terms of the probability. The considerations that guided him will be understood from the following analysis. We consider two gas systems. If we call W_1 the probability that the first system is in some specified macroscopic state, and W_2 the probability that the second system is in some other specified macroscopic state, then, according to the calculus of probabilities, the probability that the two systems will be simultaneously in the specified states is expressed by the product $W_1\,W_2$ of the two probabilities. On the other hand according to thermodynamics, the entropy of the two systems, viewed as forming a single system, is $S_1 + S_2$, where S_1 and S_2 are the entropies of the two systems considered individually. Thus, the probabilities are multiplied and the corresponding entropies are added. The foregoing correspondence between multiplication and addition is exhibited in mathematics by numbers and their logarithms. Boltzmann therefore supposed that the entropy of a gas in a given macroscopic state was proportional to the logarithm of the probability of this state.

If we call W the probability of a given macroscopic state of a given mass of gas, the entropy S of the gas in this state is determined, according to Boltzmann, by

$$S = A \log BW,$$

where A and B are two constants. The numerical value of the probability W of interest is obtained from the expression on page 395. In particular, the entropy in the state of statistical equilibrium is obtained when we replace W by its maximum value.

It can be shown that the constant A is none other than the gas constant k. As for the constant B, the kinetic theory affords no means of establishing its value, and consequently Boltzmann's formula does not yield the absolute (but only the relative) value of the entropy. This limitation was not regarded at the time as of any particular importance, because thermodynamics also, prior to the discovery of the third principle, could assign no absolute value to the entropy. But with the introduction of the third principle, the necessity of considering absolute values was recognized, and for this reason the constant B could no longer remain unspecified. In Chapter XL we shall see how an absolute value for the entropy was finally obtained.

The kinetic theory leads to an extension of the thermodynamical concept of entropy. In thermodynamics, where differences in entropy were determined in connection with reversible transformations, the concept of the entropy of a gas acquired a meaning only when the gas was in the state of statistical equilibrium. But in the kinetic theory, thanks to Boltzmann's formula, the entropy of any macroscopic state can be computed. From the standpoint of the kinetic theory, entropy serves as a measure of the randomness with which the molecules are moving—the more orderly the motion, the smaller the entropy. We may also add that the function H (mentioned on a previous page), which as Boltzmann proved, has a tendency to decrease continually as a result of molecular collisions, is none other than the entropy S with its sign reversed.

Loschmidt's Paradox—Several pitfalls are connected with Boltzmann's conception of the entropy of a gas. One of the most striking is Loschmidt's paradox; it has since been investigated by Ehrenfest. Let us consider a gas in a given microscopic state A. The positions and the momenta (or velocities) of all the individual molecules are then specified. The subsequent and also the prior evolution of our isolated mechanical gas system is rigorously determined by this initial state. Since we are dealing with a conservative mechanical system, we know that, if we leave the initial positions of the molecules unchanged but reverse all the velocities, the system will pass in inverse order through all the former states. Suppose that in the original situation the entropy of the gas is increasing during the passage of the gas through the state A. When the velocities are reversed, the entropy will be decreasing. Let us pair off in this way every microscopic state with the one in which the velocities are reversed. Obviously, there will be exactly as many situations in which the entropy is decreasing as there are situations in which an increase in the entropy is taking place. Furthermore, since according to the basic assumptions of the theory all microscopic states are equally probable, we are driven to the conclusion that the probability of the entropy decreasing at any instant is exactly the same as the probability of its increasing. How are we to reconcile this fact with the general tendency of the entropy to increase?

The paradox is easily elucidated when we examine more carefully what happens to the entropy of the gas over a long period of time. Let us plot the entropy S at any instant along a vertical, and measure the time t along a horizontal. At the initial instant we may suppose that the gas is in a highly improbable macroscopic state, *e.g.*, all the molecules may be situated in one corner of the enclosure. We know that the gas will immediately expand of its own accord and fill the enclosure.

Suppose that at the initial instant the entropy S of the gas (in this improbable macroscopic state) is represented by the point a on the graph. We now follow the change in the entropy over a long period of time and find it to be represented by a wavy curve, as in the figure. The curve starts to rise as the gas expands into the vacuum, and the state of statistical equilibrium will be attained when the entropy curve reaches its maximum elevation at b. From then on, for trillions of centuries, the entropy will hover slightly below the maximum value, touching this maximum value now and then. Eventually, at the point c, the entropy will enter upon a rapid decrease (for instance, the gas may be contracting of its own accord). Some minimum value d will be registered for the entropy,

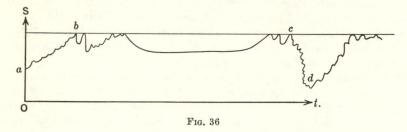

Fig. 36

and then the entropy curve will suddenly rise again, and, for another span covering trillions of centuries, it will hover just below the maximum value.

Let us now examine the situation from the standpoint of Loschmidt's paradox. If we consider the tremendously long interval of time from a to d, we are correct in saying that at any instant the curve has as much chance of pointing upwards as of slanting downwards: so we are correct in saying that the entropy has as great a chance of decreasing as of increasing at the instant of interest. But we also see that, if the velocities are reversed at any instant and the same curve is described backwards, it is very unlikely that the entropy will ever decrease to any extent. Of course, if we reversed the velocities in the state b, the gas would contract of its own accord and the entropy would fall to the value a. However, this last situation is exceptional, for the span of time from b to c covers trillions of centuries, whereas the interval of time corresponding to ab is a small fraction of a second. If, then, the velocities are reversed at any instant, this instant will most probably be comprised somewhere between b and c, and in this event a reversal of the velocities will not entail any considerable departure in the value of the entropy from the maximum value (unless we be prepared to wait for many centuries).

The error in Loschmidt's reasoning is now clear. Loschmidt was correct in claiming that the decreases in the entropy should be as frequent as the increases; but he failed to realize that when the gas is near its state of statistical equilibrium, a prolonged decrease in the entropy is exceedingly rare, and that, usually, each small decrease will be followed by a small increase as a result of which no considerable departure from the state of statistical equilibrium is to be expected.

The Viscosity of Gases—If two superposed layers of gas are moving in opposite directions, experiment shows that the motions of the two layers will slow down gradually. It would seem that a kind of friction is generated between the layers. In liquids and gases this frictional effect is called "viscosity." The kinetic theory accounts for viscosity by showing that the molecular collisions will cause some of the molecules of one layer to penetrate into the second layer, and vice versa. Thus the two layers will tend to become mixed, and eventually the layers will be moving with a common velocity.

Maxwell, by applying the kinetic theory, proved that the viscosity of a gas should increase with the density of the gas, with the mean velocity of the molecules and with the mean distance the molecules cover between collisions (mean free path). Maxwell's formula which establishes these relationships enabled him to predict that the viscosity of a gas at a given temperature should be independent of its pressure. This prediction appeared very unlikely at the time, but it was soon confirmed by experiment; and, as a result, the kinetic theory, having predicted an unknown phenomenon, registered its first important success. Maxwell's formula, linking as it does the viscosity, the mean free path, the density, and the mean velocities (and hence also the temperature) of a gas, permits us to calculate any one of these four magnitudes when three of them are known. Similar situations arise for the various formulae of the kinetic theory. A considerable amount of crosscheckings is thus possible, and the consistency in the results that have been obtained is in itself a strong argument in favor of the validity (at least approximate) of the kinetic theory.

Maxwell's Distribution Law—Suppose we are dealing with a gas maintained at constant temperature; we assume that all the molecules are of the same mass, shape and size. When the state of statistical equilibrium is reached, mathematical analysis shows that the molecules are more or less uniformly distributed throughout the container. At least this will be so if the molecules are not subjected to external forces (*e.g.*, gravitational), and if they exert no mutual attractions on one another. It can also be proved that the molecules will be moving in all directions, but that their velocities will not be the same; some of the molecules will be at rest, while others will be moving at extremely high speeds. If, however, we consider the percentages of molecules whose velocities are contained within consecutive equal intervals, we shall find that there is one interval for which the percentage of molecules is greatest. The situation is best represented by means of a graph.

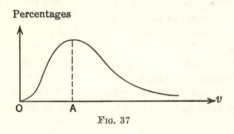

Percentages

O   A

FIG. 37

If we mark off, as abscissae, the velocities of the molecules (regardless of direction) and, as ordinates, the percentages of molecles having these velocities, we obtain the curve traced in the figure. We see that the largest percentage of molecules has velocities in the neighborhood of the velocity *OA*. The precise distribution of the velocities (which we have represented in the graph) is expressed mathematically by a law called "Maxwell's law of the velocity distribution." * Maxwell's law can be deduced from the probability considerations which form the basis of the kinetic theory. When the temperature of the gas is increased, the point *A* on the graph, corresponding to the most probable velocity, will move to the right. In other words, the velocities of the molecules will increase with the temperature. Furthermore as will be seen presently, for gases which are at the same temperature but whose molecules differ in mass, the velocities of the more massive molecules will be the smaller.

The Law of Equipartition of Energy—This important law deals with the distribution of the total kinetic energy of a gas among the various molecules when the state of statistical equilibrium is attained. A few mechanical preliminaries are here necessary.

The molecules of a gas are viewed as rigid bodies of finite size. Till now, we have implicitly assumed that the molecules undergo translations but do not rotate. We now propose to treat the problem more thoroughly and to assume that the collisions may generate rotations. Under these

---

* Or Maxwell's Distribution Law for short.

conditions, our molecules must be viewed as elastic bodies of finite size which are free to move in all possible ways. Each molecule has therefore six degrees of freedom. The presence of six degrees of freedom implies that the motion of the molecule, at any instant, may be decomposed into six independent partial motions; three of these are translational motions along three mutually perpendicular axes $Ox, Oy, Oz,$ and the three others are rotational motions round three mutually perpendicular axes. The superposition of all these partial motions, occurring simultaneously, yields the total motion of the molecule at the instant considered. We also know that kinetic energy is always associated with motion. Hence each of the six partial motions of each molecule involves a certain amount of kinetic energy. The sum of the three kinetic energies of translation of any given molecule yields the total kinetic energy of translation of the molecule. Similarly, the sum of the three rotational kinetic energies gives the total kinetic energy of rotation of the molecule. Finally, the sum of all six partial kinetic energies is the total kinetic energy of the molecule.

Let us, then, consider an aggregate of N molecules forming a gas. We shall assume that the molecules can be set into rotation as a result of collisions. We shall also suppose, for the sake of greater generality, that we are dealing with a mixture of different gases, so that the molecules are not all of the same kind. Our mixture of gases is placed in a container whose walls are impervious to heat. After a sufficient lapse of time (which in practice will be exceedingly short), the gas mixture attains its state of statistical equilibrium, say, at a temperature T. The molecules will have translational and rotational motions simultaneously. The total energy of the gas system is the sum of the energies of the various molecules. If we assume that no mutual attractions or repulsions are exerted by the molecules and that no external forces are applied (e.g., force of gravity), the energy of each individual molecule is purely kinetic and so also is the total energy of the gas. According to the principle of conservation this total energy will remain constant during the motion. We now fix our attention on the kinetic energy associated with one particular degree of freedom of one particular molecule, e.g., the translational degree of freedom along the Ox axis. As collisions arise, the magnitude of this partial kinetic energy will be subjected to sudden variations, but mathematical calculation shows that its average value over a long period of time will always be the same regardless of the particular degree of freedom or the particular molecule considered.

Now the sum total of the average kinetic energies of all the molecules is equal to the average value of the total kinetic energy of the gas; hence it is equal to the total kinetic energy of the gas (for since the total kinetic energy is constant, its average value is also its actual value). Grouping these results, we may say:

> In the state of statistical equilibrium, the total kinetic energy of a gas, or of a mixture of gases, is equipartitioned on an average among the various degrees of freedom of the various molecules.

The foregoing statement expresses the celebrated law of the "Equipartition of Energy." It is of extreme historical importance, for it was the necessity of amending this law in the case of radiation equilibrium that led to the discovery of the quantum theory. Let us observe that the equipartition of the total kinetic energy is only an average effect; it is not realized rigorously at an arbitrary instant. Indeed, any rigorous permanent distribution of the energy is obviously excluded owing to the collisions which cause sudden changes in the molecular velocities. At any given instant when the gas is in the state of statistical equilibrium, some of the molecules will have more and others less kinetic energy than would follow from rigorous equipartition. Only when we take average values over long periods of time does equipartition hold.

Thus far we have been concerned with the general law of equipartition without regard to the precise amount of energy that will be allotted to each degree of freedom. We shall now show that the average value of the kinetic energy, associated with each degree of freedom of each molecule, has the value $\dfrac{kT}{2}$, where T is the absolute temperature of the gas and k is the gas constant. To prove this point we first recall a result mentioned on a previous page. We said that, in the state of statistical equilibrium, the total kinetic energy of translation of all the N molecules of a pure gas is $\dfrac{3}{2}NkT$.* However, when we mentioned this result, we were not entertaining the possibility of the molecules being in rotation. A minor revision is required when the molecules are assumed to be rotating in addition to undergoing translations. The need for a revision is easily understood. Suppose, for instance, that two molecules which are not rotating undergo a glancing collision; the molecules will be set spinning. Whereas before the collision the energies of the two molecules were purely translational, after the collision the energies are partly rotational. Thus, although the total energy of the two molecules remains constant, the total translational energy of the two molecules and the total rotational energy may change in value as a result of the collision. Extending these con-

---

* Page 396.

siderations to the case of N molecules, we see that the total translational energy of the N molecules will not remain constant, but will fluctuate in value under the effect of the collisions. Consequently, our former statement must be revised as follows: Instead of saying that, in the state of statistical equilibrium, the total translational energy of the N molecules is measured by $\frac{3}{2} NkT$, we must say that the *average* value of the total translational energy is measured by $\frac{3}{2} NkT$. The foregoing result is valid whether we are considering a pure gas or a mixture of different gases.

We may now proceed with the proof that $\frac{1}{2} kT$ measures the kinetic energy associated (on an average) with each degree of freedom. Since there are N molecules and each of these has three translational degrees of freedom, the average value of the kinetic energy of translation, associated with any one of the translational degrees of freedom of any one of the molecules, must be $\frac{1}{3N}$ times the total average kinetic energy of translation of the entire gas. This gives the value

$$\frac{1}{3N} \times \frac{3}{2} NkT, \ i.e., \ \frac{1}{2} kT$$

for the average kinetic energy associated with each translational degree of freedom. Now according to the law of equipartition, each degree of freedom, whether translational or rotational, has the same average kinetic energy. Hence we conclude that, in the state of statistical equilibrium, $\frac{1}{2} kT$ defines the average kinetic energy associated with any one of the various degrees of freedom, whether translational or rotational, of any one of the N molecules.

As an application of the law of equipartition, consider a gas formed of a mixture of oxygen and hydrogen. When the mixture reaches the state of statistical equilibrium, the average total translational kinetic energy of each molecule will be $\frac{3}{2} kT$. Since the oxygen molecules are more massive than the hydrogen ones whereas their average translational energies are the same, the oxygen molecules will necessarily be moving more slowly than the lighter hydrogen ones. The velocities of both sets of molecules will satisfy Maxwell's distribution law; but in the Maxwell graph (Figure 37) the highest point of the curve for the hydrogen molecules will be situated to the right of the highest point of the curve corresponding to the slower oxygen molecules.

The Specific Heats of Gases—The specific heat (at constant volume) of a pure gas at temperature T is defined by the quantity of heat we must supply to some standard mass of the gas (whose volume is kept fixed) in order to increase its temperature by 1° Centigrade. It is customary to take as standard mass of gas either one gram or one gram-molecule. We shall take the latter, and we may recall that a gram-molecule of a pure gas contains N molecules, where N is Avogadro's number.

Let us, then, suppose that we are dealing with a gram-molecule of some pure gas at temperature T. We assume, as usual, that the energy of the gas is wholly kinetic. When discussing the law of equipartition, we mentioned that each molecule, on account of its possible translational and rotational motions, must be credited with six degrees of freedom. We now make the more general assumption that there are f degrees of freedom for each molecule; the reason for this generalization will be understood presently. The law of equipartition tells us that in the state of statistical equilibrium each degree of freedom of each molecule has, on an average, an amount of kinetic energy $\frac{1}{2}kT$. With our present assumptions, there are f degrees of freedom for each molecule, and as there are N molecules, there are Nf degrees of freedom in all. Consequently, the total average kinetic energy of all the molecules, and therefore also the total kinetic energy of the gas, is given by

$$Nf \times \frac{kT}{2}.$$

Since we are here considering one gram-molecule, N is Avogadro's number and Nk may be replaced by the gas constant R. The total kinetic energy of our gram-molecule of gas is thus

$$f\frac{RT}{2}.$$

When we heat the gas and increase its temperature from T to $T+1$, this total kinetic energy increases from $f\frac{RT}{2}$ to $f\frac{R(T+1)}{2}$. It therefore increases by

$$f\frac{R}{2}.$$

The increase in the kinetic energy must result from the heat we have supplied to the gas. Hence $f\frac{R}{2}$ measures by definition what we have called the specific heat of the gas. Let us observe that this expression of the specific heat does not involve T, and so our theory indicates that the specific heat must be independent of the temperature. Experiment confirms this anticipation at ordinary temperatures. We must now pay attention to f, *i.e.*, to the number of degrees of freedom of the individual molecules. It would seem that only one value can be attributed to f; namely $f = 6$. For, whatever their shapes, our molecules are assumed to be rigid bodies, and they should, in consequence, have six degrees of freedom corresponding to the translations and to the rotations that will be generated by the collisions. If we set $f = 6$ in the expression $f\dfrac{R}{2}$ of the specific heat, we obtain $6\dfrac{R}{2}$ or $3R$. Thus the kinetic theory requires that the specific heats of all perfect gases should have the same value $3R$.

But experiment contradicts this anticipation. The results actually obtained are as follows: At ordinary pressures and temperatures, the specific heat of a polyatomic gas is indeed given by $6\dfrac{R}{2}$ or $3R$, as required by the kinetic theory; but diatomic gases, under the same conditions, have specific heats measured by $5\dfrac{R}{2}$ and monoatomic gases have still smaller specific heats measured by $3\dfrac{R}{2}$. These results imply that for some reason or other the molecules of the polyatomic, diatomic, and monoatomic gases have (or behave as though they had) six, five and three degrees of freedom, respectively.

Boltzmann observed that these discrepancies might be attributed to the different shapes of the molecules. The molecule of a monoatomic gas, such as neon, consists of a single atom and may therefore be assumed spherical in shape. A diatomic molecule, such as that of hydrogen, being formed of two atoms, may be regarded as having the shape of a dumb-bell. Finally, a polyatomic molecule, *e.g.*, a molecule of SO_2, may be supposed irregular in shape. Of course, whatever the shape of the molecule, the degrees of freedom are always six in number. Nevertheless it is possible that the shape of the molecules affects their conduct during a collision. For instance, let us take the case of spherical molecules. When two billiard balls are subjected to a glancing collision, the balls rebound and at the same time start to spin. The spin, however, will be vanishingly small if the balls are perfectly smooth and frictionless. If, then, we assume that our spherical molecules are perfectly smooth, their collisions will not generate rotations and the entire kinetic energy will remain translational. To all intents and purposes, the rotational degrees of freedom will be inexistent and the spherical molecules will have only three *effective* degrees of freedom. Under these conditions the observed value of the specific heat of a monoatomic gas will be consistent with the kinetic theory. A similar situation arises for the diatomic dumb-bell molecules. If they are perfectly smooth, collisions cannot set them into rotation around the axis passing through the two knobs, though rotations will occur round the other two axes. Thus one of the rotational degrees of freedom will be inoperative, and for all practical purposes diatomic molecules will have only five degrees of freedom. Finally, polyatomic molecules, even if perfectly smooth, will (owing to their irregular shape) be set into rotation in all manner of ways by the mutual impacts. Such molecules will therefore have six effective degrees of freedom. In short, if we assume actual collisions between the molecules, we may bring the kinetic theory into agreement with the observed facts by taking into consideration the shapes of the molecules and by supposing the molecules to be perfectly smooth.

Boltzmann's unattractive hypothesis of smooth molecules may be avoided if we suppose that the molecules never actually collide but are repelled by intense forces when they come into close proximity to one another. The impossibility of rotations arising for spherical molecules may then be ascribed quite plausibly to the symmetrical distribution of the repulsive forces. However, we shall not dwell further on this matter, for, as we shall now see, at extremely low temperatures the situation becomes too baffling to be explained away so simply.

Let us first take the case of high temperatures. Experiment shows that at high temperatures the specific heat of diatomic gases tends to increase as though the molecules had acquired a greater number of degrees of freedom. This increase in itself is not surprising, for we may suppose that at high temperatures the two knobs of a dumb-bell molecule are no longer rigidly connected, and hence are free to vibrate, thereby giving rise to an increased number of degrees of freedom. But at extremely low temperatures the situation becomes incomprehensible. Experiment shows that as we near the absolute zero, the specific heats of the diatomic and polyatomic gases decrease and approximate to that of the monoatomic gases. The implication is that, for the diatomic molecules, the degrees of freedom fall from five to three. In other words, the collisions cease to generate any rotations at all. But this conclusion appears impossible, for however gentle the impacts, a collision against one of the knobs of a dumb-bell molecule must necessarily set the molecule into rotation. At least this is what is required by the laws of classical (or relativistic) mechanics; and if classical mechanics is incorrect, the kinetic theory which is based on its laws cannot itself be correct. Finally, we mentioned that, at still lower temperatures, the third principle of thermodynamics suggests the vanishing of the specific heat of a perfect gas. This additional decrease constitutes one of the aspects of "gas degeneration." In Boltzmann's time gas degeneration was unsuspected, but its existence is corroborated by the new statistics (see Chapter XL). At all events, if we accept the phenomenon of degeneration, we must assume not only that the molecules will cease to rotate, but also that some of them will be incapable of any motion at exceedingly low temperatures—and this in spite of the impacts to which they will be submitted on the part of the molecules still in motion. Such assumptions are obviously impossible from the standpoint of classical mechanics.

Closely allied with the disappearance of some or all of the degrees of freedom of the molecules, there is another difficulty which the classical kinetic theory fails to explain. The molecules in the kinetic theory are treated as bodies of so stable and permanent a nature that they are assimilable to closed worlds. Whatever energy may be stored within the atoms which form the molecules remains sealed, and whatever energy may be ceded to the molecules does not penetrate into the interior of the atoms. Even in Boltzmann's time these assumptions were known to be untenable, for energy transmissions from the interior to the exterior of an atom were known to occur when the atom radiated light. But then, if the molecules and atoms do not behave as so many closed worlds, the internal degrees of freedom of these molecules and atoms should share in the equipartition of energy with the external degrees of freedom (which relate to the spatial motions of the molecules as a whole). In short, a gas should involve a much larger number of degrees of freedom than has been supposed in our analysis. Unfortunately, if this greater number of degrees of freedom is taken into account, the kinetic theory gives far too high values for the specific heats.

We are thus led to suppose that, just as the rotational degrees of freedom of the spherical molecules are inoperative at ordinary temperatures (and those of all molecules at very low temperatures), so must the internal degrees of freedom of all molecules be inoperative at ordinary temperatures. The similarity of these various situations is obvious. All of them indicate that for some reason or other, which classical mechanics cannot explain, certain degrees of freedom may become inoperative. We are here dealing with the curious occurrences presently to be called "quantum phenomena". Thanks to the quantum theory, physicists have been able to save the general structure of the kinetic theory, while modifying it considerably in many details. But prior to the discovery of the quantum theory the situation appeared less hopeful, and in view of the difficulties encountered in connection with the specific heats, the kinetic theory was on the verge of being abandoned.

The Brownian Movement —When the kinetic theory was, so to speak, at its last gasp, it was saved or at least given a new lease of life by Einstein. In the investigations we are about to consider, Einstein did not remove any of the difficulties previously mentioned, but he opened a new field of quantitative investigation in which the anticipations of the kinetic theory were found to be in remarkable agreement with experimental measurements. We refer to the phenomenon of the Brownian movement.

Brownian movement was discovered by the botanist Brown. On peering through his microscope he perceived strange quiverings of the particles floating in a solution. Some physicists thought that these motions betrayed living matter; others contended that the illumination of the microscope generated an unequal heating of the solution and that convection currents were formed in the liquid. Gouy proved that the illumination had little or nothing to do with the matter. Similar Brownian movements were also detected in small particles suspended in gases. Those physicists who defended the kinetic theory claimed that the Brownian movement in gases was due to the random impacts of the gas molecules against the particles in suspension. A similar interpretation was then extended to the Brownian movement in liquids; and, as a result, a kinetic theory for liquids was devised. According to the kinetic theory, a particle suspended in a gas or in a liquid should every now and then be subjected to a preponderance of impacts on one or the other side. If the particle is of large mass, the unequal distribution of the impacts will not develop a cumulative action sufficient at any instant to displace the particle perceptibly. But if the particle is of sufficiently small mass, the impact of an additional molecule on one side will impart a sudden motion to the particle; and at succeeding instants the impacts predominating on one side or the other may be expected to generate a quivering of the particle. These ideas received support when it was found that the less massive the particles, the more intense the Brownian movement actually observed. But all these ideas were merely qualitative and of little value.

Einstein, in 1905, developed the mathematical theory of the Brownian movement and sought to determine how a given particle should move over a protracted period of time. At first sight it might be supposed that a particle would seldom move to any extent from the point it originally occupied. We might think that a superabundance of collisions would occur with equal probability from every direction at consecutive instants, so that the particle would merely quiver about a fixed position. But this surmise would not be correct, as may be understood from the following illustration. Suppose a man takes a step from a post in a random direction; he then takes a second step of the same length, also in a random direction; he then takes a third step, and so on. A problem, sometimes called the problem of the drunkard's walk, is to determine the probable distance of the man from the post after n steps have been taken. It can be shown that if the number n of steps is large, the probable distance from the post will be measured by $\sqrt{n}$ steps taken in the same direction. For example, if one hundred steps are taken (each step a yard in length), the probable distance from the post will be ten yards. And if one step is taken every second, the probable distance will be proportional to the square root of the number of seconds and thus to the square root of the time elapsed.

The motion of a particle suspended in a gas or liquid and hustled here and there by the random impacts of the molecules affords a very similar problem, and we may infer from the problem of the drunkard's walk that after a certain period of time the particle will have wandered away from its initial position. The following quantitative results were obtained by Einstein: If we call X^2 the mean value of the square of the particle's displacement projected on any horizontal axis, Einstein's formula is

$$X^2 = \frac{2RT}{NA}\, t,$$

where R is the gas constant, T the absolute temperature, N Avogadro's number, t the time elapsed, and A a resistance factor which increases with the viscosity of the medium and the size of the particle. On the right-hand side of this formula, all the magnitudes are known. Hence the displacement over a time t can be computed. Perrin subjected Einstein's formula to experimental verification by studying the displacements of particles suspended in a liquid. His measurements confirmed Einstein's anticipations and thereby, indirectly, the kinetic theory of gases.

From Einstein's formula, Avogadro's number N may be expressed in terms of magnitudes that can be measured. The value thus obtained for N is approximately 6.06×10^{23}. Now this value is in remarkable agreement with the value to which we are led by other trains of inquiry. Since we can scarcely attribute this agreement to mere chance, we must recognize that it affords considerable indirect support to the kinetic theory. By checks and crosschecks of this kind, the validity of the kinetic theory can be tested.

Einstein also established the formula for the rotations of the particles that arise from the molecular impacts (Brownian movements of rotation). Here also the correctness of Einstein's formula was verified by Perrin.

The rôle played by Brownian movements in commonplace phenomena is important. When a piece of sugar is placed in a glass of stagnant water, some of the sugar molecules become detached. The sugar molecules then play the part of our former particles suspended in a liquid and are subjected to the Brownian displacements just studied. As time passes, the sugar molecules move to the various portions of the water and the latter becomes uniformly sweet. The technical name for this phenomenon is diffusion, and we see that it is an effect of the Brownian movements. Einstein's formula shows that the diffusion will proceed more rapidly when the diffusing molecules are smaller and the temperature higher. Of course in practice, when we wish the water to be uniformly sweetened, we stir it, but the sweetening then arises from a transport of the sweetened water to the other parts of the liquid and is no longer a consequence of the Brownian movements.

The experimental verifications of Einstein's formula for the Brownian movement were soon extended by Maurice de Broglie to the case of particles suspended in gases. Considerable difficulties surround such experiments, for it is necessary to determine the volumes of the particles in order to apply the formula, and particles that are light enough to be suspended in the air are exceedingly minute and difficult to measure. De Broglie overcame this difficulty in a very ingenious way by utilizing tobacco smoke. The smoke is formed of tiny particles floating in the air, and these particles are found to be electrified, so that they stream in one direction when an electric field is applied. By measuring the velocity of the streaming, de Broglie obviated the necessity of measuring the dimensions of the particles. He established a relation connecting the Brownian displacement, Avogadro's number N, and the charge of the electron. If the two latter magnitudes are assumed known, Einstein's formula may be tested. Conversely, if we assume the correctness of Einstein's formula and take the accepted value of N, the charge of the electron may be determined. By following this method, de Broglie obtained one of the first rigorous determinations of the charge of the electron.

Brownian movement is interesting for many reasons. Thus particles of dust suspended in air may themselves be regarded as constituting the molecules of a gas; the air and the dust particles may then be viewed as forming a mixture of two different gases. From this standpoint the Brownian movements of the particles are none other than the random motions of the heavier gas molecules, and to this extent we may say that the random molecular motions are actually observed. Brownian movements also furnish concrete illustrations of the reversal in the growth of entropy. We mentioned previously that all the molecules under a table might simultaneously happen to be moving vertically upwards and that in this event the table would be thrown against the ceiling. Should such a phenomenon occur, the heat energy of the molecules would be transformed into mechanical energy (without an accompanying fall of heat from a higher to a lower temperature)—Carnot's principle, and hence the law of entropy, would be violated. But we also saw that this occurrence could arise only once in trillions of centuries, so that it would presumably never be observed. Yet with Brownian movements the occurrence is witnessed frequently. It arises every time our particle is thrown upwards against the force of gravity by the molecular impacts. Except for the difference in the sizes of the particle and of the table, the phenomenon is the same, and Carnot's principle is violated before our eyes.

Fluctuations—Consider a translucent medium whose structure or properties vary appreciably over microscopic distances. As may readily be understood, the local variations will appear more important to a microbe than to an observer who adopts a macroscopic viewpoint. For the latter, the heterogeneities will average out and the medium will appear homogeneous. To detect the heterogeneities, we shall therefore have to rely on some means of investigation which will be extremely sensitive to variations over tiny extensions. Waves of light, in view of their extremely short wave lengths, suggest themselves as a convenient means of detecting fine heterogeneities. The wave theory shows that if the properties of a medium vary appreciably over distances comparable to the wave lengths of visible light, and if white light is made to fall on the medium, then blue light will be scattered laterally and the transmitted light be reddened in consequence. The medium will thus exhibit a bluish opalescence. Consequently, whenever this opalescence is observed we may presume that the medium contains fine heterogeneities.

Now when white light is directed against a vapor which is about to become liquefied at the so-called critical point,* a bluish opalescence

---

* The critical point for a real gas is defined by a pressure and a temperature above either of which a condensation of the vapor into the liquid phase is impossible. The critical point arises only for real gases and not for perfect ones.

appears. Smoluchowski surmised therefore that the semi-liquefied vapor exhibited a fine heterogeneity of structure. At first sight it seems difficult to account for this heterogeneity, for we should expect a fluid in equilibrium to be homogeneous. Theory shows, however, that in the neighborhood of the critical point the compressibility of the vapor becomes infinite. Under these conditions the density of the semi-liquid vapor may quite well vary considerably over tiny distances, so that fine heterogeneities will appear even though the vapor is in equilibrium. These heterogeneities in the density of the vapor, which were predicted by Schmoluchowski, are referred to today as "fluctuations in the density." But not before Schmoluchowski had submitted his idea to mathematical treatment could he be certain that these fluctuations truly existed and were the cause of the bluish opalescence. His mathematical results, however, were found to be in good quantitative agreement with the facts observed, and so we are justified in saying that the fluctuations in density have been established by experiment.

The fluctuations we have been considering occur in a semi-liquefied vapor, but we now propose to show that, according to the kinetic theory, fluctuations of a similar kind should arise in gases. Consider a gas in the state of statistical equilibrium. In thought we imagine the volume divided into a large number of tiny cubes of equal size. The molecules fly hither and thither moving from one cube into another, but so long as the state of statistical equilibrium is realized rigorously, the number of molecules in any given cube does not vary with time and this number is the same for each cube. We know, however, that the precise state of statistical equilibrium is departed from every now and then. Consequently, the numbers of molecules in the various cubes at a given instant will not usually be exactly the same, and fluctuations in density will arise. If we could take a series of snapshot pictures of the molecular distributions at successive instants of time, we should find that a certain fluctuation (corresponding indifferently to a compression or to a rarefaction) would be more probable than any other fluctuation or than the uniform distribution of the molecules. As a result, a permanently heterogeneous condition would appear to be realized. Calculation enables us to compute the exact degree of heterogeneity that must be expected. We may add that similar fluctuations should occur for, the pressure and for the temperature.

What might be called the intensity of the heterogeneity depends on the size we credit to our tiny cubes. To see this, let us imagine that the vessel which contains the gas is divided into two parts, one large and the other very small. On certain occasions all the molecules which would normally be found in the smaller volume will pass into the larger one. The smaller volume is then empty; the density of the gas in it drops from the normal value to zero. On other occasions it will be the molecules of the larger volume that move into the smaller one. But obviously, the first occurrence will arise far more often than the second. Consequently, the fluctuations in density in the smaller volume will be more frequent and more intense than those in the larger volume. These considerations show that the heterogeneities which we may expect in gases will assume importance only within microscopic volumes.

To test by experimental means the presence of these microscopic fluctuations, we proceed, as we did before, by appealing to optical phenomena. We should expect that blue light would be scattered laterally when white light is made to fall on a mass of gas. But theory shows that the fluctuations should be much less important than they were in a vapor, with the result that the scattering will be too weak to be observed easily in a laboratory. Cumulative effects may, however, be obtained if we operate on very large masses of gas, e.g., on the atmosphere as a whole. Calculations conducted by Rayleigh, Keesom, and Einstein have shown that the blue color of the sky may be due to the fluctuations in the density of the atmosphere. These conclusions result from a comparison of the theoretical anticipations with actual measurements of the relative intensity of the blue light scattered.

Gases in a Gravitational Field—In the previous discussions we assumed that no external field was acting on the molecules. But we may also suppose that a gravitational field of force is present. This situation is important on account of its bearing on the problem of the atmosphere. Boltzmann proved that, according to the kinetic theory, when a gas is in the state of statistical equilibrium, a logarithmic relation should hold between the density of the gas at a given point and the potential of the field at this point. Boltzmann's relation shows that the higher the potential at a point, the smaller the density of the gas at this point. As applied to the problem of the atmosphere, this implies that the density will decrease when the elevation increases. This prediction of the kinetic theory is in agreement with facts.

But there is another application of the kinetic theory which yields new information. If for argument's sake we assume that the temperature of the atmosphere is uniform, the various molecules, according to the theorem of equipartition, must have the same average kinetic energy. Now the state of statistical equilibrium, to which the theorem of· equipartition applies, is departed from every now and then, and so from time to time we may expect one of the molecules to receive more than its share of kinetic energy. This molecule may speed away from the Earth with a velocity which carries it to infinity in spite of the Earth's attraction. Gradually, then, the atmosphere should become impoverished and its molecules dispersed. The rapidity with which this scattering process should be taking place can be submitted to calculation; it depends largely on the intensity of the gravitational field and hence on the mass of the planet which attracts the atmosphere. For a relatively small mass such as the moon, the impoverishment of an atmosphere should be far more rapid than it is for the Earth; this circumstance may account for the absence of a lunar atmosphere.

What we have said of atmospheres attracted by planetary masses also applies in the main to the gaseous nebulae. In the nebulae, however, the gravitational field which holds the molecules together is due to the mutual attractions of the molecules themselves. But if the nebula is of small total mass, the gravitational field it develops will be small and the molecules will escape very rapidy. Consequently, there is a minimum limit below which the mass of a nebula must not fall if the nebula is to endure for any length of time. Considerations of this sort have important applications in cosmogony.

Chemical Reactions—The methods of thermodynamics allow us, in theory, to predict the direction in which a chemical reaction should occur under specified conditions of pressure and of temperature. But, in practice, this information may be of little value owing to our ignorance of the rapidity with which the reactions take place. Thermodynamics throws no light on this point. To clarify the nature of the difficulty, let us assume that three chemical substances A, B, and C are placed in a container and that according to the principles of thermodynamics A should combine with B and also with C. Suppose that A combines very rapidly with B and very slowly with C. In this case practically all of A will have combined with B before any part of it has had time to combine with C, and the reaction between A and C will not occur. Unless, then, we have some means of anticipating the rate of a chemical reaction, our conclusions drawn from thermodynamics may turn out to be incorrect. Some advance may be made, however, when we apply statistical methods and assume that, for two atoms to combine, they must collide. If the two reacting substances are gases, we may calculate the probability, and hence the frequency, of the collisions between the atoms of the two gases. In this way the rate of the reaction may be determined.

are elastic. Each vibra⁺⁻
om and three potential ones, ⸱
hen, we consider a gram-atom of ou
is (where N is Avogadro's number).
erage energy allotted to each degree
is T, is given by $\frac{1}{2} kT$; and since
the N atoms, the total energy of the
$\times \frac{1}{2} kT$, *i.e.*, $3RT$. Consequently,
result obtained is the same as for a
of the theory are in agreement with
formed at ordinary temperatures by
that at ordinary temperatures our
ulong and Petit's empirical law.
aw is incorrect at low temperatures;
progressively as the absolute zero is
eement with the third law of thermo-

Considerations of this kind shed considerable light on the nature of chemical equilibrium. We discussed the phenomenon of chemical equilibrium in Chapter XXI, taking as an illustration the reaction between oxygen and sulphur dioxide: sulphur trioxide is formed, but the reaction does not proceed until exhaustion; instead, a condition of equilibrium is established for the three substances when a certain relationship connecting their concentrations is attained. The three substances are then said to be in chemical equilibrium. If we modify the concentrations by adding or withdrawing in part one of the substances, the reaction will proceed afresh in one direction or the other until the equilibrium relationship is reestablished.

Insofar as we can judge from thermodynamics, where the law of entropy is regarded as of absolute validity, the state of chemical equilibrium, corresponding as it does to the state of maximum entropy, is perfectly stable. Consequently when it is attained no further reactions can take place. But when we consider the matter from the standpoint of the kinetic theory, we are led to view the state of chemical equilibrium as being of the statistical type. We must then assume that fluctuations occur and that the reaction proceeds now in one direction and now in the other, quivering as it were about the state of equilibrium. We may even calculate the probability of a given fluctuation.

Boltzmann's Definition of Past and Future—Thermodynamics teaches us that the entropy of the universe is constantly increasing. Thus the growth of entropy accompanies the flow of time. Boltzmann, having identified the increase in entropy with the passage from a less probable to a more probable macroscopic state, was thus led to identify this passage with the flow of time. According to his conception, the past and future are connected respectively with the less probable and the more probable states of the universe. The fact that in the light of the kinetic theory the growth of entropy may occasionally be reversed, does not necessarily preclude its identification with the unidirectional time flow, for such reversals occur too seldom on the macroscopic level to be of any practical importance.

The interesting feature of Boltzmann's conception is that it affords a mechanical interpretation of the distinction between past and future. We recall that the evolution of a conservative dynamical system is reversible, because, if at any given instant the motions of all the parts of the system are reversed, the system passes in reverse order through the same succession of states. For this reason, mathematicians long deemed it impossible to establish, on the basis of mechanical conceptions, any intrinsic difference between a past state and a future state—a mere reversal of the motion would cause the original past state to become the new future one, and vice versa. (Indeed it was this characteristic feature of conservative mechanical systems that seemed to render illusory the hope of obtaining a mechanical model for a gas.) But Boltzmann, by investigating the mechanics of a large number of elastic spheres colliding at random, proved that, despite the fundamental reversibility of mechanical phenomena, a certain mechanical function H decreases constantly in magnitude as the system evolves. The discovery of a mechanical magnitude varying always in the same direction thus suggested the possibility of including the unidirectional flow of time within the scheme of a mechanistic universe.

In this chapter we have mentioned Boltzmann's function H only incidentally, because Boltzmann subsequently identified $-H$ with the entropy, and we have preferred to develop the theory by utilizing the thermodynamical concept of entropy.

Conclusions on the Classical Kinetic Theory of Gases—Physicists distinguish between real gases and perfect gases; the latter are mere idealizations never encountered in practice. Nevertheless, at ordinary temperatures and pressures, many gases, *e.g.*, hydrogen, behave very nearly like perfect gases, and so the earlier empirical gas laws of Boyle and Charles, though derived from experiments on real gases, happen to coincide with the laws for perfect gases. Of course, the laws for perfect gases cannot be even approximately valid for the real gases at all temperatures and pressures; indeed we know that, at sufficiently low temperatures and high pressures, real gases may be liquefied and thus cease to have the properties of gases. Now the kinetic theory, as developed to this point, is consistent with the assumption that the gas laws of Boyle and Charles are correct at all temperatures and pressures. Consequently, this theory can only apply to perfect gases and must be modified if real gases are to come within its compass.

If we glance back at the foundations of the kinetic theory, we shall observe that the volumes and the mutual attractions of the molecules were disregarded. Yet it is certain that the molecules of matter must attract one another, since it is these attractions that account for such phenomena as cohesion and surface tension. Accordingly, Van der Waals refined the kinetic theory by taking into consideration the molecular attractions and the space occupied by the molecules. In this way he obtained a new equation of state which gave a good account of the

properties of real gases, of the phenomenon of condensation, and of the existence of the critical point.

It might be supposed that Van der Waals's refinement of the theory would eliminate the difficulties concerning the specific heats at low temperatures. But such is not the case. For example, the falling off in the specific heat of a diatomic gas, such as hydrogen, at low temperatures, cannot be attributed to the fact that hydrogen ceases to constitute a perfect gas at these low temperatures. This circumstance shows that, when we confine our attention to the idealized perfect gases, the kinetic theory cannot be accepted without modification.

We may improve the kinetic theory, however, by reinterpreting it in terms of the mechanics of relativity. The foundations of the theory are not modified thereby, for Gibbs's theorems on the conservation of extension and of density in phase still hold, and the ergodic hypothesis may be accepted with the same measure of justification as before. The law of equipartition must, however, be rejected and replaced by a different law of partition. But these refinements are not of much use, for when the motions of the molecules are slow (a situation which arises at low temperatures), the relativistic theory merges into the classical one, and it is precisely at low temperatures that the discrepancies between experiment and the classical theory become notable. We are therefore certain that some much deeper modification is required than can be furnished by the relativity theory. One of the merits of the quantum theory has been to provide the necessary modifications, many of them of a fundamental nature. The "New Quantum Statistics," developed by Bose and by Fermi today supersede the classical kinetic theory. Nevertheless, the bare fact that the classical kinetic theory has proved its worth in the range of moderate temperatures and pressures shows that it is a first approximation to reality. Indeed we shall see that the new statistics of the quantum theory pass over into the classical kinetic theory when the temperature and pressure are moderate. We have here another illustration of the progressive approximations realized in the successive theories of mathematical physics; each theory is a refinement of its predecessor.

The Specific Heats of Solids—A metallic crystal (*e.g.*, a crystal of silver) is composed of atoms regularly spaced. These atoms cannot migrate as do the molecules of a gas, but they may vibrate elastically about fixed positions. Many phenomena appear to support this assumption. Now when an atom vibrates, it exchanges energy with the neighboring atoms, either by collision or by other processes. The net result is that if the crystal is isolated and one part of it is warmer than

the other, the atoms in the hotter part of the crystals, which are vibrating with greater violence, will gradually communicate their energy to the other atoms. Eventually a state of statistical equilibrium is reached, and the entire crystal is at a uniform temperature.

When we were dealing with gases, the molecules were assumed to exert no forces on one another, and their energy was purely kinetic. In the state of statistical equilibrium the total energy was then equipartitioned among the various kinetic degrees of freedom. If the situation were the same for our solid, we should have to equipartition the energy among the $3N$ translational degrees of freedom of the N atoms (since there are no rotational degrees of freedom in our present problem). But the problem of the solid differs from that of a perfect gas owing to the presence of potential energy. This potential energy is due to the forces which draw the atoms back to their centres of oscillation. We are thus led to attribute to each atom three potential degrees of freedom in addition to the three kinetic degrees previously mentioned. The total energy must now be partitioned among the $6N$ degrees of freedom (kinetic and potential) of the atoms. The law of partition will usually no longer be one of equipartition. It can be shown, however, that if the vibrations are of the amount of elastic, the potential degrees freedom partake of the same partition is energy as the kinetic one. In this case the theorem of equipartition still correct.

Let us assume that the vibrations are elastic. Each vibrating atom, having three kinetic degrees of freedom and three potential ones, has six degrees of freedom in all. Suppose, then, we consider a gram-atom of our crystal, *i.e.*, a mass containing N atoms (where N is Avogadro's number). Exactly as in the case of gases the average energy allotted to each degree of freedom, when the temperature is T, is given by $\frac{1}{2} kT$; and since there are $6N$ degrees of freedom for the N atoms, the total energy of the crystal at temperature T will be $6N \times \frac{1}{2} kT$, *i.e.*, $3RT$. Consequently, the specific heat should be $3R$. The result obtained is the same as for a polyatomic gas. These anticipations of the theory are in agreement with the experimental measurements performed at ordinary temperatures by Dulong and Petit. We conclude that at ordinary temperatures our statistical theory is in accord with Dulong and Petit's empirical law.

But experiment shows that this law is incorrect at low temperatures; the specific heat is found to decrease progressively as the absolute zero is approached. The decrease is in agreement with the third law of thermo-

dynamics, but it conflicts with the statistical theory just developed (since, according to the latter, the specific heat should always be $3R$, and hence independent of the temperature). We must therefore suppose that as the temperature is lowered, the various degrees of freedom cease to be operative; and we are thus confronted with the same situation that arose for gases. Here as elsewhere we are dealing with quantum phenomena, and an explanation of the discrepancy involves an understanding of the quantum theory. The explanation which we shall give on a later page is due to Einstein and Debye;* it is regarded as affording one of the strongest arguments in favor of the quantum theory.

The Kinetic Theory of Electronic Conduction—In view of the success of the kinetic theory for gases and solids at normal temperatures and pressures, Drude and Lorentz sought to apply the methods of the theory in other cases. According to Lorentz's electronic theory, the electric current which passes through a metallic wire, when a difference in potential is applied, is due to the streaming of the free electrons present in the wire. Suppose, then, no difference in potential is applied. There will be no general, ordered streaming of electrons, but, by analogy with the kinetic theory of gases, we may assume that the electrons are rushing hither and thither within the wire, colliding and rebounding against the vibrating atoms of the metal. In other words, the electrons are assimilated to the molecules of a gas, and the name "electron gas" was coined in consequence. All the results of the kinetic theory may tentatively be extended to this electron gas. For example, the higher the temperature, the more violent the electronic motions.

Let us examine the phenomenon of heat conduction according to this theory. In our former treatment of the specific heats of solids, we mentioned that in a metal the atoms are assumed to vibrate about fixed positions. We also mentioned that if one extremity of the metal is heated, the atoms will be vibrating in the heated part with greater amplitude, and will communicate gradually, through collisions, some of their energy to the neighboring atoms. By this means all the atoms will eventually be vibrating with the same average energy, and the temperature of the metal will become uniform. If now we accept the electronic theory, we must suppose that the transference of vibrational energy from one atom to another is accomplished not directly, but through the medium of the electrons. Except for this difference, our understanding of heat con-

---

* See page 463.

duction is the same as before. The electrons in the hotter part of a wire are moving faster, and eventually as a result of collisions all the electrons (and also all the atoms) will be moving with the same average energy. The spreading of the electronic disturbance throughout the wire, from the hotter to the colder regions, should be more rapid when the electronic motions suffer less impediment from the presence of the atoms or from other causes. Consequently, the thermal conductivity of the wire should be greater under such conditions. Since the electrical conductivity likewise increases with the ability of the electrons to move rapidly from one part of the metal to the other, we conclude that the thermal and the electrical conductivities of a metal should be proportional. Silver, for example, which is the best conductor of heat, should also be the best conductor of electricity. These general expectations are found to be verified by experiment. More important still, when the calculations of the kinetic theory were extended to the electron gas, a quantitative relation between the two conductivities was obtained, and this relation was seen to be in good agreement with the experimental results already known to physicists and expressed in the empirical law of Wiedemann-Frantz.

Despite its success in interpreting this law, the electronic theory breaks down completely when we consider the specific heats. We recall that, at ordinary temperatures, the law of Dulong and Petit gives the correct value for the specific heat of a solid. We have also seen that this law is accounted for when we assume that the total energy is equipartitioned among the degrees of freedom of the vibrating atoms. But if we accept the electronic theory, not only the degrees of freedom of the atoms but also the translational degrees of freedom of all the electrons should participate in the equipartition of energy.* This assumption leads, however, to an inordinately high value for the specific heat. If, then, we wish to retain the electronic theory, we must suppose that for some obscure reason the degrees of freedom of the electrons are inoperative and do not participate in the partitioning of the energy. We are here confronted with the same situation which we encountered in ordinary gases. Indeed, in the present case the situation is still more baffling, for the apparent freezing of the electronic degrees of freedom now occurs at ordinary temperatures instead of in the neighborhood of the mysterious absolute zero. More important still, since the electronic degrees of freedom, which are here inoperative, are exclusively translational * (instead of rotational as was the case for the molecules of diatomic and polyatomic gases), we must assume that the electrons remain motionless. But this

---

* The rotational degrees of freedom of the electrons are not considered.

assumption cannot be correct, for it is the motion of the electrons along the wire which constitutes the electric current—and the existence of such currents is a fact which cannot be denied.

The solution to the difficulty has since been given by Sommerfeld; it involves the new quantum statistics, which will be explained in Chapter XL. We shall find that the electron gas in the metal is in a state of "degeneracy." According to the new statistics (and to the third principle of thermodynamics) a similar state should occur for the usual perfect gases when the temperature approaches the absolute zero, or when the density of packing of the particles becomes enormous. Under such conditions quantum phenomena are no longer negligible and they must be taken into account.

Statistical Mechanics in Astronomy—All the applications of statistical mechanics we have thus far considered were necessarily of a speculative nature: firstly, because we had to start from some more or less hypothetical model of the physical substance under investigation; and secondly, because we were assuming the validity of classical mechanics in the domain of the microscopic. These objections do not arise when we apply the theorems of statistical mechanics to the motions of the stars of our galaxy, for here the model is before our eyes, and in addition the laws of classical mechanics are known to hold with a high measure of accuracy. For this reason statistical mechanics may be applied with comparative safety in many astronomical problems. (The necessity of accepting the ergodic hypothesis still introduces an element of uncertainty, however.)

Statistical mechanics has furnished information on the age of our galaxy. Thus the various stars are viewed as the molecules of a gigantic bubble of gas. The stars rarely enter into direct collision, but they nevertheless exchange energy because of their mutual gravitational actions. The effect of these actions and the deflections to which they give rise are in all respects analogous to those generated by the hypothetical elastic collisions between the gas molecules of the kinetic theory. After a sufficient period of time we must expect a state of statistical equilibrium to occur, the total energy being equipartitioned among the various degrees of freedom of the different stars. The stars will then be moving in random directions, the less massive ones with the faster motions. But whereas in the case of swiftly-moving gas molecules, statistical equilibrium is attained in a fraction of a second, many thousands of years will elapse before it can be established in the star system. The required period of time may be calculated. Astronomical observation shows that the present translational motions of the stars seem to be in accord with the requirements of the theorem of equipartition, but that, in the case of double stars circling around each other, statistical equilibrium has not yet been attained. The fact, however, that the translational energies are equipartitioned shows that our galaxy must have existed for at least the time necessary to bring about this situation; we are thus able to assign a lower limit to the age of the stellar system.*

Finally, we may mention that statistical mechanics played an important part in suggesting to Einstein his hypothesis of the spatially closed universe. We saw, when discussing the problem of gaseous nebulae, that molecules must always be escaping, moving away to infinity. If, then, we view all the stellar masses of the entire universe as so many molecules of a gigantic nebula, we must expect the material universe to be losing matter incessantly. To avoid this conclusion which was displeasing to him, Einstein postulated a closed universe whence nothing could escape.

---

* The discussion on the age of our galaxy given in the text is intended as a mere illustration of a possible application of statistical mechanics in astronomy. Many astronomers dispute the claim that a state of equipartition holds in our galaxy; they point out that the stellar motions manifest definite drifts and are not distributed at random. Furthermore the situation is complicated by the rotational motion of the galaxy as a whole.

AUGUSTIN FRESNEL (1788-1827)

LORD RAYLEIGH (1842-1919)

VIKTOR MEYER (1848-1897)

SADI CARNOT (1796-1832)

RUDOLF JULIUS EMANUEL CLAUSIUS (1822-1888)

J. WILLARD GIBBS (1839-1903)

WALTER HERMANN NERNST (1864-1941)

JAMES CLERK MAXWELL (1831-1879)

LUDWIG BOLTZMANN (1844-1906)

A CATALOGUE OF SELECTED DOVER BOOKS
IN ALL FIELDS OF INTEREST

A CATALOGUE OF SELECTED DOVER BOOKS IN ALL FIELDS OF INTEREST

CELESTIAL OBJECTS FOR COMMON TELESCOPES, T. W. Webb. The most used book in amateur astronomy: inestimable aid for locating and identifying nearly 4,000 celestial objects. Edited, updated by Margaret W. Mayall. 77 illustrations. Total of 645pp. 5⅜ x 8½.
20917-2, 20918-0 Pa., Two-vol. set $9.00

HISTORICAL STUDIES IN THE LANGUAGE OF CHEMISTRY, M. P. Crosland. The important part language has played in the development of chemistry from the symbolism of alchemy to the adoption of systematic nomenclature in 1892. ". . . wholeheartedly recommended,"—Science. 15 illustrations. 416pp. of text. 5⅝ x 8¼. 63702-6 Pa. $6.00

BURNHAM'S CELESTIAL HANDBOOK, Robert Burnham, Jr. Thorough, readable guide to the stars beyond our solar system. Exhaustive treatment, fully illustrated. Breakdown is alphabetical by constellation: Andromeda to Cetus in Vol. 1; Chamaeleon to Orion in Vol. 2; and Pavo to Vulpecula in Vol. 3. Hundreds of illustrations. Total of about 2000pp. 6⅛ x 9¼.
23567-X, 23568-8, 23673-0 Pa., Three-vol. set $27.85

THEORY OF WING SECTIONS: INCLUDING A SUMMARY OF AIRFOIL DATA, Ira H. Abbott and A. E. von Doenhoff. Concise compilation of subatomic aerodynamic characteristics of modern NASA wing sections, plus description of theory. 350pp. of tables. 693pp. 5⅜ x 8½.
60586-8 Pa. $8.50

DE RE METALLICA, Georgius Agricola. Translated by Herbert C. Hoover and Lou H. Hoover. The famous Hoover translation of greatest treatise on technological chemistry, engineering, geology, mining of early modern times (1556). All 289 original woodcuts. 638pp. 6¾ x 11.
60006-8 Clothbd. $17.95

THE ORIGIN OF CONTINENTS AND OCEANS, Alfred Wegener. One of the most influential, most controversial books in science, the classic statement for continental drift. Full 1966 translation of Wegener's final (1929) version. 64 illustrations. 246pp. 5⅜ x 8½. 61708-4 Pa. $4.50

THE PRINCIPLES OF PSYCHOLOGY, William James. Famous long course complete, unabridged. Stream of thought, time perception, memory, experimental methods; great work decades ahead of its time. Still valid, useful; read in many classes. 94 figures. Total of 1391pp. 5⅜ x 8½.
20381-6, 20382-4 Pa., Two-vol. set $13.00

ALIGN LAST LINE WITH FACING PAGE

DRAWINGS OF WILLIAM BLAKE, William Blake. 92 plates from Book of Job, *Divine Comedy, Paradise Lost,* visionary heads, mythological figures, Laocoon, etc. Selection, introduction, commentary by Sir Geoffrey Keynes. 178pp. 8⅛ x 11. 22303-5 Pa. $4.00

ENGRAVINGS OF HOGARTH, William Hogarth. 101 of Hogarth's greatest works: *Rake's Progress, Harlot's Progress, Illustrations for Hudibras, Before and After, Beer Street and Gin Lane,* many more. Full commentary. 256pp. 11 x 13¾. 22479-1 Pa. $12.95

DAUMIER: 120 GREAT LITHOGRAPHS, Honore Daumier. Wide-ranging collection of lithographs by the greatest caricaturist of the 19th century. Concentrates on eternally popular series on lawyers, on married life, on liberated women, etc. Selection, introduction, and notes on plates by Charles F. Ramus. Total of 158pp. 9⅜ x 12¼. 23512-2 Pa. $6.00

DRAWINGS OF MUCHA, Alphonse Maria Mucha. Work reveals draftsman of highest caliber: studies for famous posters and paintings, renderings for book illustrations and ads, etc. 70 works, 9 in color; including 6 items not drawings. Introduction. List of illustrations. 72pp. 9⅜ x 12¼. (Available in U.S. only) 23672-2 Pa. $4.00

GIOVANNI BATTISTA PIRANESI: DRAWINGS IN THE PIERPONT MORGAN LIBRARY, Giovanni Battista Piranesi. For first time ever all of Morgan Library's collection, world's largest. 167 illustrations of rare Piranesi drawings—archeological, architectural, decorative and visionary. Essay, detailed list of drawings, chronology, captions. Edited by Felice Stampfle. 144pp. 9⅜ x 12¼. 23714-1 Pa. $7.50

NEW YORK ETCHINGS (1905-1949), John Sloan. All of important American artist's N.Y. life etchings. 67 works include some of his best art; also lively historical record—Greenwich Village, tenement scenes. Edited by Sloan's widow. Introduction and captions. 79pp. 8⅜ x 11¼. 23651-X Pa. $4.00

CHINESE PAINTING AND CALLIGRAPHY: A PICTORIAL SURVEY, Wan-go Weng. 69 fine examples from John M. Crawford's matchless private collection: landscapes, birds, flowers, human figures, etc., plus calligraphy. Every basic form included: hanging scrolls, handscrolls, album leaves, fans, etc. 109 illustrations. Introduction. Captions. 192pp. 8⅞ x 11¾. 23707-9 Pa. $7.95

DRAWINGS OF REMBRANDT, edited by Seymour Slive. Updated Lippmann, Hofstede de Groot edition, with definitive scholarly apparatus. All portraits, biblical sketches, landscapes, nudes, Oriental figures, classical studies, together with selection of work by followers. 550 illustrations. Total of 630pp. 9⅛ x 12¼. 21485-0, 21486-9 Pa., Two-vol. set $15.00

THE DISASTERS OF WAR, Francisco Goya. 83 etchings record horrors of Napoleonic wars in Spain and war in general. Reprint of 1st edition, plus 3 additional plates. Introduction by Philip Hofer. 97pp. 9⅜ x 8¼. 21872-4 Pa. $4.00

THE SENSE OF BEAUTY, George Santayana. Masterfully written discussion of nature of beauty, materials of beauty, form, expression; art, literature, social sciences all involved. 168pp. 5⅜ x 8½.   20238-0 Pa. $3.00

ON THE IMPROVEMENT OF THE UNDERSTANDING, Benedict Spinoza. Also contains *Ethics, Correspondence,* all in excellent R. Elwes translation. Basic works on entry to philosophy, pantheism, exchange of ideas with great contemporaries. 402pp. 5⅜ x 8½.   20250-X Pa. $4.50

THE TRAGIC SENSE OF LIFE, Miguel de Unamuno. Acknowledged masterpiece of existential literature, one of most important books of 20th century. Introduction by Madariaga. 367pp. 5⅜ x 8½.
                                                     20257-7 Pa. $4.50

THE GUIDE FOR THE PERPLEXED, Moses Maimonides. Great classic of medieval Judaism attempts to reconcile revealed religion (Pentateuch, commentaries) with Aristotelian philosophy. Important historically, still relevant in problems. Unabridged Friedlander translation. Total of 473pp. 5⅜ x 8½.   20351-4 Pa. $6.00

THE I CHING (THE BOOK OF CHANGES), translated by James Legge. Complete translation of basic text plus appendices by Confucius, and Chinese commentary of most penetrating divination manual ever prepared. Indispensable to study of early Oriental civilizations, to modern inquiring reader. 448pp. 5⅜ x 8½.   21062-6 Pa. $5.00

THE EGYPTIAN BOOK OF THE DEAD, E. A. Wallis Budge. Complete reproduction of Ani's papyrus, finest ever found. Full hieroglyphic text, interlinear transliteration, word for word translation, smooth translation. Basic work, for Egyptology, for modern study of psychic matters. Total of 533pp. 6½ x 9¼. (Available in U.S. only)   21866-X Pa. $5.95

THE GODS OF THE EGYPTIANS, E. A. Wallis Budge. Never excelled for richness, fullness: all gods, goddesses, demons, mythical figures of Ancient Egypt; their legends, rites, incarnations, variations, powers, etc. Many hieroglyphic texts cited. Over 225 illustrations, plus 6 color plates. Total of 988pp. 6⅛ x 9¼. (Available in U.S. only)
                                      22055-9, 22056-7 Pa., Two-vol. set $16.00

THE STANDARD BOOK OF QUILT MAKING AND COLLECTING, Marguerite Ickis. Full information, full-sized patterns for making 46 traditional quilts, also 150 other patterns. Quilted cloths, lame, satin quilts, etc. 483 illustrations. 273pp. 6⅞ x 9⅝.   20582-7 Pa. $4.95

CORAL GARDENS AND THEIR MAGIC, Bronsilaw Malinowski. Classic study of the methods of tilling the soil and of agricultural rites in the Trobriand Islands of Melanesia. Author is one of the most important figures in the field of modern social anthropology. 143 illustrations. Indexes. Total of 911pp. of text. 5⅝ x 8¼. (Available in U.S. only)
                                                     23597-1 Pa. $12.95

THE PHILOSOPHY OF HISTORY, Georg W. Hegel. Great classic of Western thought develops concept that history is not chance but a rational process, the evolution of freedom. 457pp. 5⅜ x 8½.   20112-0 Pa. $4.50

LANGUAGE, TRUTH AND LOGIC, Alfred J. Ayer. Famous, clear introduction to Vienna, Cambridge schools of Logical Positivism. Role of philosophy, elimination of metaphysics, nature of analysis, etc. 160pp. 5⅜ x 8½. (Available in U.S. only)   20010-8 Pa. $2.00

A PREFACE TO LOGIC, Morris R. Cohen. Great City College teacher in renowned, easily followed exposition of formal logic, probability, values, logic and world order and similar topics; no previous background needed. 209pp. 5⅜ x 8½.   23517-3 Pa. $3.50

REASON AND NATURE, Morris R. Cohen. Brilliant analysis of reason and its multitudinous ramifications by charismatic teacher. Interdisciplinary, synthesizing work widely praised when it first appeared in 1931. Second (1953) edition. Indexes. 496pp. 5⅜ x 8½.   23633-1 Pa. $6.50

AN ESSAY CONCERNING HUMAN UNDERSTANDING, John Locke. The only complete edition of enormously important classic, with authoritative editorial material by A. C. Fraser. Total of 1176pp. 5⅜ x 8½.
20530-4, 20531-2 Pa., Two-vol. set $16.00

HANDBOOK OF MATHEMATICAL FUNCTIONS WITH FORMULAS, GRAPHS, AND MATHEMATICAL TABLES, edited by Milton Abramowitz and Irene A. Stegun. Vast compendium: 29 sets of tables, some to as high as 20 places. 1,046pp. 8 x 10½.   61272-4 Pa. $14.95

MATHEMATICS FOR THE PHYSICAL SCIENCES, Herbert S. Wilf. Highly acclaimed work offers clear presentations of vector spaces and matrices, orthogonal functions, roots of polynomial equations, conformal mapping, calculus of variations, etc. Knowledge of theory of functions of real and complex variables is assumed. Exercises and solutions. Index. 284pp. 5⅝ x 8¼.   63635-6 Pa. $5.00

THE PRINCIPLE OF RELATIVITY, Albert Einstein et al. Eleven most important original papers on special and general theories. Seven by Einstein, two by Lorentz, one each by Minkowski and Weyl. All translated, unabridged. 216pp. 5⅜ x 8½.   60081-5 Pa. $3.50

THERMODYNAMICS, Enrico Fermi. A classic of modern science. Clear, organized treatment of systems, first and second laws, entropy, thermodynamic potentials, gaseous reactions, dilute solutions, entropy constant. No math beyond calculus required. Problems. 160pp. 5⅜ x 8½.
60361-X Pa. $3.00

ELEMENTARY MECHANICS OF FLUIDS, Hunter Rouse. Classic undergraduate text widely considered to be far better than many later books. Ranges from fluid velocity and acceleration to role of compressibility in fluid motion. Numerous examples, questions, problems. 224 illustrations. 376pp. 5⅝ x 8¼.   63699-2 Pa. $5.00

THE AMERICAN SENATOR, Anthony Trollope. Little known, long unavailable Trollope novel on a grand scale. Here are humorous comment on American vs. English culture, and stunning portrayal of a heroine/villainess. Superb evocation of Victorian village life. 561pp. 5⅜ x 8½.
23801-6 Pa. $6.00

WAS IT MURDER? James Hilton. The author of *Lost Horizon* and *Goodbye, Mr. Chips* wrote one detective novel (under a pen-name) which was quickly forgotten and virtually lost, even at the height of Hilton's fame. This edition brings it back—a finely crafted public school puzzle resplendent with Hilton's stylish atmosphere. A thoroughly English thriller by the creator of Shangri-la. 252pp. 5⅜ x 8. (Available in U.S. only)
23774-5 Pa. $3.00

CENTRAL PARK: A PHOTOGRAPHIC GUIDE, Victor Laredo and Henry Hope Reed. 121 superb photographs show dramatic views of Central Park: Bethesda Fountain, Cleopatra's Needle, Sheep Meadow, the Blockhouse, plus people engaged in many park activities: ice skating, bike riding, etc. Captions by former Curator of Central Park, Henry Hope Reed, provide historical view, changes, etc. Also photos of N.Y. landmarks on park's periphery. 96pp. 8½ x 11.
23750-8 Pa. $4.50

NANTUCKET IN THE NINETEENTH CENTURY, Clay Lancaster. 180 rare photographs, stereographs, maps, drawings and floor plans recreate unique American island society. Authentic scenes of shipwreck, lighthouses, streets, homes are arranged in geographic sequence to provide walking-tour guide to old Nantucket existing today. Introduction, captions. 160pp. 8⅞ x 11¾.
23747-8 Pa. $6.95

STONE AND MAN: A PHOTOGRAPHIC EXPLORATION, Andreas Feininger. 106 photographs by *Life* photographer Feininger portray man's deep passion for stone through the ages. Stonehenge-like megaliths, fortified towns, sculpted marble and crumbling tenements show textures, beauties, fascination. 128pp. 9¼ x 10¾.
23756-7 Pa. $5.95

CIRCLES, A MATHEMATICAL VIEW, D. Pedoe. Fundamental aspects of college geometry, non-Euclidean geometry, and other branches of mathematics: representing circle by point. Poincare model, isoperimetric property, etc. Stimulating recreational reading. 66 figures. 96pp. 5⅝ x 8¼.
63698-4 Pa. $2.75

THE DISCOVERY OF NEPTUNE, Morton Grosser. Dramatic scientific history of the investigations leading up to the actual discovery of the eighth planet of our solar system. Lucid, well-researched book by well-known historian of science. 172pp. 5⅜ x 8½. 23726-5 Pa. $3.50

THE DEVIL'S DICTIONARY. Ambrose Bierce. Barbed, bitter, brilliant witticisms in the form of a dictionary. Best, most ferocious satire America has produced. 145pp. 5⅜ x 8½. 20487-1 Pa. $2.25

THE CURVES OF LIFE, Theodore A. Cook. Examination of shells, leaves, horns, human body, art, etc., in *"the* classic reference on how the golden ratio applies to spirals and helices in nature . . . . "—Martin Gardner. 426 illustrations. Total of 512pp. 5⅜ x 8½.   23701-X Pa. $5.95

AN ILLUSTRATED FLORA OF THE NORTHERN UNITED STATES AND CANADA, Nathaniel L. Britton, Addison Brown. Encyclopedic work covers 4666 species, ferns on up. Everything. Full botanical information, illustration for each. This earlier edition is preferred by many to more recent revisions. 1913 edition. Over 4000 illustrations, total of 2087pp. 6⅛ x 9¼.   22642-5, 22643-3, 22644-1 Pa., Three-vol. set $25.50

MANUAL OF THE GRASSES OF THE UNITED STATES, A. S. Hitchcock, U.S. Dept. of Agriculture. The basic study of American grasses, both indigenous and escapes, cultivated and wild. Over 1400 species. Full descriptions, information. Over 1100 maps, illustrations. Total of 1051pp. 5⅜ x 8½.   22717-0, 22718-9 Pa., Two-vol. set $15.00

THE CACTACEAE,, Nathaniel L. Britton, John N. Rose. Exhaustive, definitive. Every cactus in the world. Full botanical descriptions. Thorough statement of nomenclatures, habitat, detailed finding keys. The one book needed by every cactus enthusiast. Over 1275 illustrations. Total of 1080pp. 8 x 10¼.   21191-6, 21192-4 Clothbd., Two-vol. set $35.00

AMERICAN MEDICINAL PLANTS, Charles F. Millspaugh. Full descriptions, 180 plants covered: history; physical description; methods of preparation with all chemical constituents extracted; all claimed curative or adverse effects. 180 full-page plates. Classification table. 804pp. 6½ x 9¼.
23034-1 Pa. $12.95

A MODERN HERBAL, Margaret Grieve. Much the fullest, most exact, most useful compilation of herbal material. Gigantic alphabetical encyclopedia, from aconite to zedoary, gives botanical information, medical properties, folklore, economic uses, and much else. Indispensable to serious reader. 161 illustrations. 888pp. 6½ x 9¼. (Available in U.S. only)
22798-7, 22799-5 Pa., Two-vol. set $13.00

THE HERBAL or GENERAL HISTORY OF PLANTS, John Gerard. The 1633 edition revised and enlarged by Thomas Johnson. Containing almost 2850 plant descriptions and 2705 superb illustrations, Gerard's *Herbal* is a monumental work, the book all modern English herbals are derived from, the one herbal every serious enthusiast should have in its entirety. Original editions are worth perhaps $750. 1678pp. 8½ x 12¼.
23147-X Clothbd. $50.00

MANUAL OF THE TREES OF NORTH AMERICA, Charles S. Sargent. The basic survey of every native tree and tree-like shrub, 717 species in all. Extremely full descriptions, information on habitat, growth, locales, economics, etc. Necessary to every serious tree lover. Over 100 finding keys. 783 illustrations. Total of 986pp. 5⅜ x 8½.
20277-1, 20278-X Pa., Two-vol. set $11.00

AMERICAN BIRD ENGRAVINGS, Alexander Wilson et al. All 76 plates. from Wilson's *American Ornithology* (1808-14), most important ornithological work before Audubon, plus 27 plates from the supplement (1825-33) by Charles Bonaparte. Over 250 birds portrayed. 8 plates also reproduced in full color. 111pp. 9⅜ x 12½. 23195-X Pa. $6.00

CRUICKSHANK'S PHOTOGRAPHS OF BIRDS OF AMERICA, Allan D. Cruickshank. Great ornithologist, photographer presents 177 closeups, groupings, panoramas, flightings, etc., of about 150 different birds. Expanded *Wings in the Wilderness*. Introduction by Helen G. Cruickshank. 191pp. 8¼ x 11. 23497-5 Pa. $6.00

AMERICAN WILDLIFE AND PLANTS, A. C. Martin, et al. Describes food habits of more than 1000 species of mammals, birds, fish. Special treatment of important food plants. Over 300 illustrations. 500pp. 5⅜ x 8½. 20793-5 Pa. $4.95

THE PEOPLE CALLED SHAKERS, Edward D. Andrews. Lifetime of research, definitive study of Shakers: origins, beliefs, practices, dances, social organization, furniture and crafts, impact on 19th-century USA, present heritage. Indispensable to student of American history, collector. 33 illustrations. 351pp. 5⅜ x 8½. 21081-2 Pa. $4.50

OLD NEW YORK IN EARLY PHOTOGRAPHS, Mary Black. New York City as it was in 1853-1901, through 196 wonderful photographs from N.-Y. Historical Society. Great Blizzard, Lincoln's funeral procession, great buildings. 228pp. 9 x 12. 22907-6 Pa. $8.95

MR. LINCOLN'S CAMERA MAN: MATHEW BRADY, Roy Meredith. Over 300 Brady photos reproduced directly from original negatives, photos. Jackson, Webster, Grant, Lee, Carnegie, Barnum; Lincoln; Battle Smoke, Death of Rebel Sniper, Atlanta Just After Capture. Lively commentary. 368pp. 8⅜ x 11¼. 23021-X Pa. $8.95

TRAVELS OF WILLIAM BARTRAM, William Bartram. From 1773-8, Bartram explored Northern Florida, Georgia, Carolinas, and reported on wild life, plants, Indians, early settlers. Basic account for period, entertaining reading. Edited by Mark Van Doren. 13 illustrations. 141pp. 5⅜ x 8½. 20013-2 Pa. $5.00

THE GENTLEMAN AND CABINET MAKER'S DIRECTOR, Thomas Chippendale. Full reprint, 1762 style book, most influential of all time; chairs, tables, sofas, mirrors, cabinets, etc. 200 plates, plus 24 photographs of surviving pieces. 249pp. 9⅞ x 12¾. 21601-2 Pa. $7.95

AMERICAN CARRIAGES, SLEIGHS, SULKIES AND CARTS, edited by Don H. Berkebile. 168 Victorian illustrations from catalogues, trade journals, fully captioned. Useful for artists. Author is Assoc. Curator, Div. of Transportation of Smithsonian Institution. 168pp. 8½ x 9½. 23328-6 Pa. $5.00

THE EARLY WORK OF AUBREY BEARDSLEY, Aubrey Beardsley. 157 plates, 2 in color: *Manon Lescaut, Madame Bovary, Morte Darthur, Salome,* other. Introduction by H. Marillier. 182pp. 8⅛ x 11. 21816-3 Pa. $4.50

THE LATER WORK OF AUBREY BEARDSLEY, Aubrey Beardsley. Exotic masterpieces of full maturity: *Venus and Tannhauser, Lysistrata, Rape of the Lock, Volpone,* Savoy material, etc. 174 plates, 2 in color. 186pp. 8⅛ x 11. 21817-1 Pa. $5.95

THOMAS NAST'S CHRISTMAS DRAWINGS, Thomas Nast. Almost all Christmas drawings by creator of image of Santa Claus as we know it, and one of America's foremost illustrators and political cartoonists. 66 illustrations. 3 illustrations in color on covers. 96pp. 8⅜ x 11¼. 23660-9 Pa. $3.50

THE DORÉ ILLUSTRATIONS FOR DANTE'S DIVINE COMEDY, Gustave Doré. All 135 plates from Inferno, Purgatory, Paradise; fantastic tortures, infernal landscapes, celestial wonders. Each plate with appropriate (translated) verses. 141pp. 9 x 12. 23231-X Pa. $4.50

DORÉ'S ILLUSTRATIONS FOR RABELAIS, Gustave Doré. 252 striking illustrations of *Gargantua and Pantagruel* books by foremost 19th-century illustrator. Including 60 plates, 192 delightful smaller illustrations. 153pp. 9 x 12. 23656-0 Pa. $5.00

LONDON: A PILGRIMAGE, Gustave Doré, Blanchard Jerrold. Squalor, riches, misery, beauty of mid-Victorian metropolis; 55 wonderful plates, 125 other illustrations, full social, cultural text by Jerrold. 191pp. of text. 9⅜ x 12¼. 22306-X Pa. $7.00

THE RIME OF THE ANCIENT MARINER, Gustave Doré, S. T. Coleridge. Dore's finest work, 34 plates capture moods, subtleties of poem. Full text. Introduction by Millicent Rose. 77pp. 9¼ x 12. 22305-1 Pa. $3.50

THE DORE BIBLE ILLUSTRATIONS, Gustave Doré. All wonderful, detailed plates: Adam and Eve, Flood, Babylon, Life of Jesus, etc. Brief King James text with each plate. Introduction by Millicent Rose. 241 plates. 241pp. 9 x 12. 23004-X Pa. $6.00

THE COMPLETE ENGRAVINGS, ETCHINGS AND DRYPOINTS OF ALBRECHT DURER. "Knight, Death and Devil"; "Melencolia," and more—all Dürer's known works in all three media, including 6 works formerly attributed to him. 120 plates. 235pp. 8⅜ x 11¼. 22851-7 Pa. $6.50

MECHANICK EXERCISES ON THE WHOLE ART OF PRINTING, Joseph Moxon. First complete book (1683-4) ever written about typography, a compendium of everything known about printing at the latter part of 17th century. Reprint of 2nd (1962) Oxford Univ. Press edition. 74 illustrations. Total of 550pp. 6⅛ x 9¼. 23617-X Pa. $7.95

THE ANATOMY OF THE HORSE, George Stubbs. Often considered the great masterpiece of animal anatomy. Full reproduction of 1766 edition, plus prospectus; original text and modernized text. 36 plates. Introduction by Eleanor Garvey. 121pp. 11 x 14¾. 23402-9 Pa. $6.00

BRIDGMAN'S LIFE DRAWING, George B. Bridgman. More than 500 illustrative drawings and text teach you to abstract the body into its major masses, use light and shade, proportion; as well as specific areas of anatomy, of which Bridgman is master. 192pp. 6½ x 9¼. (Available in U.S. only) 22710-3 Pa. $3.50

ART NOUVEAU DESIGNS IN COLOR, Alphonse Mucha, Maurice Verneuil, Georges Auriol. Full-color reproduction of *Combinaisons orne-mentales* (c. 1900) by Art Nouveau masters. Floral, animal, geometric, interlacings, swashes—borders, frames, spots—all incredibly beautiful. 60 plates, hundreds of designs. 9⅜ x 8-1/16. 22885-1 Pa. $4.00

FULL-COLOR FLORAL DESIGNS IN THE ART NOUVEAU STYLE, E. A. Seguy. 166 motifs, on 40 plates, from *Les fleurs et leurs applications decoratives* (1902): borders, circular designs, repeats, allovers, "spots." All in authentic Art Nouveau colors. 48pp. 9⅜ x 12¼. 23439-8 Pa. $5.00

A DIDEROT PICTORIAL ENCYCLOPEDIA OF TRADES AND IN-DUSTRY, edited by Charles C. Gillispie. 485 most interesting plates from the great French Encyclopedia of the 18th century show hundreds of working figures, artifacts, process, land and cityscapes; glassmaking, paper-making, metal extraction, construction, weaving, making furniture, clothing, wigs, dozens of other activities. Plates fully explained. 920pp. 9 x 12. 22284-5, 22285-3 Clothbd., Two-vol. set $40.00

HANDBOOK OF EARLY ADVERTISING ART, Clarence P. Hornung. Largest collection of copyright-free early and antique advertising art ever compiled. Over 6,000 illustrations, from Franklin's time to the 1890's for special effects, novelty. Valuable source, almost inexhaustible.
Pictorial Volume. Agriculture, the zodiac, animals, autos, birds, Christmas, fire engines, flowers, trees, musical instruments, ships, games and sports, much more. Arranged by subject matter and use. 237 plates. 288pp. 9 x 12. 20122-8 Clothbd. $14.50

Typographical Volume. Roman and Gothic faces ranging from 10 point to 300 point, "Barnum," German and Old English faces, script, logotypes, scrolls and flourishes, 1115 ornamental initials, 67 complete alphabets, more. 310 plates. 320pp. 9 x 12. 20123-6 Clothbd. $15.00

CALLIGRAPHY (CALLIGRAPHIA LATINA), J. G. Schwandner. High point of 18th-century ornamental calligraphy. Very ornate initials, scrolls, borders, cherubs, birds, lettered examples. 172pp. 9 x 13. 20475-8 Pa. $7.00

THE DEPRESSION YEARS AS PHOTOGRAPHED BY ARTHUR ROTH-STEIN, Arthur Rothstein. First collection devoted entirely to the work of outstanding 1930s photographer: famous dust storm photo, ragged children, unemployed, etc. 120 photographs. Captions. 119pp. 9¼ x 10¾.
23590-4 Pa. $5.00

CAMERA WORK: A PICTORIAL GUIDE, Alfred Stieglitz. All 559 illustrations and plates from the most important periodical in the history of art photography, Camera Work (1903-17). Presented four to a page, reduced in size but still clear, in strict chronological order, with complete captions. Three indexes. Glossary. Bibliography. 176pp. 8⅜ x 11¼.
23591-2 Pa. $6.95

ALVIN LANGDON COBURN, PHOTOGRAPHER, Alvin L. Coburn. Revealing autobiography by one of greatest photographers of 20th century gives insider's version of Photo-Secession, plus comments on his own work. 77 photographs by Coburn. Edited by Helmut and Alison Gernsheim. 160pp. 8⅛ x 11.
23685-4 Pa. $6.00

NEW YORK IN THE FORTIES, Andreas Feininger. 162 brilliant photographs by the well-known photographer, formerly with Life magazine, show commuters, shoppers, Times Square at night, Harlem nightclub, Lower East Side, etc. Introduction and full captions by John von Hartz. 181pp. 9¼ x 10¾.
23585-8 Pa. $6.95

GREAT NEWS PHOTOS AND THE STORIES BEHIND THEM, John Faber. Dramatic volume of 140 great news photos, 1855 through 1976, and revealing stories behind them, with both historical and technical information. Hindenburg disaster, shooting of Oswald, nomination of Jimmy Carter, etc. 160pp. 8¼ x 11.
23667-6 Pa. $5.00

THE ART OF THE CINEMATOGRAPHER, Leonard Maltin. Survey of American cinematography history and anecdotal interviews with 5 masters—Arthur Miller, Hal Mohr, Hal Rosson, Lucien Ballard, and Conrad Hall. Very large selection of behind-the-scenes production photos. 105 photographs. Filmographies. Index. Originally Behind the Camera. 144pp. 8¼ x 11.
23686-2 Pa. $5.00

DESIGNS FOR THE THREE-CORNERED HAT (LE TRICORNE), Pablo Picasso. 32 fabulously rare drawings—including 31 color illustrations of costumes and accessories—for 1919 production of famous ballet. Edited by Parmenia Migel, who has written new introduction. 48pp. 9⅜ x 12¼. (Available in U.S. only)
23709-5 Pa. $5.00

NOTES OF A FILM DIRECTOR, Sergei Eisenstein. Greatest Russian filmmaker explains montage, making of Alexander Nevsky, aesthetics; comments on self, associates, great rivals (Chaplin), similar material. 78 illustrations. 240pp. 5⅜ x 8½.
22392-2 Pa. $4.50

A MAYA GRAMMAR, Alfred M. Tozzer. Practical, useful English-language grammar by the Harvard anthropologist who was one of the three greatest American scholars in the area of Maya culture. Phonetics, grammatical processes, syntax, more. 301pp. 5⅜ x 8½. 23465-7 Pa. $4.00

THE JOURNAL OF HENRY D. THOREAU, edited by Bradford Torrey, F. H. Allen. Complete reprinting of 14 volumes, 1837-61, over two million words; the sourcebooks for *Walden*, etc. Definitive. All original sketches, plus 75 photographs. Introduction by Walter Harding. Total of 1804pp. 8½ x 12¼. 20312-3, 20313-1 Clothbd., Two-vol. set $70.00

CLASSIC GHOST STORIES, Charles Dickens and others. 18 wonderful stories you've wanted to reread: "The Monkey's Paw," "The House and the Brain," "The Upper Berth," "The Signalman," "Dracula's Guest," "The Tapestried Chamber," etc. Dickens, Scott, Mary Shelley, Stoker, etc. 330pp. 5⅜ x 8½. 20735-8 Pa. $4.50

SEVEN SCIENCE FICTION NOVELS, H. G. Wells. Full novels. *First Men in the Moon, Island of Dr. Moreau, War of the Worlds, Food of the Gods, Invisible Man, Time Machine, In the Days of the Comet.* A basic science-fiction library. 1015pp. 5⅜ x 8½. (Available in U.S. only)
20264-X Clothbd. $8.95

ARMADALE, Wilkie Collins. Third great mystery novel by the author of *The Woman in White* and *The Moonstone*. Ingeniously plotted narrative shows an exceptional command of character, incident and mood. Original magazine version with 40 illustrations. 597pp. 5⅜ x 8½.
23429-0 Pa. $6.00

MASTERS OF MYSTERY, H. Douglas Thomson. The first book in English (1931) devoted to history and aesthetics of detective story. Poe, Doyle, LeFanu, Dickens, many others, up to 1930. New introduction and notes by E. F. Bleiler. 288pp. 5⅜ x 8½. (Available in U.S. only)
23606-4 Pa. $4.00

FLATLAND, E. A. Abbott. Science-fiction classic explores life of 2-D being in 3-D world. Read also as introduction to thought about hyperspace. Introduction by Banesh Hoffmann. 16 illustrations. 103pp. 5⅜ x 8½.
20001-9 Pa. $2.00

THREE SUPERNATURAL NOVELS OF THE VICTORIAN PERIOD, edited, with an introduction, by E. F. Bleiler. Reprinted complete and unabridged, three great classics of the supernatural: *The Haunted Hotel* by Wilkie Collins, *The Haunted House at Latchford* by Mrs. J. H. Riddell, and *The Lost Stradivarious* by J. Meade Falkner. 325pp. 5⅜ x 8½.
22571-2 Pa. $4.00

AYESHA: THE RETURN OF "SHE," H. Rider Haggard. Virtuoso sequel featuring the great mythic creation, Ayesha, in an adventure that is fully as good as the first book, *She*. Original magazine version, with 47 original illustrations by Maurice Greiffenhagen. 189pp. 6½ x 9¼.
23649-8 Pa. $3.50

HOUSEHOLD STORIES BY THE BROTHERS GRIMM. All the great Grimm stories: "Rumpelstiltskin," "Snow White," "Hansel and Gretel," etc., with 114 illustrations by Walter Crane. 269pp. 5⅜ x 8½.
21080-4 Pa. $3.50

SLEEPING BEAUTY, illustrated by Arthur Rackham. Perhaps the fullest, most delightful version ever, told by C. S. Evans. Rackham's best work. 49 illustrations. 110pp. 7⅞ x 10¾.
22756-1 Pa. $2.50

AMERICAN FAIRY TALES, L. Frank Baum. Young cowboy lassoes Father Time; dummy in Mr. Floman's department store window comes to life; and 10 other fairy tales. 41 illustrations by N. P. Hall, Harry Kennedy, Ike Morgan, and Ralph Gardner. 209pp. 5⅜ x 8½.
23643-9 Pa. $3.00

THE WONDERFUL WIZARD OF OZ, L. Frank Baum. Facsimile in full color of America's finest children's classic. Introduction by Martin Gardner. 143 illustrations by W. W. Denslow. 267pp. 5⅜ x 8½.
20691-2 Pa. $3.50

THE TALE OF PETER RABBIT, Beatrix Potter. The inimitable Peter's terrifying adventure in Mr. McGregor's garden, with all 27 wonderful, full-color Potter illustrations. 55pp. 4¼ x 5½. (Available in U.S. only)
22827-4 Pa. $1.25

THE STORY OF KING ARTHUR AND HIS KNIGHTS, Howard Pyle. Finest children's version of life of King Arthur. 48 illustrations by Pyle. 131pp. 6⅛ x 9¼.
21445-1 Pa. $4.95

CARUSO'S CARICATURES, Enrico Caruso. Great tenor's remarkable caricatures of self, fellow musicians, composers, others. Toscanini, Puccini, Farrar, etc. Impish, cutting, insightful. 473 illustrations. Preface by M. Sisca. 217pp. 8⅜ x 11¼.
23528-9 Pa. $6.95

PERSONAL NARRATIVE OF A PILGRIMAGE TO ALMADINAH AND MECCAH, Richard Burton. Great travel classic by remarkably colorful personality. Burton, disguised as a Moroccan, visited sacred shrines of Islam, narrowly escaping death. Wonderful observations of Islamic life, customs, personalities. 47 illustrations. Total of 959pp. 5⅜ x 8½.
21217-3, 21218-1 Pa., Two-vol. set $12.00

INCIDENTS OF TRAVEL IN YUCATAN, John L. Stephens. Classic (1843) exploration of jungles of Yucatan, looking for evidences of Maya civilization. Travel adventures, Mexican and Indian culture, etc. Total of 669pp. 5⅜ x 8½.
20926-1, 20927-X Pa., Two-vol. set $7.90

AMERICAN LITERARY AUTOGRAPHS FROM WASHINGTON IRVING TO HENRY JAMES, Herbert Cahoon, et al. Letters, poems, manuscripts of Hawthorne, Thoreau, Twain, Alcott, Whitman, 67 other prominent American authors. Reproductions, full transcripts and commentary. Plus checklist of all American Literary Autographs in The Pierpont Morgan Library. Printed on exceptionally high-quality paper. 136 illustrations. 212pp. 9⅛ x 12¼.
23548-3 Pa. $12.50

CATALOGUE OF DOVER BOOKS

AN AUTOBIOGRAPHY, Margaret Sanger. Exciting personal account of hard-fought battle for woman's right to birth control, against prejudice, church, law. Foremost feminist document. 504pp. 5⅜ x 8½.
20470-7 Pa. $5.50

MY BONDAGE AND MY FREEDOM, Frederick Douglass. Born as a slave, Douglass became outspoken force in antislavery movement. The best of Douglass's autobiographies. Graphic description of slave life. Introduction by P. Foner. 464pp. 5⅜ x 8½. 22457-0 Pa. $5.50

LIVING MY LIFE, Emma Goldman. Candid, no holds barred account by foremost American anarchist: her own life, anarchist movement, famous contemporaries, ideas and their impact. Struggles and confrontations in America, plus deportation to U.S.S.R. Shocking inside account of persecution of anarchists under Lenin. 13 plates. Total of 944pp. 5⅜ x 8½.
22543-7, 22544-5 Pa., Two-vol. set $12.00

LETTERS AND NOTES ON THE MANNERS, CUSTOMS AND CONDITIONS OF THE NORTH AMERICAN INDIANS, George Catlin. Classic account of life among Plains Indians: ceremonies, hunt, warfare, etc. Dover edition reproduces for first time all original paintings. 312 plates. 572pp. of text. 6⅛ x 9¼. 22118-0, 22119-9 Pa.. Two-vol. set $12.00

THE MAYA AND THEIR NEIGHBORS, edited by Clarence L. Hay, others. Synoptic view of Maya civilization in broadest sense, together with Northern, Southern neighbors. Integrates much background, valuable detail not elsewhere. Prepared by greatest scholars: Kroeber, Morley, Thompson, Spinden, Vaillant, many others. Sometimes called Tozzer Memorial Volume. 60 illustrations, linguistic map. 634pp. 5⅜ x 8½.
23510-6 Pa. $10.00

HANDBOOK OF THE INDIANS OF CALIFORNIA, A. L. Kroeber. Foremost American anthropologist offers complete ethnographic study of each group. Monumental classic. 459 illustrations, maps. 995pp. 5⅜ x 8½.
23368-5 Pa. $13.00

SHAKTI AND SHAKTA, Arthur Avalon. First book to give clear, cohesive analysis of Shakta doctrine, Shakta ritual and Kundalini Shakti (yoga). Important work by one of world's foremost students of Shaktic and Tantric thought. 732pp. 5⅜ x 8½. (Available in U.S. only)
23645-5 Pa. $7.95

AN INTRODUCTION TO THE STUDY OF THE MAYA HIEROGLYPHS, Syvanus Griswold Morley. Classic study by one of the truly great figures in hieroglyph research. Still the best introduction for the student for reading Maya hieroglyphs. New introduction by J. Eric S. Thompson. 117 illustrations. 284pp. 5⅜ x 8½. 23108-9 Pa. $4.00

A STUDY OF MAYA ART, Herbert J. Spinden. Landmark classic interprets Maya symbolism, estimates styles, covers ceramics, architecture, murals, stone carvings as artforms. Still a basic book in area. New introduction by J. Eric Thompson. Over 750 illustrations. 341pp. 8⅜ x 11¼.
21235-1 Pa. $6.95

AMERICAN ANTIQUE FURNITURE, Edgar G. Miller, Jr. The basic coverage of all American furniture before 1840: chapters per item chronologically cover all types of furniture, with more than 2100 photos. Total of 1106pp. 7⅞ x 10¾.   21599-7, 21600-4 Pa., Two-vol. set $17.90

ILLUSTRATED GUIDE TO SHAKER FURNITURE, Robert Meader. Director, Shaker Museum, Old Chatham, presents up-to-date coverage of all furniture and appurtenances, with much on local styles not available elsewhere. 235 photos. 146pp. 9 x 12.   22819-3 Pa. $6.00

ORIENTAL RUGS, ANTIQUE AND MODERN, Walter A. Hawley. Persia, Turkey, Caucasus, Central Asia, China, other traditions. Best general survey of all aspects: styles and periods, manufacture, uses, symbols and their interpretation, and identification. 96 illustrations, 11 in color. 320pp. 6⅛ x 9¼.   22366-3 Pa. $6.95

CHINESE POTTERY AND PORCELAIN, R. L. Hobson. Detailed descriptions and analyses by former Keeper of the Department of Oriental Antiquities and Ethnography at the British Museum. Covers hundreds of pieces from primitive times to 1915. Still the standard text for most periods. 136 plates, 40 in full color. Total of 750pp. 5⅜ x 8½.
          23253-0 Pa. $10.00

THE WARES OF THE MING DYNASTY, R. L. Hobson. Foremost scholar examines and illustrates many varieties of Ming (1368-1644). Famous blue and white, polychrome, lesser-known styles and shapes. 117 illustrations, 9 full color, of outstanding pieces. Total of 263pp. 6⅛ x 9¼. (Available in U.S. only)   23652-8 Pa. $6.00

Prices subject to change without notice.

Available at your book dealer or write for free catalogue to Dept. GI, Dover Publications, Inc., 31 East Second Street, Mineola, N.Y. 11501. Dover publishes more than 175 books each year on science, elementary and advanced mathematics, biology, music, art, literary history, social sciences and other areas.